T0180735

FROM SUNS TO LIFE: A CHRONOLOGICAL APPROACH TO THE HISTORY OF LIFE ON EARTH

From Suns to Life: A Chronological Approach to the History of Life on Earth

Edited by

MURIEL GARGAUD
PHILIPPE CLAEYS
PURIFICACIÓN LÓPEZ-GARCÍA
HERVÉ MARTIN
THIERRY MONTMERLE
ROBERT PASCAL
JACQUES REISSE

Reprinted from *Earth Moon, and Planets*
Volume 98, Nos. 1–4, 2006

 Springer

A C.I.P catalogue record for this book is available from the library of Congress

ISBN 1-4939-3885-1
ISBN 978-1-4939-3885-8
DOI 10.1007/978-0-387-45083-4

Published by Springer,
P.O. Box 17, 3300 AA, Dordrecht, The Netherlands

www.springer.com

Printed on acid-free paper

Cover image: Persistence of Memory, 1931, © Salvadore Dali, Fundactión Gala-Salvador Dali, c/o Beeldrecht Amsterdam 2006

A C.I.P catalogue record for this book is available from the library of Congress

Published by Springer,
P.O. Box 17, 3300 AA, Dordrecht, The Netherlands
www.springer.com

Printed on acid-free paper

Cover image: Persistence of Memory, 1931, © Salvadore Dali, Fundactión Gala-Salvador Dali, c/o Beeldrecht Amsterdam 2006

Printed in the Netherlands

Table of Contents

Earth, Moon, and Planets (2006) 98: 1–9
DOI 10.1007/s11038-006-9085-7

1. From the Arrow of Time to the Arrow of Life

MURIEL GARGAUD

Observatoire Aquitain des Sciences de l'Univers, Université Bordeaux1, Bordeaux, France
(E-mail: gargaud@obs.u-bordeaux1.fr)

JACQUES REISSE

Faculté des Sciences Appliquées (CP 165/64), Université Libre de Bruxelles, Brussels, Belgium
(E-mail: jreisse@ulb.ac.be)

(Accepted 4 April 2006)

Abstract. Astrobiology, like many (but not all) sciences, must take into account questions of the "Why?", "Where?", "How?" and "When?" type. In this introductory chapter, we explain why, in this book, we will only consider two of these questions that are, moreover, deeply interrelated. Chronology is by definition related to the "when?" question but as soon as we are interested in the history of Earth or the history of life, it is impossible to treat these questions and their answers without explicit references to the "how?" questions. We also present in this chapter the genesis and the aim of the book.

Keywords: Time, dating

1.1. The Notion of Time in Astrobiology

Scientists are trained to ask questions about Nature. As is well known in dialectics, the quality of the question determines the quality of the answer. Questions that remain unanswered for a long time are probably questions that must be formulated in a different way.

Each chapter (from the "Dating methods and corresponding chronometers in astrobiology" chapter 2 to the final "Life on Earth and elsewhere" chapter 9) has been submitted to a very severe internal refereeing, each author having read many, if not all, contributions to make comments, criticisms and advice. As a general rule these comments/remarks were discussed amongst the authors and were included in the original text. Ideally, all the authors could have signed together all the contributions but we have preferred to "render to Caesar the things which are Caesar's" and the specialist who first wrote it signed each subchapter. For further information on the contributing authors, please refer to CVs', included at the end of this volume. All subchapters devoted to the same general field have been gathered together as a chapter under a general title. Consequently, each chapter is alphabetically co-signed by all the authors of the subchapters, and the first author is the coordinator who supervised its homogeneity and completeness. In all cases, each author did his (or her) best to give to the reader the most accurate and recent data, along with the evidence, but also assumptions, on which the data is founded and, when necessary, the caution required for its interpretation. A glossary of terminology used in all chapters is available at the end of this volume.

Among the numerous questions scientists are interested in, those related to time are probably the most fascinating because time is a very peculiar dimension. From Einstein's work, we learnt that, for physics, time is "just" the fourth dimension but, on the other hand, we are also aware that the time dimension, as we perceive it, is qualitatively different from the space dimensions. Time "flows"; time is irreversible and associated with past, present and future.

"Why is time different from the other dimensions?" remains an open question, and this is one major reason why many scientists are fascinated by problems related to time. Of course, not all sciences are historical, i.e. not all sciences focus their study on similarities or differences occurring between past, present and future events. For example, a chemist studying the evolution of a reaction as a function of time certainly knows that if pressure, temperature and all other experimental conditions are kept the same, the reaction will evolve tomorrow in a deterministic way, exactly as it does today. For this chemist, time is a parameter easily measurable with a chronometer. Even if the reaction under study is irreversible, chemists (as all scientists) know that the physical laws do not change with time. As Noether showed, this time translation independence is related to the energy conservation law (Zee, 1986)

The situation is completely different for a geologist or a biologist interested in evolutionary problems. They must take into account the historical time and, therefore, the irreversible flow of time, the so-called "arrow of time" (e.g. Klein and Spiro, 1997). For these scientists, a chronometer is useless: they need to measure time with respect to a conventional reference time. Their situation can be compared to that of a historian who, in Western countries, uses as reference time the birth of Christ, even though its date is still debated among historians, being uncertain by several years. All of them, historians, geologists or biologists have in common their need to apply to a time reference chosen by convention.

Cosmologists are the only scientists who could claim to use a time scale based on an absolute zero even if, today, it is discussed whether the Big Bang itself could be considered as the origin of our time. Planetologists and geologists use different reference times depending on the problem they are interested in. They know from cosmologists and astronomers that the Universe is probably 13.7 billion years old, but they do not use the Big Bang as reference time: they use the age of the oldest meteorites instead. Frequently, but not always, they use a time scale that takes as reference time the "present time" defined as 1950 AD[1], which is the reference used for ^{14}C dating. Any time is thus expressed in "years before present" (yr BP). On this scale and just as examples, the accretion of the solar system took place approximately

---

[1] AD = Anno Domini = After Jesus-Christ.

4.6×10^9 yrs BP while the extinction of dinosaurs took place approximately 65×10^6 yrs BP. As explained in Chapter 2 on chronometers, for various reasons different communities have to use different time scales, both forward and backwards, even for events that are a common subject of study.

As mentioned above, not all sciences are historical. Yet many scientists, if not all of them, are concerned by the history of the solar system, of the Earth, and of life on it. Naturally, many of them also wonder about the existence of extraterrestrial living systems, whose emergence and survival would depend on various physico-chemical parameters, including time. Some of these scientists investigate specific aspects of these broad topics and, from their collaboration, a new scientific field, called Astrobiology, has emerged. Interestingly enough, although Astrobiology is a scientific field in itself, scientists tend to avoid describing themselves as astrobiologists. In any case, Astrobiology (exobiology or bioastronomy are also used as synonyms) is probably one of the best examples of a truly interdisciplinary field. By putting together pieces of a huge puzzle, astronomers, geologists, physicists, chemists, and biologists try to bring to the fore scenarios that led to the emergence of life on Earth and, eventually, to see if these scenarios could apply to other planets. The collaboration between these scientists from different horizons does certainly contribute to our understanding of the Earth's remote past and to better apprehend the conditions that allowed the emergence of life. Astrobiology is, today, a well-identified science. Of course, we are still far from having definitive answers to the countless questions related to events that took place billions of years ago. We should even consider the possibility that some of these answers will never be found. This may be perturbing for an experimental scientist who is able to test his hypotheses by performing experiments in his laboratory. However, this situation is absolutely normal if we consider the nature of all the historical sciences, including history itself. Nobody knows with absolute certainty how and when language did originate in human populations. A few hypotheses exist, but compelling evidence proving one of them while disproving the others is still missing. It might be found tomorrow, in one century or never!

If one considers Astrobiology as a good example of a historical natural science, the types of questions that can be formulated may then be classified into four families: "why", "when", "where" and "how"

1.2. Why, Where, When and How: Here are the Questions

1.2.1. WHY?

Although we will not discuss here the qualitative differences between all these families, we would like to highlight the peculiarity of any question of the but its occurrence at a given time point does not preclude the existence of life hundreds, thousands or millions of years before. Strictly speaking, it is possible to determine the age of the oldest sediment containing microorganisms but it is impossible to ascertain the age of the first microorganisms. A similar situation applies to the age of the Earth or that of the first oceans. We can establish the age of the oldest refractory inclusions in a particular type of chondrite or the age of the oldest zircons, but we are unable to measure directly the age of the Earth or that of the first oceans. Nevertheless, by fascinating "language shifts", we tend to speak about the age of the Earth or about the age of the first oceans. Moreover, these ages are considered to be known with great accuracy whereas, in fact, what we know accurately is the age of what was measured i.e. the age of pieces of rocks or even of single minerals within rocks (e.g. Jack Hills zircons are 4.4 billion years old, but the rocks they were extracted from are much younger). All these ages are strongly dependent on the validity of theoretical models describing the accretion of the solar system or the formation of a hydrosphere on the surface of the young Earth as very rapid processes. Therefore, "when" questions are often dependent on answers to "how" questions.

1.3. Our Modest Contributions to the "When" (and "How"?) Answers

1.3.1. THE "WHEN AND WHERE" MAKING-OF

This project started in September 2003 during an exobiology summer school (Exobio'03) we (M. Gargaud and D. Despois) organise in Propriano (Corsica) every two years since 1999 and where 80 researchers working in astronomy, geology, chemistry and biology try, year after year, to reconstitute the story of the emergence and evolution of life on Earth and its possible distribution elsewhere in the Universe.

1.3.2. THE "WHY" MAKING-OF

At that time, two of us (M. Gargaud and D. Despois) had the modest ambition to put on a sheet of paper some chronological data relevant to the origins of life and to understand what was the exact meaning of numbers in sentences like "the solar system is 4.569 Ga old", "the first proto-ocean appeared around 4.4 Ga", "the first undisputable evidence of life is dated around 2.7 Ga, but oldest traces of life could be dated as soon as 3.5 Ga". In brief, their goal was simply to understand:

– What is the exact meaning of data read in the literature?

This aspect is particularly important for people, even scientist, who are not specialists of the field and have no other choice than to take for granted what specialists say. If specialists all agree, it's not really a problem but of course it's not often the case. The important secondary questions are then:

– What are the error bars associated with these numbers?
– What sort of chronometer each scientific field uses and what is the reliability of each of them?
– Which hypotheses (models, observations, experiments) are assumed or taken for granted to validate these data?

On the other hand, a number, as precise as could be, can rarely reflect a sudden event (the transition from "non-living" to "living" can certainly not be represented by a Dirac function) and some questions relative so these numbers are even more important. For instance:

– What has been exactly the duration of that event and how rigorous is it to speak about a "beginning" and an "end" of a given one (planetary disk formation, Moon formation, ocean formation, late heavy bombardment, etc... but also and much more difficult, prebiotic chemistry, early biochemistry, etc...)
– What is the time reference used by astronomers, chemists, geologists or biologists?

1.3.3. THE "HOW" MAKING-OF

Of course the questions previously listed were not obvious ones, but we were enthusiastic enough to hope that one day of brainstorming with the whole community would allow us to order, at least sequentially if not absolutely, the main events having led to the emergence of life. At the time of Exobio'03 (September 2003), we didn't even think to publish anything on these topics; we just wanted to clarify our own ideas.

In fact, we rapidly realized that we wouldn't leave Propriano with the answers we were looking for, but we were far from thinking that we had put a finger in a terrible set of gear – wheels from which we could only escape more than two years later (and the story is probably not finished...), after two others specialized workshops on the subject and hours and hours of vivid and passionate discussions...

Of course we were aware of the difficulties for an astronomer (geologist, chemist, biologist) to think like a biologist (astronomer, geologist, chemist) but having organised several conferences and summer schools devoted to various aspects of astrobiology, we thought that a common language was now more or less acquired. In a sense, we were right: the language was common, but the culture and the ways of thinking were (still are?) completely different. Indeed, the nature of the problem each discipline can solve, the difficulties encountered, the tools used to solve them, the interpretation of theory/modelling/observations/experiments are by nature absolutely different. For example, physicists, and even more so astrophysicists, are used to extensive logical and deterministic constructions (mathematical models) spanning a bridge between facts and deduction. Indeed, and just as an example, many details of the nuclear reactions inside stars rely, *in fine*, on the measurements of only T and L on their surface, many thousands kilometres away from the centre. ...Of course it's far from being intellectually satisfying, but as nobody will never go inside a star nucleus to measure physical parameters, (even though, in the case of the Sun, which is the only star we know with great precision, we have access to deep layers via neutrinos and helioseismology), the only choice astrophysicists have is to built models not at variance with observations.

Concerning the chronometers each discipline can rely on, here again the situation is completely different from one field to another. For example, astronomers can collect a lot of information but the chronometers they have in hand are indirect and only of statistical nature (chapter 2.1). Geochemists have very efficient radioactivity chronometers (chapters 2.2 and 3.2) but the difficulty for them is to determine what is exactly dated and what they can infer from these data (the latest measurements of Calcium Aluminium Inclusion in Allende meteorite gives a very precise age of $4.5685 \pm 0.0004 \times 10^9$ yrs, but on what reliable hypotheses can we deduct that this gives also the age of the solar system with an error bar of less than 10^6 yrs?). Chemists have no chronometers (chapters 2.3 and 5.1) and it's even an impossible mission for them to reconstruct the prebiotic chemistry and biochemistry period (between 4.4 and 2.7 billion years) when absolute and even to some extents relative chronology remains totally unknown. Hopefully the situation gets better in biology where molecular clocks are invaluable tools for reconstructing evolutionary timescales (chapter 2.4), but biochemical and biological problems are so complex that of course the reliability of chronometers helps but doesn't solve everything in a definitive way...

Another problem we had to face was the relevance of an absolute t_0 time of reference. Of course it does not really matter to know that Ramses II lived 1250 years BC or 3200 years BP, but if historians want to compare the reign of Ramses II with those of Amenophys III or Cleopatra, they have first to agree on a unique reference time. And what is important in fact for Egyptian history is not so much that Ramses II reigned 1250 years C, but that he reigned during 66 years, after Amenophys III and before Cleopatra. Once again, duration is a very important parameter, and duration of course imply "a" beginning and "an" end.

As indicated in the previous paragraph we could have decided to choose the Big Bang as an absolute reference time. Nevertheless as we were interested in the history of life on Earth, we started our study from the formation of the solar system (we could have started from the formation of the Earth as well) and we defined an absolute and arbitrary time t_0^* which corresponds to the start of the collapse of the molecular core cloud which precedes a time $t_0 = 4568.5 \times 10^6$ yrs corresponding to the oldest dated solids formed in the proto-solar nebula. This allowed us to describe the first million years by reference to t_0^* (see section 2.5) and to introduce the different stages of protostar and T-Tauri star, necessary to form the Sun, which occurred before t_0. But, as explained in chapter 2, t_0 is the age of calcium aluminium inclusions in Allende meteorite, not the age of the solar system for which we'll never know "exactly" when it started to form. We could also have chosen different relative times (t_1 = impact of Theia on Earth, t_2 = formation of proto-ocean, t_3 = end of the Late Heavy Bombardment, etc...) and described the different following events by reference to these relative times. For practical reasons we finally agreed to take by convention a unique t_0 (whatever could it be) and to introduce sometimes time elapsed since t_0. Indeed it's by far easier to remember that Moon formed between 10 and 70 million years after the formation of the first solid in the solar system than remembering that Moon formed between 4.558 and 4.498 billion years ago.

After some very enriching discussions on the determination of t_0 (which we stress again is without real importance but is somewhere the emerged part of the time-iceberg problem), we had to choose "when" to stop this reconstitution of the history of life on Earth. Here again, and depending on the field of competence of each of us, opinions were different. We decided finally to stop approximately 0.5 billion years BP and we all agreed that between 100 millions years before the accretion of the Earth and 0.5 Ga BP, when multicellular life exploded at the beginning of the Cambrian, all kind of processes involved in the evolution of our planet took place. Of course macroscopic life diversified spectacularly during the last 0.5 billion years (and the literature on this period is very abundant) and the Earth itself continued to evolve, but the most important steps for the emergence of life had already occurred.

In the following six chapters astronomers, geologists, chemists and biologists will review what is known about the chronology of some key events (formation of the solar system, accretion and differentiation of the Earth, formation of the first oceans, late heavy bombardment, plate tectonics, appearance of prokaryotic life, evolution of the Earth atmosphere, origin of eukaryotic life) and how this chronology has been established. A general chronological frieze bringing to the fore the most important events relevant (at least to our opinion) to the emergence of life is presented in chapter 8, and final conclusions (chapter highlights and questions about the ubiquity of life in the Universe) are presented in Chapter 9.

Acknowledgements

Writing a collegial paper with 25 researchers belonging to very different scientific fields is a challenge comparable to a "frog race"[2]. As we finally, all together, reach the finish, the editors would really like to thank all the authors who accepted to participate in this adventure and who accepted to stay on the "chronology wheelbarrow" whatever the difficulties could have been and the time needed to reach the goal. We would also like to warmly thank the «Château Monlot Capet», St Emilion, France http://www.belair-monlot.com/anglais/bienvenue.htm and the «Fondation Antoine d'Abbadie» of the French Academy of Sciences, Hendaye, France http://www.academie-sciences.fr/Abbadia.htm for their kind hospitality during the May and December 2004 «chronology workshops». The vivid and controversial ideas discussed during these workshops were at the origin of this review. Finally we are very grateful to Centre National de la Recherche Scientifique (CNRS), and especially to the Aquitaine and Limousin Delegation, to Centre National d'Etude Spatiale (CNES), to the GDR (Groupement de Recherche) Exobio, to the Conseil Régional d'Aquitaine, to the Université Bordeaux 1 and to the Observatoire Aquitain des Sciences de l'Univers, near Bordeaux, for their financial support in the course of this project.

References

Klein, E. and Spiro, M.: 1996, Le temps et sa flèche. Champs-Flammarion (Paris).
Mayr, E.: 2004. *What makes Biology Unique: Consideration on the Autonomy of a Scientific Discipline*, Cambridge University Press, Cambridge, USA.
Zee, A.: 1986. *Fearful Symmetry: the Search for Beauty in Modern Physics*, Princeton University Press, Princeton, USA.

[2] The game consists in starting from a point A with N frogs you have to put on a wheelbarrow, and to reach a point B as quickly as possible with all the frogs still on the wheelbarrow. The story goes that you never reach the finish of the race because you spend all your time looking for frogs which, one after one, (and sometimes all together), jump outside the wheelbarrow.

Earth, Moon, and Planets (2006) 98: 11–38
DOI 10.1007/s11038-006-9086-6

© Springer 2006

2. Dating Methods and Corresponding Chronometers in Astrobiology

MURIEL GARGAUD
Observatoire Aquitain des Sciences de l'Univers, Université Bordeaux 1, Bordeaux, France
(E-mail: gargaud@obs.u-bordeaux1.fr)

FRANCIS ALBARÈDE
Ecole Normale Supérieure, Lyon, France
(E-mail: albarede@ens-lyon.fr)

LAURENT BOITEAU
Départment de Chimie, Université Montpellier II, Montpellier, France
(E-mail: laurent.bioteau@unive-montp2.fr)

MARC CHAUSSIDON
Centre de Recherches Pétrographiques et Géochimiques (CRPG), Nancy, France
(E-mail: chocho@crpg.cnrs-nancy.fr)

EMMANUEL DOUZERY
Institut des Sciences de l'Evolution, Université Montpellier II, Montpellier, France
(E-mail: douzery@isem.univ-montp2.fr)

THIERRY MONTMERLE
Laboratoire d'Astrophysique de Grenoble, Université Joseph Fourier, Grenoble, France
(E-mail: montmerle@obs.ujf-grenoble.fr)

(Received 1 February 2006; Accepted 4 April 2006)

Abstract. This chapter concerns the tools with which time or durations are measured in the various disciplines contributing to the chronology of the solar system until the emergence of life. These disciplines and their tools are successively: astronomy (use of the Herzsprung–Russell diagram), geochemistry (radioactive dating), chemistry (no clocks!), and biology (molecular clocks, based on rates of molecular evolution over phylogenetic trees). A final section puts these tools in perspective, showing the impossibility of using a unique clock to describe the evolution of the solar system and of life until today.

Keywords: Dating methods, chronometers, Herzsprung-Russel diagram, radioactive dating, molecular clocks

2.1. Astronomy: Dating Stellar Ages with the "Herzsprung–Russell Diagram"

THIERRY MONTMERLE

Since the beginning of the 20th century, astronomers have been using the "Herzsprung–Russell Diagram" (after the name of its discoverers; "HRD" for short) to classify stars and understand their evolution.

Observationally, astronomers first determine the magnitude and spectral type of the stars. These numbers are then transformed into luminosity L_* and temperature T_{eff}, which are physical quantities that can be compared with models. This assumes (i) knowing the distance (to convert magnitudes into luminosities), which is determined by various methods (to an accuracy of ~20–30% in the case of young stars), and (ii) converting from spectral types to temperatures, which can be done with models of stellar photospheres. For "simple" stars like the Sun, the temperature determination is very precise (<1%), but for more complex spectra, like Young stars (T Tauri stars) which have a circumstellar disk, the uncertainty may reach 10–20% or more.

In the course of their evolution, stars live two fundamentally different lives. As is now well understood, stars like the Sun are in a quiet stage, lasting billions of years, in which hydrogen is slowly converted into helium: this is known as the "main sequence". The evolution continues after the main sequence in a more complex way, but the energy output is always *thermonuclear* in origin, with successive nuclear reaction networks driving important changes in the overall stellar structure (like the formidable expansion phase of solar-type stars, known as "red giants", in which the stellar radius becomes larger than the size of the solar system, ending in spectacular "planetary nebulae"). This evolution is strongly dependent on mass: the most massive stars (>10 $M_\odot$[1]) end their lives in catastrophic explosive events known as supernovae that entirely disrupt them. At the other end of the mass spectrum, low-mass stars (<0.7 $M_\odot$) are essentially eternal: their lifetimes are longer than the age of the universe!

Figure 2.1.1 summarizes two important factors that crucially depend on stellar mass (adapted from Montmerle and Prantzos, 1988): the stellar luminosity (left) and the stellar lifetime (right). One can see that stellar luminosities (*on the main sequence*) span 9 orders of magnitude (L_* from 10^{-3} $L_\odot$[2] to 10^6 $L_\odot$), for masses M_* between 0.1 and 100 $M_\odot$. This is an expression of the well-known law $L_* \propto M_*^3$, which can be demonstrated when the stellar energy is derived only from the conversion of hydrogen into helium. Correspondingly, massive stars burn more hydrogen per unit time than lower-mass stars, and above 20 $M_\odot$ live only a few million years.

Figure 2.1.2 (also adapted from Montmerle and Prantzos, 1988) summarizes the fate of stars, depending on their mass. In brief, low-mass stars, including the Sun ($M_* < 6$–7 $M_\odot$), become "red giants" and lose mass to expand as "planetary nebulae" after 10^8–10^9 yrs, leaving behind an Earth-sized, very hot (10^5 K) compact star: a "white dwarf". More massive stars evolve faster (Figure 2.1.1) and end their lives exploding as supernovae,

---

[1] $M_\odot$ = 1 Solar mass = 1.989×10^{30} kg

[2] $L_\odot$ = 1 Solar luminosity = 3.826×10^{26} w

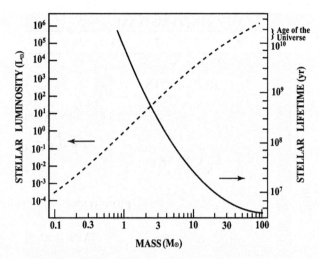

Figure 2.1.1. Stellar luminosity on the main sequence (dotted line, left-hand scale) and stellar lifetimes (continuous line, right-hand scale), as a function of mass. The most massive stars radiate an enormous power (up to 1 million suns), but have extremely brief lifetimes (a few million years), whereas low-mass stars ($M_* < 0.7\ M_\odot$) are extremely faint (less than 10^{-4} suns) and are essentially "eternal", i.e., have lifetimes longer than the age of the universe (currently accepted value: 13.7 billion years).

themselves leaving behind even more compact, city-sized stars (neutron stars, black holes). Since the most massive stars live only a few million years, they explode as supernovae in the same region as where they were born, and interact violently with their parent molecular cloud, possibly triggering new generations of stars, and "polluting" the cloud with freshly synthesized elements such as the short-lived radioactive nuclei ^{26}Al and ^{60}Fe (see Section 3.2.2.3).

Before the main sequence, however, the situation is entirely different, since the stars slowly shrink as the radiation generated by gravitational contraction is evacuated in the form of light. In other words, the energy of "pre-main sequence" stars (protostars and T Tauri stars) in *not nuclear* in origin, but is drawn only from *gravitation*. This "simplification" explains that, as early as 1966 (when only low-power computers were available!), the first theoretical model of pre-main sequence evolution could be devised by Hayashi and his collaborators in Japan (Hayashi, 1966).

This early work on "Hayashi" evolutionary tracks in the HRD (luminosity as a function of temperature) distinguished two main phases which are still used as a reference today:

(i) The "convective" phase, in which the stars are fully convective, and evolve essentially isothermally (i.e., at the same temperature, so that the theoretical tracks as a function of mass are all approximately vertical);

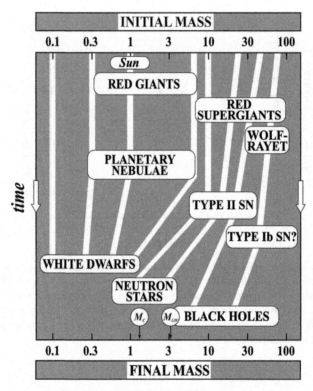

Figure 2.1.2. Stellar evolution in a nutshell: the fate of stars as a function of their initial mass (in $M_\odot$). The mass stays constant on the main sequence (vertical line as a function of time), but as a result of nuclear reactions and changes of structure, all stars start to lose mass, at a rate which increases with the mass (evolutionary line bent to the left). They all end in "compact objects", when reaching a final, critical mass in their core: Earth-sized "white dwarfs", up to ~1 $M_\odot$, the "Chandrasekhar mass" ($M_C = 1.4\ M_\odot$) for neutron stars, and the "Landau–Oppenheimer–Volkoff" mass ($M_{LOV} = 3.1\ M_\odot$) for black holes.

(ii) The "radiative" phase, in which a radiative core develops inside the star, which then becomes hotter but evolves at an almost constant luminosity (following the contraction in radius).

For a star like the Sun, the convective phase lasts about 10 million years, during which the temperature is held almost constant, around 4600 K, and the luminosity drops by a factor of ~20 with respect to the early T Tauri stage (when the young Sun starts to be optically visible). Then the radiative phase lasts from ~10 million to ~100 million years at a roughly constant luminosity (~1 L_{sun}), until the central regions become hot enough (15 MK) that thermonuclear reactions transforming hydrogen into helium start. This marks the beginning of the main sequence, on which the Sun has been for ~4.5 billion

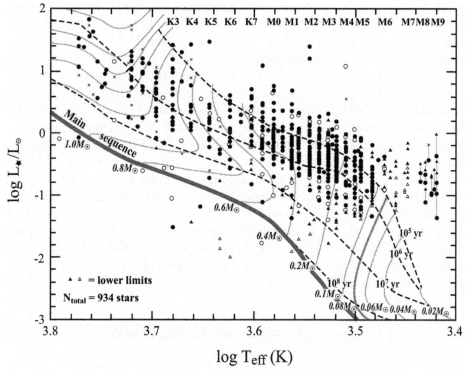

$$\log T_{eff} (K)$$

Figure 2.1.3. Herzsprung–Russell diagram of 934 low-mass members of the Orion Nebula Cluster (adapted from Hillenbrand, 1997). The spectral types corresponding to surface temperatures (T_{eff}) are also indicated at the top of the diagram. The theoretical grid is labeled in masses, from 0.02 to 5 M$_\odot$ (dotted lines; the upper left-hand dotted lines for higher masses are highly uncertain) and ages, from 10^5 to 10^8 yrs (dashed lines). The main sequence is also indicated. The thick dotted line to the right indicates the theoretical evolution of stars of mass < 0.08 M$_\odot$ (brown dwarfs), which never ignite nuclear reactions, hence never reach the main sequence.

years and will be for the next ~5 billion years, before it becomes, as mentioned above, a solar-system-sized red giant and then a planetary nebula.

The pre-main sequence phase is illustrated more generally in Figure 2.1.3 by the HRD of the Orion Nebula Cluster ("ONC"), the classic cluster associated with the Orion nebula M42, adapted from Hillenbrand (1997). The reason for this choice is far from arbitrary: we now believe (see Chapter 3) that the Sun was probably formed in a similar star cluster. The ONC itself comprises about 2000 stars, of which the best characterized 934 low-mass stars are plotted in the diagram (dots). The observational points lie in a grid of theoretical lines: (i) the dotted lines label stellar masses, from 0.02 M$_\odot$, to 5 M$_\odot$, with a thick dotted line marking the mass boundary below which stars never ignite thermonuclear reactions (the so-called "brown dwarfs"); (ii) the oblique dashed lines correspond to stellar ages, labelled by factors of 10, from

10^5 yr (top) to 10^8 yr (bottom, main sequence). As they evolve, the stars follow the "Hayashi evolutionary tracks" (dotted lines), successively convective and radiative as explained above for the Sun (1 $M_\odot$ by definition), until they reach the main sequence.

Then in such a diagram each (L_*, T_{eff}) observational point is converted, via theoretical model grids, into (mass, age) estimates (see Hillenbrand, 1997 for details and references; new theoretical models have been constructed since that paper, especially covering the high-mass and very low-mass ends: e.g., Chabrier and Baraffe, 2000; Palla and Stahler, 2001). For the ONC, one sees that the observed masses for the majority of stars run from ~0.1 $M_\odot$ to 5 $M_\odot$ and above, and the ages from less than (i.e., younger than) 10^5 yrs to a few million years. Ages smaller than 10^5 yrs are very uncertain, and all the stars shown should be understood as being "very young" only. The age of older stars ($>10^6$ yrs) is more reliable (uncertainty ~20–30%). One major conclusion from this diagram is that star formation in Orion has not been instantaneous, but is spread over a few million years and still continues today (see Section 3.2.2.3).

2.2. Geochemistry: Principles of Radioactive Dating

FRANCIS ALBARÈDE AND MARC CHAUSSIDON

A number of radioactive elements were present in the solar system when it formed. As a consequence, geochemical ages can be obtained for a given rock from the amount of daughter isotopes that have been accumulated in its different minerals via radioactive decay of the parent isotopes. It must be noted that the proportions of parent isotopes remaining in rocks have no relationship with the age of the rock, but are a direct function of the age of the considered chemical element, i.e., the average time elapsed since the nucleosynthesis of this element. The strength of isotopic dating is that radioactive decay is a nuclear process so that the rate of decay is constant and thus independent of the history of a rock.

Radioactivity is a memoryless process (atoms do not age) and is therefore a nuclear event whose probability of occurrence per unit of time, noted λ, is independent of time. This probability, termed the *decay constant*, is specific to each radioactive nuclide. Radioactive decay is a Poisson process, where the number of events is proportional to the time over which the observation is made. In the absence of any other loss or gain, the proportion of parent atoms (or radioactive nuclides) disappearing per unit of time t is constant:

$$\frac{dP}{Pdt} = -\lambda. \tag{1}$$

Occasionally, the notions of half-life or of mean life are used instead of λ. The half-life ($T_{1/2}$) is the time required for the decay of half the number of radioactive atoms originally present in the system:

$$T_{1/2} = \frac{\ln 2}{\lambda}. \tag{2}$$

The mean life (τ) is the inverse of the decay constant ($\tau = 1/\lambda$), it differs from the half-life $T_{1/2}$ by a factor ln2. The values of λ and $T_{1/2}$ of the main radiometric chronometers are given in Table 2.1.

For a number of parent atoms $P = P_0$ at time $t = 0$, Equation (1) integrates as:

$$P = P_0 e^{-\lambda t}. \tag{3}$$

It is therefore possible to determine the age of a system by measuring the number P of parent atoms that it contains today. However, this requires P_0 to be known and therefore, in this form, Equation (3) is in general not a chronometer (a notable exception is the ^{14}C method). For each parent atom, a daughter atom (or radiogenic nuclide) is created, usually of a single element, whose amount can be noted D. In a closed system and for a stable daughter nuclide D, the number of parent and daughter atoms is constant, therefore:

$$D = D_0 + P_0 - P = D_0 + P(e^{\lambda t} - 1). \tag{4}$$

The term $P(e^{\lambda t} - 1)$ is a measure of the radiogenic nuclides accumulated during time t, D_0 being the amount of isotope D at $t = 0$, therefore:

$$t = \frac{1}{\lambda} \ln\left(1 + \frac{D - D_0}{P}\right). \tag{5}$$

Even if D and P are measured, this equation is no more a timing device than Equation (3), unless the number of daughter atoms D_0 at time $t = 0$ is known.

Two different types of radioactive isotopes can be used to constrain the timescales of the formation of the solar system:

- Radioactive elements with a long half-life are useful to determine absolute ages for the different components of meteorites. The most commonly used long-lived nuclides are ^{235}U ($T_{1/2} = 0.704$ Ga), ^{238}U ($T_{1/2} = 4.47$ Ga), ^{87}Rb ($T_{1/2} = 48.81$ Ga), ^{147}Sm ($T_{1/2} = 106$ Ga) and ^{176}Lu ($T_{1/2} = 35.9$ Ga); see Table 2.1 for details.

TABLE 2.1
Decay constant λ and half life ($T_{(1/2)}$) for main radiometric chronometer

Parent nuclide	Daughter nuclide	λ = Decay constant	$T_{1/2}$ = Half life
^{7}Be	^{7}Li	4,7735 yr^{-1}	53 d
^{228}Th	^{224}Ra	3.63×10^{-1} yr^{-1}	1.91 yr
^{210}Pb	^{210}Bi	3.11×10^{-2} yr^{-1}	22.3 yr
^{32}Si	^{32}P	2.1×10^{-3} yr^{-1}	330 yr
^{226}Ra	^{222}Rn	4.33×10^{-4} yr^{-1}	1.60×10^{3} yr
^{14}C	^{14}N	1.245×10^{-4} yr^{-1}	5.59×10^{3} yr
^{231}Pa	^{227}Ac	2.11×10^{-5} yr^{-1}	3.29×10^{4} yr
^{230}Th	^{226}Ra	9.21×10^{-6} yr^{-1}	7.53×10^{4} yr
^{59}Ni	^{59}Co	9.12×10^{-6} yr^{-1}	7.60×10^{4} yr
^{41}Ca	^{41}K	6.93×10^{-6} yr^{-1}	1.00×10^{5} yr
^{81}Kr	^{81}Br	3.03×10^{-6} yr^{-1}	2.29×10^{5} yr
^{234}U	^{230}Th	2.83×10^{-6} yr^{-1}	2.45×10^{5} yr
^{36}Cl	^{36}Ar	2.30×10^{-6} yr^{-1}	3.01×10^{5} yr
^{26}Al	^{26}Mg	9.80×10^{-7} yr^{-1}	7.07×10^{5} yr
^{107}Pd	^{107}Ag	6.5×10^{-7} yr^{-1}	1.07×10^{6} yr
^{60}Fe	^{60}Ni	4.62×10^{-7} yr^{-1}	1.50×10^{6} yr
^{10}Be	^{10}B	4.59×10^{-7} yr^{-1}	1.51×10^{6} yr
^{53}Mn	^{53}Cr	1.87×10^{-7} yr^{-1}	3.71×10^{6} yr
^{182}Hf	^{182}W	7.7×10^{-8} yr^{-1}	9.00×10^{6} yr
^{129}I	^{129}Xe	4.3×10^{-8} yr^{-1}	1.61×10^{7} yr
^{92}Nb	^{92}Zr	1.93×10^{-8} yr^{-1}	3.59×10^{7} yr
^{244}Pu	$^{131-136}$Xe	8.66×10^{-9} yr^{-1}	8.00×10^{7} yr
^{235}U	^{207}Pb	9.849×10^{-10} yr^{-1}	7.04×10^{8} yr
^{146}Sm	^{142}Nd	6.73×10^{-10} yr^{-1}	1.03×10^{9} yr
^{40}K	^{40}Ar	5.50×10^{-10} yr^{-1}	1.26×10^{9} yr
^{40}K	^{40}Ca	4.96×10^{-10} yr^{-1}	1.40×10^{9} yr
^{187}Re	^{187}Os	1.64×10^{-11} yr^{-1}	4.23×10^{10} yr
^{238}U	^{206}Pb	1.551×10^{-10} yr^{-1}	4.47×10^{9} yr
^{87}Rb	^{87}Sr	1.42×10^{-11} yr^{-1}	4.88×10^{10} yr
^{40}K	^{40}Ar	5.81×10^{-11} yr^{-1}	1.19×10^{10} yr
^{232}Th	^{208}Pb	4.95×10^{-11} yr^{-1}	1.40×10^{10} yr
^{176}Lu	^{176}Hf	1.93×10^{-11} yr^{-1}	3.59×10^{10} yr
^{147}Sm	^{143}Nd	6.54×10^{-12} yr^{-1}	1.06×10^{11} yr
^{138}La	^{138}Ce	2.24×10^{-12} yr^{-1}	3.09×10^{11} yr
^{130}Te	^{130}Xe	8.66×10^{-23} yr^{-1}	8.00×10^{21} yr

- Radioactive elements with a short half-life are useful to build a relative chronology with a sharp time resolution (on the order of the half-life or shorter) for early solar system processes. These short-lived radioactive elements are also called *extinct radioactivities* or *extinct radioactive nuclides*, as per today, 4.56 Ga after the formation of the solar system, they have totally decayed and are no more present in meteorites, yet their former presence can be inferred by the presence of their daughter products. It is for this type of radioactive elements that a so-called "last minute origin" (see Chapter 3) is required to explain their presence in the early Solar system. The short-lived nuclides detected so far are ^{7}Be ($T_{1/2}$ = 53 days), ^{41}Ca ($T_{1/2}$ = 0.1 Ma), ^{36}Cl ($T_{1/2}$ = 0.301 Ma), ^{26}Al ($T_{1/2}$ = 0.707 Ma), ^{10}Be ($T_{1/2}$ = 1.51 Ma), ^{60}Fe ($T_{1/2}$ = 1.5 Ma) and ^{53}Mn ($T_{1/2}$ = 3.71 Ma); see Table 2.1 for details.

2.2.1. LONG-LIVED CHRONOMETERS

2.2.1.1. *"Rich" chronometers: $D_0 \ll D$*
The condition $D_0 \ll D$ applies for instance to the U–Pb dating of zircons, in which the amounts of initial ^{206}Pb and ^{207}Pb are negligible when compared to ^{206}Pb and ^{207}Pb produced by the radioactivity of ^{238}U and ^{235}U, respectively (radiogenic ingrowth). Equation (5) may then be written:

$$t = \frac{1}{\lambda_{238U}} \ln \left(1 + \frac{^{206}\text{Pb}_t}{^{238}\text{U}_t} \right)$$

This condition is also met for the K–Ar dating method.

These ages date the isolation of the analysed mineral, and consequently, they can be different from the age of the host rock.

2.2.1.2. *"Poor" chronometers: the isochron method*
When the condition $D_0 \ll D$ does not apply, it is replaced by the principle of isotopic homogenisation. When mineral phases, melts and fluids, separate from each other, such as during melting, vaporisation, or metamorphic alteration, it is safely assumed that these processes do not selectively separate the radiogenic from stable nuclides (other isotope fractionation processes, either natural or instrumental, are corrected using a different pair of stable isotopes from the same element). Equation (4) is transformed by dividing each member by the number D' of atoms of a stable isotope (i.e., neither radioactive nor radiogenic) of D. For a closed system, D' remains constant and therefore:

$$\left(\frac{D}{D'} \right)_t = \left(\frac{D}{D'} \right)_0 + \left(\frac{P}{D'} \right)_t (e^{\lambda t} - 1). \tag{6}$$

For the ^{87}Rb–^{87}Sr chronometer $P = {}^{87}$Rb, $D = {}^{87}$Sr, and $D' = {}^{86}$Sr, and therefore:

$$\left(\frac{^{87}\text{Sr}}{^{86}\text{Sr}}\right)_t = \left(\frac{^{87}\text{Sr}}{^{86}\text{Sr}}\right)_0 + \left(\frac{^{87}\text{Rb}}{^{86}\text{Sr}}\right)_t \left(e^{\lambda_{87\text{Rb}}t} - 1\right).$$

In this equation, D/D' represents the ratio of the radiogenic nuclide to its stable isotope (e.g., ^{87}Sr/^{86}Sr) and P/D' is the "parent/daughter" ratio, called this way as in practice it is proportional to an elemental ratio (here Rb/Sr). In a diagram (^{87}Sr/^{86}Sr) versus (^{87}Rb/^{86}Sr), several samples formed at the same time from a well-mixed reservoir (meteorites from the nebula, rocks from a magma) define a straight-line called "isochron" and the slope a of this line, which simply is $e^{\lambda_{87\text{Rb}}t} - 1$, gives the age t of the rock as:

$$t = \frac{1}{\lambda_{87\text{Rb}}} \ln a.$$

This isochron equation can be graphically solved if, in the sample to be dated, several fractions having different parent/daughter ratios (^{87}Rb/^{86}Sr) can be analysed (Figure 2.2.1).

This age dates the time at which the two samples last shared a same ^{87}Sr/^{86}Sr ratio. This method is commonly used for parent–daughter systems with a long half-life, typically ^{143}Nd–^{144}Nd, ^{176}Lu–^{176}Hf and ^{187}Re–^{187}Os.

A particular application combines the two chronometers ^{238}U–^{206}Pb and ^{235}U–^{207}Pb, in which the parent isotopes (^{238}U and ^{235}U) are not explicitly considered but only the daughter isotopes (^{206}Pb–^{207}Pb); this method is known as the Pb–Pb method.

2.2.2. SHORT-LIVED CHRONOMETERS: EXTINCT RADIOACTIVITIES

The so-called extinct radioactivities (see Section 3.2.2) have a short half-life ($T_{1/2}$) and therefore a large λ. For large values of λt, P becomes vanishingly small and therefore the closed system condition reads:

$$D_{\text{today}} = (D + P)_t \tag{7}$$

for any sample formed at any time t after the isolation or the solar system from the nucleosynthetic processes. Let us write this equation for a sample (spl) formed from the solar nebula (SN), which we suppose to be isotopically homogenous, and divide it by D':

$$\left(\frac{D}{D'}\right)^{\text{spl}}_{\text{today}} = \left(\frac{D}{D'}\right)^{\text{spl}=\text{SN}}_t + \left(\frac{P}{P'}\right)^{\text{spl}=\text{SN}}_t \left(\frac{P'}{D'}\right)^{\text{spl}}_{\text{today}} \tag{8}$$

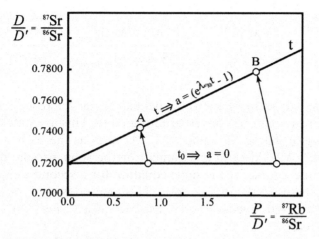

Figure 2.2.1. D/D' vs. P/D' isochron diagram showing how the slope a of the "isochron" is a time dependent parameter, whose knowledge allows to calculate time t. See other examples in Chapter 3.2: Figures 3.2.7, 3.2.8.

in which, as above, P' is an isotope of the parent nuclide P, and the closed system assumption $(P')_t = (P')_0$ holds. This equation is equivalent to:

$$\left(\frac{D}{D'}\right)^{\text{spl}}_{\text{today}} = \left(\frac{D}{D'}\right)^{\text{SN}}_{\text{today}} + \left(\frac{P}{P'}\right)^{\text{sple=SN}}_{t} \left[\left(\frac{P'}{D'}\right)^{\text{spl}}_{\text{today}} - \left(\frac{P'}{D'}\right)^{\text{SN}}_{\text{today}}\right] \tag{9}$$

in which we used a transformation of Equation (7) as:

$$\left(\frac{D}{D'}\right)^{\text{SN}}_{t} = \left(\frac{D}{D'}\right)^{\text{SN}}_{\text{today}} - \left(\frac{P}{D'}\right)^{\text{SN}}_{t} = \left(\frac{D}{D'}\right)^{\text{SN}}_{\text{today}} - \left(\frac{P}{P'}\right)^{\text{SN}}_{t} \left(\frac{P'}{D'}\right)^{\text{SN}}_{\text{today}} \tag{10}$$

with the usual closed-system constraint on both D' and P'.

In the case of the ^{26}Al–^{26}Mg chronometer, Equation (8) reads:

$$\left(\frac{^{26}\text{Mg}}{^{24}\text{Mg}}\right)^{\text{spl}}_{\text{today}} = \left(\frac{^{26}\text{Mg}}{^{24}\text{Mg}}\right)^{\text{spl=SN}}_{t} + \left(\frac{^{26}\text{Al}}{^{27}\text{Al}}\right)^{\text{spl=SN}}_{t} \left(\frac{^{27}\text{Al}}{^{24}\text{Mg}}\right)^{\text{spl}}_{\text{today}}$$

which, for a set of samples formed at time t from a homogeneous nebula, is the equation of an isochron in a ^{26}Mg/^{24}Mg vs. ^{27}Al/^{24}Mg plot. Both the slope and the intercept of the isochron (which revolves around the point representing the solar nebula) are time-dependent.

The ^{26}Al/^{27}Al ratio of the solar nebula at the time a particular sample formed is obtained by isolating the $(P/P')_t$ ratio from equation (9), here for ^{26}Al:

$$\left(\frac{^{26}\text{Al}}{^{27}\text{Al}}\right)_t^{\text{spl}=\text{SN}} = \left(\frac{^{26}\text{Al}}{^{27}\text{Al}}\right)_0^{\text{SN}} e^{-\lambda t} = \frac{\left(^{26}\text{Mg}/^{24}\text{Mg}\right)_{\text{today}}^{\text{spl}} - \left(^{26}\text{Mg}/^{24}\text{Mg}\right)_{\text{today}}^{\text{SN}}}{\left(^{27}\text{Al}/^{24}\text{Mg}\right)_{\text{today}}^{\text{spl}} - \left(^{27}\text{Al}/^{24}\text{Mg}\right)_{\text{today}}^{\text{SN}}}.$$

If the $^{26}\text{Al}/^{27}\text{Al}$ ratio of the solar nebula at the reference time $t = 0$ is assumed, the result can be converted into an age. This age dates the time at which the sample shared the same $^{26}\text{Al}/^{27}\text{Al}$ ratio as the solar nebula. If no history of the $^{26}\text{Al}/^{27}\text{Al}$ ratio is assumed for the solar nebula, dividing this equation for one sample by the same equation for a second sample gives the age difference between the two samples. This method is used for a number of "extinct" short-lived nuclides, such as ^{41}K–^{41}Ca, ^{60}Fe–^{60}Ni, ^{53}Mn–^{53}Cr, ^{146}Sm–^{142}Nd, etc.

The isochron equation can also be graphically solved if, in the same sample, several fractions having different parent/daughter ratios $(^{27}\text{Al}/^{24}\text{Mg})$ can be analysed. In this case, and if the system remained closed after its formation, the daughter isotopic ratios plot on a line (isochron) as function of the parent/daughter elemental ratios. The slope $(^{26}\text{Al}/^{27}\text{Al})_0$ and the zero-intercept give the isotopic composition of the parent and of the daughter elements, respectively, at the time the sample was formed. Since the considered parent isotope is a short-period radioactive nuclide, its isotopic composition rapidly changes with time. The $^{26}\text{Al}/^{27}\text{Al}$ ratio decreases, for instance, by a factor of 2 in 0.7 Ma. Thus, two samples 1 and 2 formed in the same original reservoir at different times will show a formation age difference $\Delta t = t_1 - t_2$, which can be written as function of the isotopic ratios of the parent nuclides according to, in the case of ^{26}Al for instance:

$$\frac{(^{26}\text{Al}/^{27}\text{Al})_{t_1}}{(^{26}\text{Al}/^{27}\text{Al})_{t_2}} = e^{-\lambda \cdot \Delta t}.$$

2.2.3. THE LIMITS OF THE METHOD

In theory, variations in short-lived radioactive nuclides isotopic compositions should allow time differences between several samples formed from the same reservoir to be measured with a good precision (<1 Myr). Nevertheless, relative chronologies should be anchored with absolute ages derived from long-lived radioactivities. This condition is difficult to meet and remains a major limitation. Ongoing efforts to calibrate the ^{26}Al chronology of calcium, aluminium-rich inclusions (CAIs) of primitive meteorites, the oldest solar system condensates against the U/Pb age are very promising (U/Pb age of 4567.2 ± 0.6 Ga; see discussion and references in Section 3.2.2.2).

Other limitations exist in using radioactive isotopes to date rocks. The system to be dated is assumed to have formed very quickly and to have subsequently remained closed to exchanges with gases, fluid phases, and adjacent minerals for both the parent and daughter isotopes. For shocked meteorites, this condition is usually not met and perturbations of the isotopic systems are often clearly visible. In addition, minerals do not cool instanta-neously, simply because of the thermal inertia of the host planetary body. Most meteorite dates therefore reflect a cooling age, i.e., the time at which the host rock in the parent planetesimal cooled down below the so-called blocking temperature. This temperature marks the point when solid state diffusion becomes too slow to allow a redistribution of the parent and daughter isotopes. Cooling rates are not an issue for CAIs and chondrules that rapidly cooled in the nebular gas. However, the chondrite parent bodies kept accreting long after the formation of the CAIs and chondrules they host, and a protracted thermal history of the parent planetesimals heated by the decay of ^{26}Al and ^{60}Fe is expected.

Dating minerals and rocks in meteorites therefore entails more than producing isotopic "ages". It requires a deep understanding of what these ages mean with respect to the processes that lead to the isolation of the chronometers, the cooling history of their carrier, and any perturbation invalidating the basic dating premises.

Additional material about the principles of isotopic rock dating can be found in Albarède, 2001, 2003; Allègre, 2005; Faure, 1986; Vidal, 1998.

2.3. Chemistry: The Impossible use of Chemical Clocks in a Prebiotic Scope

LAURENT BOITEAU

The use of chemistry as a clock ("chemical chronology") basically relies on the quantification of molecular compounds involved in known chemical reactions (either as reactants or as products). Although in most astrophysical contexts the term "chemistry" strictly refers to element/isotope quantification, we shall not deal with these latter items since they are rather relevant to nucleosynthesis and/or radioactivity (nuclear physics). This also excludes the quantification of given chemical elements as time markers in e.g. geological stratigraphy, since the dating is not provided by chemistry itself, but rather by other physical methods. The most popular example is the anomalously high abundance of iridium and other siderophilic elements in the K/T layer, mostly considered to be directly connected to the fall of an asteroid ca. 65 Myr ago.

A preliminary requisite for using chemistry as a clock is a concern of analytical chemistry. Indeed it is necessary to be able to quantify – both accurately and precisely – the targeted molecular compound(s) from its – usually solid – matrix. Considering we have to deal with "natural" samples (geological or archaeological), such a problem is far from trivial: targeted organic analytes are likely to be present at trace level, in mixture with many other compounds (either similar or different), often included in a mineralized matrix, which complicates the extraction and analysis process (for a review of the complexity of this issue see Vandenabeele et al., 2005).

However it would be misleading to consider "chemical chronology" as just a concern of analytical chemistry, although far from negligible. The most fundamental element is chemical kinetics. In theory any set of chemical reactions could be considered, provided that the following elements are known:

- The set of reactions involving the given analyte (including catalytic processes);
- The kinetic law of these reactions;
- The boundary conditions: amount of reactants and products at time $t = 0$, as well as temperature, pressure etc.

It must be mentioned in addition that conversely to radioisotope decay (which is strictly first-order) the kinetics of most chemical reactions are dependent on pressure and – especially – on temperature (a temperature increase of 10 °C often involves a doubling of the reaction rate and probably much more for many slow reactions that may be useful for dating, see Wolfenden et al., 1999). Therefore, when these parameters are not constant the knowledge of their historicity is also necessary. An implicit condition is that the system is closed (no exchange of matter with the surrounding environment), otherwise the historicity of input/output of reactants/products must also be known. With the knowledge of the above elements, the building of a kinetic model (predicting the time-dependence of involved compounds) for given boundary conditions is possible through (numerically) solving a set of differential equations. In many cases the inverse problem can also be solved, i.e. retrieving the set of kinetic equations from monitoring the involved analytes, mostly through numerical simulations and fitting with experimental data.

In most cases however – especially in a geological/archaeological context – the problem is too open, with lack of information for instance about the historicity of temperature or the boundary conditions. Moreover, in many cases a given set of (analyte) measured values can correspond to several possible sets of boundary conditions, especially when reactions other than first-order are involved. Thus practically almost only *first-order* reactions can efficiently serve as clocks, what mostly means *unimolecular* reactions, for instance degradation of macromolecules or epimerisation of asymmetric centres.

2.3.1. Some chemical clocks and their limits

2.3.1.1. *Temperature dependence*
While the variations of pressure have usually a slow influence on reaction kinetics in condensed phase (unless reaching very high values), temperature variations strongly affect chemical kinetics (according to Arrhenius' or Eyring's laws). Since the historicity of temperature is mostly unknown over geological timescales, it must be assumed to be constant, a condition very rarely fulfilled over long extents of time. This is a major drawback against the use of chemical reactions as geochronometers. Conversely however, measurement of the extent of given chemical degradations for otherwise well-dated samples can provide useful information on temperature historicity (Schroeder and Bada, 1976).

2.3.1.2. *Epimerisation of amino acids*
An example is given with the most documented reaction so far in this field, namely acid epimerisation through diagenesis of remains of dead organisms (Schoeder and Bada, 1976; Section 8.1 in Geyh and Schleicher, 1990). The validity range of such a method (on the condition of additional information on temperature historicity) has been estimated to be of the order of 10^6 yrs. Moreover, the use of this reaction as a clock entirely relies on the homo-chirality of protein residues in alive organic matter, thus being probably useless in a prebiotic context where the boundary conditions (initial enantiomeric excess) are unknown.

2.3.1.3. *Hydration of obsidian and silicate glass*
The adsorption of water at the surface of glass induces a diffusion-controlled hydration reaction, resulting into the slow growth of a hydrated layer. Due to the compactness of the glass material, the reaction front can remain very sharp over ages, being detectable through quite simple optical observations. Modelling of glass hydration kinetics allowed to make it a reliable chrono-metric method for samples aged up to 10^6 yr (Section 8.6 in Geyh and Schleicher, 1990).

Other chemically-based chronometric methods are mentioned in the literature (Sections 8.2–8.3 in Geyh and Schleicher, 1990), not suitable to dates earlier than the quaternary era: the degradation of amino acids (from proteins) in fossilised shells (ca. 2×10^6 yr); the measurement of nitrogen and/or collagen content in bones (ca. 10^5 yr). In addition, so-called "molecular clocks" actually based on the comparison of protein or DNA sequences of living organisms (in order to determine their "evolutive" age), are not really relevant to chemistry, but rather to molecular biology and will thus be discussed in part 2.4.

2.3.1.4. *Stable isotope fractionation*
To a certain extent, stable isotope fractionation used as a chronological marker can be relevant to chemistry, since in many cases the fractionation is the consequence of slight differences in stability and/or reactivity of isotopomers (compounds of same molecular structure varying only by their isotope composition). However in such a case chemical factors are almost indissociable from physical factors such as specific gravity and/or vapour pressure. An example is oxygen isotopic fractionation ($^{18}O/^{16}O$) during seawater evaporation (measured through oxygen isotopic ratio deviation in Antarctic ice mantle or in sediments), used as a marker of Earth temperature historicity in tertiary/quaternary eras (Gat, 1996; Section 7.2 in Geyh and Schleicher, 1990). Another chronostratigraphic time scale covering the complete phanerozoic era (0–650 Ma) relies on isotopic ratio deviation of sulphur ($\delta^{34}S$), carbon ($\delta^{13}C$) and/or strontium ($^{87}Sr/^{86}Sr$ ratio), and on correlations between theses values, such isotope fractionation having occurred upon biochemical or geochemical processes (Section 7.3 in Geyh and Schleicher, 1990).

2.3.1.5. *Time scales extent*
Even considering that the above-mentioned limitations could be overcome, the time scales of chemical processes are far from adequate for prebiotic chemistry. For instance the epimerisation of α-hydrogenated amino acids is estimated to be complete within at most 10^9 yrs (Schroeder and Bada, 1976). Polypeptide (protein) degradation is likely to be faster, while the survivability of fossil DNA (much more stable than RNA) is of the order of 10^7 yrs under geological conditions (Paabo and Wilson, 1991). Time scales based on stable isotope fractionation do not extend beyond the phanerozoic era. Applications would thus be mostly limited to "recent" palaeontology or archaeology, while time scales concerned by prebiotic events are several orders of magnitude older.

2.3.2. CONCLUSIONS

The chemical clocks mentioned above, are far from applicable to the prebiotic scope mostly because of too-short operative time scales, and because of the lack of information on boundary and environment conditions. In spite of good expectations expressed a few decades ago, long-period chemical clocks have been quite overwhelmed by the important progress meanwhile accomplished in the sensitivity of (either stable or radioactive) isotope analysis, see part 2.2.

2.4. Biology: The Molecular Clocks

EMMANUEL DOUZERY

With the mining of eukaryotic genome sequences, it is possible to assemble various sets of homologous genes or proteins, i.e., nucleotide or amino acid sequences that share common ancestry. From aligned homologous sequences, phylogenetic, evolutionary trees are reconstructed and provide two kinds of information. First, the topology of trees depicts the sisterhood of species. For example, Fungi are phylogenetically closer to animals than they are to plants (Philippe et al., 2004). Second, the branch lengths of trees depict the amount of evolution elapsed between two nodes (i.e., bifurcations, or speciation events), or between one node and a terminal taxon. On the Figure 2.4-1, taxon F did accumulate more molecular divergence than E since both last shared a common ancestor. To estimate phylogenies from molecular characters, probabilistic approaches are commonly used. For example, the maximum likelihood criterion identifies the topology and the interconnected branches that maximize the probability of exactly retrieving the input sequences under a given model of DNA or protein evolution. Each branch of a topology is characterized by a length that is the product of two quantities: the evolutionary rate of DNA or protein along that branch, and its time duration. Provided that there is one way to know the rates, and keeping in mind that branch lengths are estimated by probabilistic methods, then divergence times of species may be deduced. In the following section, different biological chronometers – the molecular evolutionary clocks – are briefly described to illustrate how it is possible to measure evolutionary rates of genomes in order to deduce species divergence times.

2.4.1. HISTORICAL PERSPECTIVE ON THE MOLECULAR CLOCK

Forty years ago, a linear, increasing relationship between the number of amino acid differences among proteins of vertebrate species, e.g. globins, and the age of the common ancestor of these species as measured by paleontology was evidenced (Zuckerkandl and Pauling, 1965). Molecular evolutionary clocks were born. Protein and DNA clocks were unexpected, and questioned, as researchers thought that morphological and molecular evolution proceeded in the same way, i.e., with large rate variations over time and among species. However, even if each protein seemed to exhibit a constant rate through time, different proteins had different absolute rates, e.g., exceedingly slow for histones, slow for cytochrome c, intermediate for globins, and faster for fibrinopeptides (Dickerson, 1971).

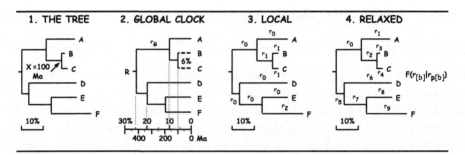

Figure 2.4-1 to −4. Phylogenetic tree reconstructed from homologous molecular sequences and depicting the evolutionary relationships among the six taxa A–F. The time line is horizontal, and branch lengths are proportional to the amount of accumulated molecular divergence. The scale is 10 nucleotide or amino acid substitutions per 100 sites compared (%). Let X be the most recent common ancestor of B and C. X is here the external fossil calibration, with a paleontological age of 100 million years (Ma). 2.- The previous tree is converted into a clocklike tree, where all taxa are equidistant from the origin of the tree (R, root). Relative rate tests evaluate the degree of distortion between the initial and clocklike trees. The maximum likelihood estimate of branch lengths XB = XC corresponds to 6% (dashed lines). The absolute substitution rate per branch is therefore $r_G = 0.06\%/Ma$. Extrapolation of this global clock allows the calculation of divergence times (see the time scale expressed in Ma). For example, RA = 25%, indicating that the age of R is $RA/r_G = 417$ Ma. 3.- The initial tree is converted into a maximum likelihood local clock tree, with three different rates: r_0, assigned to 6 branches, r_1 to three branches, and r_2 to one branch, with $r_1 < r_0 < r_2$. Knowing the local clocks $r_0 - r_2$, divergence times are calculated. 4.- Rates are allowed to vary along branches, and one distinct rate is estimated per branch. The rate of a given branch b ($r_{[b]}$) is linked to the rate of its parental branch ($r_{p[b]}$). For example, $r_{p[1]} = r_0$. In the relaxed clock approaches, the distributions of rates $r_0 - r_9$ and divergence times are estimated under $F(r_{[b]}|r_{p[b]})$, a penalty function that reduces too large rate variations between daughter and parental branches.

The neutral theory of molecular evolution provided an explanation to the quite regular ticking of the molecular chronometers (Kimura and Ohta, 1971). Mutations in genomes are either neutral (or quasi-neutral) – i.e., without effect (or nearly so) on the fitness of organisms – or under natural selection (positively or negatively selected). In natural populations, neutral substitutions accumulate at a rate that is only influenced by the mutation rate. As long as a gene contains significantly more neutral positions than selected, and as the mutation rate remains unchanged, then the evolution of DNA will be clocklike (Bromham and Penny, 2003).

The behaviour of biological chronometers, DNA and protein clocks, is a discrete, probabilistic process. In molecular evolution, a Poisson distribution is commonly used to model the time intervals between independent nucleotide or amino acid substitution events – the "ticks" of the clock. This distribution is characterized by the fact that its variance is equal to its mean, indicating that the ticks of the molecular clock are regular and random. Biological chronometers should therefore be seen as stochastic clocks rather than perfect, deterministic metronomes. Similarly, geological chronometers based on radioactivity decay follow the same stochastic behaviours (see part 2.2).

2.4.2. THE GLOBAL MOLECULAR CLOCK: A SINGLE RATE APPROACH

A constant rate of molecular evolution over the whole phylogenetic tree will be called a *global molecular clock*. Its calibration by an external date based on taxa with a rich fossil record is used to estimate divergence times for other living organisms (Figure 2.4-2). However, with the growing number of studies using the molecular clock to estimate the divergence age of organisms, different and independent problems appeared (Graur and Martin, 2004).

A first, important issue of the molecular clock approach is the fact that a global clocklike behaviour of the sequences is certainly not the rule. Several empirical studies evidenced variations in the rate of molecular evolution among taxa, both at the nucleotide and amino acid levels. These trends are not restricted to a few genes and proteins. For example, a sample of 129 proteins from an animal like a drosophila and a plant like the rice accumulated the same amount of differences, whereas trypanosomes (flagellate parasites) evolve at least twice faster than humans with respect to the amino acid replacement rate through time (Philippe et al., 2004). Statistical tests were therefore developed to measure the degree of departure of sequence data from the clock hypothesis. The most famous one is the relative rate test (Sarich and Wilson, 1973). Let AB and AC be the genetic distances from the two compared species (B and C) relative to a third, external one (A: Figure 2.4–1). If AB and AC are nearly equal, then the evolution of B and C is considered as clocklike. Now, if we consider the external species D relative to E and F, then DF > DE, and F is considered to evolve faster than E: their evolution is not clocklike. An other test is built in a maximum likelihood framework. The significance of the loss of likelihood between a clocklike set of branch lengths relative to the one of the same set of branches without the global clock assumption is compared by a likelihood ratio test (Felsenstein, 1988). However, these tests display a low resolving power, even if this may be partially corrected through the increase of the number of species and nucleotide or amino acid sites analysed (Philippe et al., 1994; Bromham et al., 2000; Robinson-Rechavi and Huchon, 2000). Though questionable when rate variation is the rule rather than the exception, the systematic elimination of erratic molecular rates across taxa and/or genes has been proposed, in order to restrict molecular dating analyses to the more clocklike data sets (Ayala et al., 1998; Kumar and Hedges, 1998). For example, the use of 39 constant rate proteins and the paleontological reference of a mammals/

birds split at 310 ± 0 Ma was extrapolated to estimate that the animals/fungi split may have occurred 1,532 ± 75 Ma ago (Wang et al., 1999).

Interestingly, multiple substitutions on the same DNA or protein position across the taxa compared can yield a saturation phenomenon, leading to observe a virtually similar amount of genetic divergence among the sequences compared. When the noisiest sites are discarded, changes in the topology of the phylogenetic tree and in the evolutionary rates may be revealed (Brinkmann and Philippe, 1999; Burleigh and Mathews, 2004): sequences initially thought to evolve clocklike actually display differences of substitution rates.

A second problem of the global clock approach is the use of a unique calibration point. As a fixed time point, it will ignore the inherent uncertainty of the fossil record. Moreover, when the calibration is chosen within a slow-evolving lineage (or conversely, in a fast-evolving lineage), the inferred rate of substitution will overestimate (or conversely, underestimate) the true divergence times. Actually, to explain the length of a given branch, e.g. 10% of DNA substitutions, a faster rate of evolution (e.g., 1%/Ma) will involve a shorter time duration (10 Ma), whereas a slower rate (0.1%/Ma) will involve a longer time (100 Ma). In numerous recent studies, vertebrates have been taken as the single calibrating fossils (e.g., Wang et al., 1999). However, vertebrates apparently display a slow rate of genomic evolution, at least for the proteins usually sampled (Philippe et al., 2005). There is therefore a concern that the deep divergence times observed between the major eukaryotic kingdoms, e.g., animals or fungi, would reflect the use of a single paleontological reference to calibrate the slow-evolving vertebrates (Douzery et al., 2004). To circumvent the above-mentioned problems, alternative molecular clock approaches have been designed which are based on the estimate of a few or several rates rather than a single one.

2.4.3. THE LOCAL MOLECULAR CLOCKS: A FEW RATE APPROACH

When a phylogeny is reconstructed under maximum likelihood, two extreme models of substitution rates among branches may be used. The first assumes one independent rate for each branch (Figure 2.4-1). The second assumes a single rate for all branches, a situation called the global molecular clock (see above). An intermediate approach has been suggested: the *local molecular clock model* assumes that some branches are characterized by a first rate–for example those connecting the most closely related species–whereas other branches display a second, distinct rate (Rambaut and Bromham, 1998; Yoder and Yang, 2000). In other words, local constancy of rates is tolerated despite potential greater variation at larger phylogenetic scales in the tree (Figure 2.4-3). Application of the local clocks to empirical data suggests that this approach is a reasonable compromise between too few and too much evolutionary rate classes per phylogenetic tree (Douzery et al., 2003). However, a subsequent difficulty is to identify the set of branches which will share the same rate, as well as the number of rate classes to assign on the whole tree.

Recent progresses in the local clock framework involve (i) the possibility to deal with multiple genes and multiple calibration points (Yang and Yoder, 2003), and (ii) the development of an improved algorithm – combining likelihood, Bayesian, and rate-smoothing procedures – to automatically assign branches to rate groups during local molecular clocks analyses (Yang, 2004).

2.4.4. THE RELAXED MOLECULAR CLOCKS: A MULTIPLE RATE APPROACH

As a solution to circumvent the confounding effect of non-clocklike behaviour of mutations, dating methods have been developed to relax the molecular clock assumption by allowing discrete or continuous variations of the rate of molecular evolution along branches of a phylogenetic tree (Sanderson, 1997, 2002; Thorne et al., 1998; Huelsenbeck et al., 2000; Kishino et al., 2001; Thorne and Kishino, 2002; Aris-Brosou and Yang, 2002, 2003). Changes in the rate from a parental branch to a daughter branch are governed by a penalty function (Welch and Bromham, 2005), which reduces too large variations (Figure 2.4-4). Penalties are either simple expressions, like the quadratic one – $(r_{[b]}-r_{p[b]})^2$ – which is time independent (Sanderson, 1997, 2002), or more complex functions which depend upon time (stationary lognormal: Kishino et al., 2001; exponential: Aris-Brosou and Yang, 2002).

Some of these approaches are implemented in a Bayesian framework, with estimation of the distributions of rates and divergence times, recapitulated as values with associated uncertainty (credibility) intervals. Moreover, the uncertainty of paleontological estimates is explicitly incorporated in the form of prior constraints on divergence times provided by fossil information (Kishino et al., 2001). Calibration time intervals for several independent nodes in the tree may as well be used simultaneously to reduce the impact of choosing a particular calibration reference. This method has been convincingly applied to understand the chronological evolution of very diverse taxonomic groups, like viruses (Korber et al., 2000), plants (Bell et al., 2005), mammals (Springer et al., 2003), and eukaryotes (Douzery et al., 2004). In the latter study, the relaxed clock on 129 proteins is calibrated by six fossil references, and suggests that the animals/fungi split may have occurred 984 ± 65 Ma ago. This estimate, markedly smaller than the 1,532 ± 75 Ma of Wang et al. (1999), is likely to be more accurate, owing to the use of a flexible molecular clock, and a greater number of proteins, species, and calibrations. It also illustrates how different data, calibrations and methods can provide contrasted timeline estimates.

2.4.5. THE FUTURE MOLECULAR CLOCKS: ACCURACY AND PRECISION

During the past four decades, molecular clocks have became an invaluable tool for reconstructing evolutionary timescales, opening new views on our understanding of the temporal origin and diversification of species (Kumar, 2005). However, to facilitate in the near future the comparison between rocks and clocks, critical paleontological references should be developed (Müller and Reisz, 2005), and molecular dating should become more accurate and more precise. Enhanced accuracy of molecular clocks might be achieved through the analysis of more genes for more species, but also through the development of realistic models of rate change, with discrete as well as continuous variations of evolutionary rate through time (Ho et al., 2005). Greater precision is however not warranted due to the inherent, stochastic nature of biological chronometers.

2.5. The Triple Clock of Life in the Solar System

THIERRY MONTMERLE AND MURIEL GARGAUD

2.5.1. IS THERE A BEGINING? THE PROBLEM OF "TIME ZERO" AND THE LOGARITHMIC CLOCK

As discussed in Chapter 1, the solar system must *a priori* have a "beginning". In other words, there must exist a "time zero" t_0^* to mark the origin of the solar system, so that by definition the Sun is formed at $t = t_0^*$, and then everything proceeds and can be dated according to a regular time scale: the formation of the Sun, disk evolution, condensation of the first solids, planet formation, existence of oceans, emergence of life, etc...But as a matter of fact, for very fundamental reasons similar to those of the universe itself, to define t_0^* as the age of the solar system is astrophysically impossible. Indeed, even if the universe is expanding, and a protostar is contracting, the structure of the equations that govern these evolutions is such that time flows *logarithmically*, not linearly. In other words, looking for t_0^* is like going backwards in time in units of fractions of an arbitrary reference time, t_r, say: $t_r/10$, $t_r/100$, $t_r/1000$, etc. As for Zenon's paradox, the ultimate origin, t_0^*, can never be reached, only approximated to a predefined level. At least as long as the equations remain valid: in the case of star formation, for instance, it is clear that, starting from a uniform medium in gravitational equilibrium (which is essentially the state of molecular clouds), stars should never form. So some external disturbance, not present in the equations, has to trigger the gravitational collapse (see section 3.1.1). Then this "initial" disturbance is forgotten very fast, and the "astronomical", logarithmic clock starts ticking: time just flows, and only relative times (time scales, or time intervals), corresponding to successive phases dominated by specific physical mechanisms, are meaningful.

We have illustrated this for the solar system in Figure 2.5.1. The first stage of the formation of the Sun is free-fall gravitational collapse of a molecular core, which, for solar-type stars, takes about 10^4 yrs. The protosun is thus essentially a rapidly evolving dense and extended envelope, but since this envelope is made of dust grains and gas with heavy elements, at some point it becomes opaque to its own radiation. Then the gravitational energy becomes trapped: a slower evolution ensues, regulated by radiative transfer (cooling) of its outer layers. The next timescale for evolution is about 10^5 yrs, when the formation of the Sun is eventually completed (Chapter 3.1). Indeed, at the centre, a dense protosolar embryo has formed, and here also the evolution is

Figure 2.5.1. A very simplified "astronomical history" of the solar system, using a logarithmic clock, from the time of the formation of the Sun as a protostar, until its end as a planetary nebula. The detailed history is the subject of Chapter 3.

regulated by the opacity, but inside a much denser and hotter body. A quasi-static gravitational equilibrium becomes established, which will last until nuclear reactions start 10^8 yrs later when the Sun reaches the main sequence. Meanwhile, and because of rotation, a circumstellar accretion disk forms. This disk has its own evolution, and the growth of dust grains under the influence of dynamical interactions has a characteristic timescale of 10^6 yrs. Then planetesimals collide and eventually build giant planets on timescales of 10^7 yrs. (Chapter 3.2). Finally, terrestrial planets form in 10^8 yrs (Chapter 3.3), which, by coincidence in the case of the solar system, is also the time-scale for the Sun to arrive on the main sequence.

This global (albeit very simplified) chronology of the so-called "pre-main sequence" phases of evolution of (solar-like) stars is also apparent in the "Herzsprung–Russell Diagram" introduced in Section 2.1 (except for the earliest stages which, being embedded and going through optically invisible phases, cannot be plotted on such a diagram). So even though t_0^* cannot be defined strictly speaking, the succession of phases, with its clock scaled (in yrs) in powers of 10, is well-defined.

The history of the solar system can be continued using the same clock. On this scale, life emerges in 10^9 yrs but must end at 10^{10} yrs because of the evolution of the Sun, which will become a red giant and engulf the whole solar system before becoming a planetary nebula (Section 2.1). These are the two only important events of the solar system at this stage! (Note that the characteristic timescale of 10^{10} yrs is also valid for stars of mass < 0.7 $M_\odot$, which live longer than the age of the universe, itself of the order of 10^{10} yrs.)

As Figure 2.5.1 illustrates, the "powers of 10" stages described above also have a certain duration, and at a given time some stages end while others have begun: for instance, circumstellar disks already exist at the protostellar stage (Section 3.2.1), and similarly the embryos of terrestrial planets (Section 3.2.4 and 3.3.1) are already present in the course of disk evolution. Strictly speaking, the "start" point of each phase is as impossible to define as t_0^* itself, since phase n already includes some evolution of phase $n + 1$. Figure 2.5.1 is only meant to give schematically a very synthetic idea of the astronomical chronology described in Chapter 3. What is certainly noteworthy, however, is that since the existence of the Earth is a prerequisite for the emergence of life as we know it, the slope (rate of growth) appears much steeper for life as it is for all the previous phases: paradoxically (in this chronology) life develops much faster that the Sun itself! And certainly the emergence of life is short (10%) compared with the lifetime of the Sun. We will come back to some important consequences of this fact in the next paragraph.

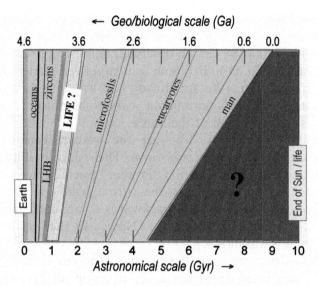

Figure 2.5.2. The astronomical, biological and geological histories of the solar system, using linear clocks. While time proceeds *forward* in astronomy (conventional unit: Gyr) since the formation of the Sun and planets (which are very short compared with the lifetime of the Sun, hence look almost instantaneous), time is measured *backwards* from 1950 AD (the "Present") in geology and biology (conventional unit: Ga). The main geological and biological events of these histories, as described in the text, are indicated in both scales.

2.5.2. ASTRONOMY VS. GEOLOGICAL AND BIOLOGICAL CHRONOLOGIES: THE LINEAR CLOCKS

Once the Earth is formed, we become interested in the chronology of events that have shaped its evolution, with the ultimate goal of understanding how life emerged, and its subsequent evolution into organized, self-replicating and increasingly complex structures. The essence of dating events comes from geology, and it is clear that the numbers obtained by all methods (radioactive decay, geological layers, etc.) are linearly ordered. In other words, to describe the corresponding chronology we have to adopt a perhaps more familiar *linear* clock. This necessary switch from the previous clock offers a paradox quite similar to that mentioned previously about life: as illustrated in Figure 2.5.2, on a linear astronomical scale, the Sun and the Earth are formed essentially instantaneously (1% of the lifetime of the Sun).

The "linear clock" can then be used in two different ways. Because when using this clock the formation of the Sun and of the Earth are almost simultaneous, the exact definition of $t_0{}^*$ is not important: it does not really matter to know whether the age of the oldest meteorites is 4.567 Gyr or even 3 Gyr – what is important is that this age is much larger than the duration of the formation of the Sun and terrestrial planets. However the astronomical scale is graduated with *increasing* values of time, which is the natural way to follow the evolution of the Sun (for instance, small, but possibly important fluctuations in its luminosity can be observed) and incorporate the *future of the Earth*, at least, again, until its doom as a vaporized planet *inside* the red giant Sun, within 5.5 Gyr. (We are already half-way into this evolution, of which the last steps will take a few million years, so here also will look on this scale as brief as the formation of the Sun.)

Correspondingly, geologists, who are the watchmakers of biologists, naturally measure the time going *backwards*. Their clock is still linear, but is graduated starting from 1950 AD: the "Present" (see Chapter 1 and Section 2.2). So while for solar system studies the astronomical time flows forward,[3] by necessity geologists have to look backwards so are concerned only with the past, not the future, of the Earth. (Note also that on all scales the emergence of man is instantaneous.)

The correspondence between the linear astronomical and "geobiological" clocks is illustrated in Figure 2.5.2, covering all known periods of the past (geology and biology) and of the future (astronomy). Also shown are the main events that occurred in the remote history of the Earth, as described in detail in the following chapters, from the formation of the first oceans, the so-called Late Heavy Bombardment (LHB), to the appearance of eukaryotes, bracketing the still controversial period where life is thought to have appeared. A summary of the main geological, chemical and biological events covering this period is presented in Chapters 8 and 9.

Acknowledgments

We would like to thank Frédéric Delsuc, Purificaciòn Lopez-Garcia, Hervé Martin, David Moreira, Hervé Philippe, Vincent Ranwez and Jacques Reisse for their contributions in this work.

References

Albarède, F.: 2001, *La géochimie*, Paris, 190 pp.
Albarède, F.: 2003. *Geochemistry: An Introduction*, Cambridge University Press, Cambridge, 262 pp.
Allègre, C. J.: 2005. *Géologie Isotopique*, Belin, Paris, 495 pp.

---

[3] Note that this is not true in cosmology, where astronomers can explore the past of the universe, since the more distant a galaxy (say) is, the younger it looks to us because of the finite time light takes to travel.

Aris-Brosou, S. and Yang, Z.: 2002, *Syst. Biol.* **51**, 703–714.
Aris-Brosou, S. and Yang, Z.: 2003, *Mol. Biol. Evol.* **20**, 1947–1954.
Ayala, F. J., Rzhetsky, A. and Ayala, F. J.: 1998, *Proc. Natl. Acad. Sci. USA* **95**, 606–611.
Bell, C. D., Soltis, D. E. and Soltis, P. S.: 2005, *Evolution* **59**, 1245–1258.
Brinkmann, H. and Philippe, H.: 1999, *Mol. Biol. Evol.* **16**, 817–825.
Bromham, L. and Penny, D.: 2003, *Nature Rev. Genet.* **4**, 216–224.
Bromham, L., Penny, D., Rambaut, A. and Hendy, M. D.: 2000, *J. Mol. Evol.* **50**, 296–301.
Burleigh, J. G. and Mathews, S.: 2004, *Am. J. Bot.* **91**, 1599–1613.
Chabrier, G. and Baraffe, I.: 2000, *Ann. Rev. Astr. Ap.* **38**, 337–377.
Dickerson, R. E.: 1971, *J. Mol. Evol.* **1**, 26–45.
Douzery, E. J. P., Delsuc, F., Stanhope, M. J. and Huchon, D.: 2003, *J. Mol. Evol.* **57**, S201–S213.
Douzery, E. J. P., Snell, E. A., Bapteste, E., Delsuc, F. and Philippe, H.: 2004, *Proc Natl. Acad. Sci. USA* **101**, 15386–15391.
Faure, G.: 1986. *Principles of Isotope Geology*, John Wiley and sons, New York, 589 pp.
Felsenstein, J.: 1988, *Ann. Rev. Genet.* **22**, 521–565.
Gat, J. R.: 1996, *Annu. Rev. Earth Planet. Sci.* **24**, 225–262.
Geyh, M. A. and Schleicher, H.: 1990, in *Absolute Age Determination*, Springer-Verlag, Berlin, 503 pp.
Graur, D. and Martin, W.: 2004, *Trends Genet.* **20**, 80–86.
Hayashi, C.: 1966, *Ann. Rev. Astron. Astrophys.* **4**, 171–192.
Hillenbrand, L. A.: 1997, *Astron. J.* **113**, 1733–1768.
Ho, S. Y. W., Phillips, M. J., Drummond, A. J. and Cooper, A.: 2005, *Mol. Biol. Evol.* **22**, 1355–1363.
Huelsenbeck, J. P., Larget, B. and Swofford, D.: 2000, *Genetics* **154**, 1879–1892.
Kimura, M. and Ohta, T.: 1971, *J. Mol. Evol.* **1**, 1–17.
Kishino, H., Thorne, J. L. and Bruno, W. J.: 2001, *Mol. Biol. Evol.* **18**, 352–361.
Korber, B., Muldoon, M., Theiler, J., Gao, F., Gupta, R., Lapedes, A., Hahn, B. H., Wolinsky, S. and Bhattacharya, T.: 2000, *Science* **288**, 1789–1796.
Kumar, S.: 2005, *Nature Rev. Genet.* **6**, 654–662.
Kumar, S. and Hedges, S. B.: 1998, *Nature* **392**, 917–920.
Montmerle, T. and Prantzos, N.: 1988, *Soleils Eclatés*, Presses du CNRS/CEA, 160 pp.
Müller, J. and Reisz, R. R.: 2005, *BioEssays* **27**, 1069–1075.
Paabo, S. and Wilson, A.: 1991, *Curr. Biol.* **1**, 45–46.
Palla, F. and Stahler, S. W.: 2001, *Astrophys. J.* **553**, 299–306.
Philippe, H., Lartillot, N. and Brinkmann, H.: 2005, *Mol. Biol. Evol.* **22**, 1246–1253.
Philippe, H., Snell, E. A., Bapteste, E., Lopez, P., Holland, P. W. H. and Casane, D.: 2004, *Mol. Biol. Evol.* **9**, 1740–1752.
Philippe, H., Sörhannus, U., Baroin, A., Perasso, R., Gasse, F. and Adoutte, A.: 1994, *J. Evol. Biol.* **7**, 247–265.
Rambaut, A. and Bromham, L.: 1998, *Mol. Biol. Evol.* **15**, 442–448.
Robinson-Rechavi, M. and Huchon, D.: 2000, *Bioinformatics* **16**, 296–297.
Sanderson, M. J.: 1997, *Mol. Biol. Evol.* **14**, 1218–1231.
Sanderson, M. J.: 2002, *Mol. Biol. Evol.* **19**, 101–109.
Sarich, V. M. and Wilson, A. C.: 1973, *Science* **179**, 1144–1147.
Schroeder, R. A. and Bada, J. L.: 1976, *Earth Sci. Rev.* **12**, 347–391.
Springer, M. S., Murphy, W. J., Eizirik, E. and O'Brien, S. J.: 2003, *Proc. Natl. Acad. Sci. USA* **100**, 1056–1061.
Thorne, J. L. and Kishino, H.: 2002, *Syst. Biol.* **51**, 689–702.
Thorne, J. L., Kishino, H. and Painter, I. S.: 1998, *Mol. Biol. Evol.* **15**, 1647–1657.

Vandenabeele-Trambouze, O., Garrelly, L. and Dobrijevic, M.: 2005, in M. Gargaud, P. Claeys and H. Martin (eds.), *Des atomes aux planètes habitables*, Presses Universitaires de Bordeaux, Bordeaux, pp. 323–356.

Vidal, P.: 1998. *Géochimie*, Dunod, Paris, 190 pp.

Wang, D. Y., Kumar, S. and Hedges, S. B.: 1999, *Proc. R. Soc. Lond. B, Biol. Sci.* **266**, 163–171.

Welch, J. J. and Bromham, L.: 2005, *Trends Ecol. Evol.* **20**, 320–327.

Wolfenden, R., Snider, M., Ridgway, C. and Miller, B.: 1999, *J. Am. Chem. Soc.* **121**, 7419–7420.

Yang, Z.: 2004, *Act. Zool. Sin.* **50**, 645–656.

Yang, Z. H. and Yoder, A. D.: 2003, *Syst. Biol.* **52**, 705–716.

Yoder, A. D. and Yang, Z.: 2000, *Mol. Biol. Evol.* **17**, 1081–1190.

Zuckerkandl, E. and Pauling, L.: 1965, in V. Bryson and H. J. Vogel (eds.), *Evolving Genes and Proteins*, Academic Press, New York, pp. 97–166.

Earth, Moon, and Planets (2006) 98: 39–95
DOI 10.1007/s11038-006-9087-5

3. Solar System Formation and Early Evolution: the First 100 Million Years

THIERRY MONTMERLE and JEAN-CHARLES AUGEREAU

Laboratoire d'Astrophysique de Grenoble, Université Joseph Fourier, Grenoble, France
(E-mails: montmerle@obs.ujf-grenoble.fr; augereau@obs.ujf-grenoble.fr)

MARC CHAUSSIDON

Centre de Recherches Pétrographiques et Géochimiques (CRPG), Nancy, France
(E-mail: chocho@crpg.cnrs-nancy.fr)

MATTHIEU GOUNELLE[1,2]

[1]Muséum National d'Histoire Naturelle, Paris, France; [2]Natural History Museum, London, UK
(E-mail: gounelle@mnhn.fr)

BERNARD MARTY

Ecole Nationale Supérieure de Géologie, Nancy, France
(E-mail: bmarty@crpg.cnrs-nancy.fr)

ALESSANDRO MORBIDELLI

Observatoire de la Côte d'Azur, Nice, France
(E-mail: morby@obs-nice.fr)

(Received 1 February 2006; Accepted 4 April 2006)

Abstract. The solar system, as we know it today, is about 4.5 billion years old. It is widely believed that it was essentially completed 100 million years after the formation of the Sun, which itself took less than 1 million years, although the exact chronology remains highly uncertain. For instance: which, of the giant planets or the terrestrial planets, formed first, and how? How did they acquire their mass? What was the early evolution of the "primitive solar nebula" (solar nebula for short)? What is its relation with the circumstellar disks that are ubiquitous around young low-mass stars today? Is it possible to define a "time zero" (t_0), the epoch of the formation of the solar system? Is the solar system exceptional or common? This astronomical chapter focuses on the early stages, which determine in large part the subsequent evolution of the proto-solar system. This evolution is logarithmic, being very fast initially, then gradually slowing down. The chapter is thus divided in three parts: (1) The first million years: the stellar era. The dominant phase is the formation of the Sun in a stellar cluster, via accretion of material from a circumstellar disk, itself fed by a progressively vanishing circumstellar envelope. (2) The first 10 million years: the disk era. The dominant phase is the evolution and progressive disappearance of circumstellar disks around evolved young stars; planets will start to form at this stage. Important constraints on the solar nebula and on planet formation are drawn from the most primitive objects in the solar system, i.e., meteorites. (3) The first 100 million years: the "telluric" era. This phase is dominated by terrestrial (rocky) planet formation and differentiation, and the appearance of oceans and atmospheres.

Keywords: Star formation: stellar clusters, circumstellar disks, circumstellar dust, jets and outflows; solar nebula: high-energy irradiation, meteorites, short-lived radionuclides, extinct radioactivities, supernovae; planet formation: planetary embryos, runaway growth, giant planets, migration, asteroid belt, formation of the Moon; early Earth: atmosphere, core differentiation, magnetic field

3.1. The First Million Years: The "Stellar Era"[1]

THIERRY MONTMERLE

3.1.1. THE SUN'S BIRTHPLACE

Stars are not born in isolation, but in clusters. This is what astronomical observations of our galaxy (the Milky Way) and other galaxies tell us. The birthplace of stars are the so-called "molecular clouds", i.e., vast, cold volumes of gas (mostly molecular hydrogen and helium, and also complex organic molecules, with so far up to 11 C atoms: see Ehrenfreund and Charnley, 2000 for a review). These clouds also contain dust grains (which include heavy elements in the form of silicates, hydrocarbons, and various ices). The masses of molecular clouds typically range from 10^6 to 10^8 $M_\odot$: in principle, molecular clouds are sufficiently massive to form millions of stars. However, somewhat paradoxically, molecular clouds do not naturally tend to form stars: gravitation, which would tend to generate the "free-fall" collapse of molecular clouds in less than 1 Myr,[2] appears to be balanced by an internal source of pressure which keeps them in gravitational equilibrium. The basic answer lies in the study, in the radio range, of the velocity distribution in the gas. It can be shown that this distribution corresponds to a state of *turbulence*, i.e., gaseous eddies that exchange energy from the large scale (the size of the cloud) to the small scales ("cloudlets" of size ~0.1 pc); smaller scales may be present but are currently beyond the spatial capabilities of existing radiotelescopes. However, on a large scale, it can be seen from mid- to far-IR observations (which have a better spatial resolution than in the mm range), that molecular clouds are in fact *filamentary*, but these filaments are constantly moving, as attested by their velocity distribution. Figure 3.1 shows a 100-micron image by the *IRAS* satellite of the Orion molecular cloud complex, where dense and cold filaments are conspicuous. According to turbulence theories (and dedicated laboratory experiments), energy is transferred from the large scales to the small scales. So the question becomes: what drives the turbulence? Or, in other words, where does the supporting energy come from? Current explanations are still debated. They focus either on an external energy source like a neighboring supernova, or on an internal feedback mechanism: as we shall see below (Section 3.1.2), young stars drive powerful outflows of matter (Reipurth and Bally, 2001), in such a way

---

[1] How astronomers determine stellar ages is described in Chapter 2 on "Chronometers" (Section 2.1).

[2] Here we adopt the usual astronomical convention: 1 "Myr" = 10^6 years. This is exactly synonymous to 1 "Ma", as used for instance by geologists elsewhere in the article (where "a" stands for "annum").

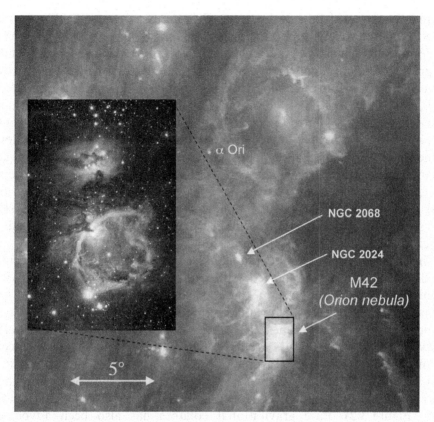

Figure 3.1. The Orion complex. *Left*: image of the Orion nebula M42 in the visible domain (© Anglo-Australian Telescope). *Background*: far-IR image (100 microns) of the Orion complex, by the IRAS satellite (1986), covering a very wide area (the angular scale is given). Note the widespread filamentary structure of the "giant molecular cloud". The bright spots are several star-forming regions belonging to the same complex, the most active one being M42 (box).

that they "inflate" the turbulence cells to keep the cloud from collapsing. Nevertheless, energy is dissipated at the smallest scales, so that some form of collapse is inevitable: the idea is that the smallest cloud structures, "cloudlets" or "prestellar cores", eventually collapse to form stars (e.g., Bate and Bonnell, 2004; Goodwin et al., 2004; Padoan et al., 2004).

Other arguments point to an important role of the *interstellar magnetic field*. In principle, a molecular cloud is by definition cold and entirely neutral, thus cannot be influenced by the presence of magnetic fields. But in practice, a minute fraction of the gas (roughly 10^{-7}) is ionized (electrically charged) by ambient cosmic rays and also by hard radiation from young stars (UV and X-rays, see Section 3.2.1.3). The charged particles are tied to the magnetic field, and thus, through collisions with them, neutral particles are in turn influenced by it: this is called *ambipolar diffusion*. In a way, ambipolar diffusion acts as a dragnet through which neutral particles flow across magnetic field lines. This effect is quantitatively important: measurements of magnetic field intensity inside molecular clouds (via the Zeeman effect on molecular lines) show that the gas pressure and the magnetic pressure are just about equal, with a difference of at most a factor of 2 in either direction, depending on the clouds (e.g., Padoan et al., 2004; Crutcher, 2005). This means that in reality we are probably dealing with *magnetically regulated turbulence*: the flow of gas in turbulent cells is not free, but is slowed down by magnetic fields and preferentially proceeds along filaments (e.g., Pety and Falgarone, 2003; Falgarone et al., 2005). In particular, this means that, at the small scales, gravitational collapse proceeds either on a short, free-fall time scale (typically $\sim 10^4$ years) if the magnetic field is on the weak side (magnetic pressure < gas pressure), or on a long, ambipolar diffusion time scale (which can reach several 10^5 years or more) if the magnetic fields is sufficiently strong (magnetic pressure > gas pressure).

This picture, at least qualitatively, leads to the idea that molecular clouds are stable, self-supported structures, but on the verge of gravitational collapse. Depending on the intensity of the magnetic field, star formation may occur at many places in the cloud, perhaps in sequence, within a relatively short timescale, of the order of a few 10^4 years locally, a few 10^5 years to 10^6 years globally. One such global theoretical mechanism is called "competitive accretion": stars form out of shocks within a pool of colliding gaseous filaments, where they compete to acquire their mass as they move through the cloud (Bate and Bonnell, 2004, Clark et al. 2005; Figure 3.2). Pure gravitational collapse has also been advocated (Krumholz et al., 2005).

Whatever the details of the various star formation mechanisms, the net result is star formation *in clusters*. Molecular observations (Motte et al., 1998), in the mass range between ~ 0.1 and $1\ M_{\odot}$, show that above $0.5\ M_{\odot}$ the core mass distribution and the observed stellar "initial mass function" (IMF) are the same, which strongly suggests (but does not prove) that the stellar mass distribution directly derives from the core mass distribution, itself linked with the turbulent structure of molecular clouds. (The IMF is the distribution of stellar masses at formation: it is observed to be a universal law, expressed as $dN_*/d\log M_* \propto M_*^{-1.5}$ for $M_* \geq 0.5\ M_{\odot}$, where N_* is the number of stars in the mass range M_*, $M_* + dM_*$; explaining it is one of the hardest challenges for star formation theories: see Kroupa, 2002 for a review.) Depending on a number of external conditions, such as the total molecular cloud mass, the passage of a shock wave of a nearby supernova explosion from the most massive stars (see below), etc., the high-mass end of the IMF is observed to be cut-off at some value $M_{\max}$. Some clusters have massive to very massive stars ($M_{\max}$ up to several tens of $M_{\odot}$), others have only intermediate-mass stars ($M_{\max} = $ a few $M_{\odot}$ at most), all the way to very small masses ("brown dwarfs", that are not massive enough to eventually

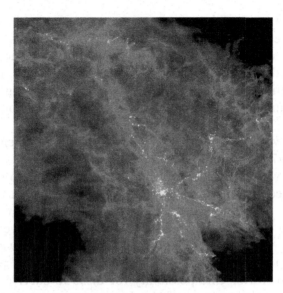

Figure 3.2. Numerical three-dimensional simulation of star formation in a 10,000 M$_\odot$ cloud, ~600,000 yrs after the initial collapse (P.C. Clark, private communication; simulations done at the UK Astrophysical Fluids Facility in Exeter). The figure is 5 pc on a side. Note the similarity of the cloud structure with that of the Orion complex shown in the previous figure. The simulation eventually leads to the formation of ~500 stars. This is less than observed in Orion (~2000 stars in M42). Complicated effects such as feedback on turbulence from stellar outflows and ionizing radiation have not been included. A better agreement is expected in the future when these effects are taken into account.

trigger nuclear reactions, $M_* < 0.08$ M$_\odot$). Because of the observed universality of the IMF, a large M_{max} implies a large number of stars, a small M_{max} implies a small number of stars. For example, star-forming regions like Orion display stars up to 20 M$_\odot$ or more, and contain altogether several thousand stars, while others like Ophiuchus, Taurus, etc., do not go beyond a few M$_\odot$ and harbour only a few tens to a few hundred stars.

Once the stars are formed, what remains of the parental cloud, not yet condensed into stars, is eventually dispersed, and the stars become optically visible. At this point, the stellar cluster becomes free from its parent cloud, and its evolution is regulated by dynamical effects in its own gravitational potential, leading after a few tens of Myr to "open clusters", then to a broad dispersal of the stars in the galaxy (at typical velocities on the order of a few km/s), much like a beehive, and thus to a loss of "memory" of how and where they were formed individually.

The Sun probably has been one of such stars. The statistics of "field stars" (like the Sun today) vs. the number of stars in star-forming regions leads to a probability argument drawn from observations: *nearly 90% of solar-like stars must have been born in clusters*, of a few tens to a few thousand stars (Adams and Myers, 2001).

It is therefore impossible at present to know *a priori* in which type of stellar cluster the Sun was born: our star is about 50 times older than the oldest open clusters. But the issue is fundamental for planet formation in general and for the origin of the solar system in particular, because of the short time evolution of the circumstellar disks around young stars (see Section 3.2.1). To simplify, there are two extreme possibilities for the birthplace of the Sun:

(i)   The Sun was born in a "rich", Orion-like environment. (see Hillen-brand 1997; Figure 3.3) The most massive stars (the Trapezium-like stars) are very hot, and thus emit UV photons able to strongly ionize their immediate environment. The disks of the less massive stars then tend also to be ionized and evaporate, as shown in Figure 3.4 According to most calculations, the disks, which have masses in the range $10^{-2} - 10^{-4}$ M_{Sun} (i.e., 0.1–10 Jupiter masses), will disappear in a few million years only, likely before any terrestrial planet has had the time to form. One possibility for disks to survive is not to stay too long in the vicinity of the hottest stars, so as to escape, via dynamical effects, the original "beehive". We observe that most disks around young stars are typically 10 times larger that the size of the present-day solar sytem; some disks may be cut-off to the size of the Kuiper Belt (i.e., the radius of the solar sytem, ~50 AU),[3] mainly because of evaporation processes (Adams et al. 2004, 2006). On the other hand, the discovery of the existence of Sedna, a relatively massive, high-eccentricity solar system "planet", implies that the protosolar disk did not suffer stellar encounters closer than 1,000 AU (Morbidelli and Levison, 2004), so retained its initial large size for some time, or else managed to be not entirely vaporized. This would be possible in the outskirts of the nebula, or in a less rich cluster like NGC1333 (Adams et al., 2006). In any case, disk survival in a Orion-like environment is certainly not easy.[4]

(ii) The Sun was born in a "poor", Ophiuchus-like environment (Figure 3.5). With no massive star around, stars form deeply embedded inside the molecular cloud, and once formed they stay protected from external disturbances. All disks survive, and can remain very large. We know of some examples of stars with large disks in the immediate vicinity of molecular clouds (Figure 3.6). It is not clear how in this case

---

[3] 1 UA = 1 Astronomical Unit = Sun-Earth distance = 150 million km.

[4] To be complete, one should mention the recent work by Throop and Bally (2005), who argue that dust grains actually grow into planetesimals under the coagulating effect of UV radiation, hence that planet formation (see below, Section 3.2.1), is favored by evaporation. But even in this case, the exposure to UV radiation must be fine-tuned for the whole system to survive.

Figure 3.3. Near-IR (2 microns) image of the center of M42, revealing the stars of the rich Orion nebula cluster, and in particular the four central hot stars called the "Trapezium", which excite the nebula. The nebula is 450 pc away. The image is about 10′ on a side, which is 250 000 AU. (© ESO, VLT-ISAAC, by M. McCaughrean.).

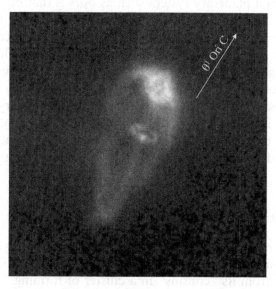

Figure 3.4. A "tear" in Orion. This is an evaporating circumstellar disk, 500 AU in diameter. The central star is clearly visible. The bright spot is oriented towards θ^1 Ori C, the hottest star of the Trapezium. The evaporating gas is shaped by the wind from this star (© NASA Hubble Space Telescope: J. Bally, H. Throop, & C.R. O'Dell).

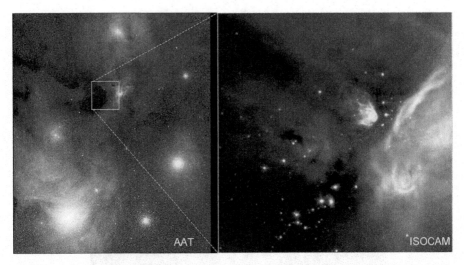

Figure 3.5. Left: The "poor" ρ Oph cluster in the optical range, hidden in a dark cloud of gas and dust. It is located 150 pc from the Sun. (The bright stars at the bottom left is the foreground star Antares, and the fuzzy spot to its right is the distant globular cluster M15. (© Anglo-Australian Telescope.) *Right*: Mid-IR image of the cloud core, revealing embedded young stars and protostars invisible in the optical range (ISOCAM, Abergel et al. 1996).

planet-forming disks would shrink by a factor of 10 to be confined within the size of the solar system. In the end, perhaps the issue of the existence of a planetary system such as ours is that of the "survival of the fittest": to suffer disk truncation in a dense, Orion-like cluster, or at least to avoid complete evaporation by way of dynamical effects, i.e. to be ejected from the original cluster quickly enough. The "quiet" environment of the Ophiuchus-like, looser clusters, would provide a "safe" evolution, but, likely, at the cost of carrying along a large, massive disk which would lead to planetary systems very different from our own. In this scenario, the solar system would be born in rare, though not uncommon (10% of the stars), environment, with a key role played by infrequent dynamical interactions, cutting off its original disk, very early (10^5 years?) after its birth.

3.1.2. THE SUN AS A FORMING STAR

Let us now zoom on the Sun as a forming star, which we shall assume isolated for simplicity (knowing from above that it *has* to be isolated, at some very early stage, from its "cousins" in a cluster of forming stars).

The knowledge of what we believe must have been the first stages of formation and evolution of the Sun, is drawn from a wealth of observations of a multitude of star-forming regions that astronomers have been able to

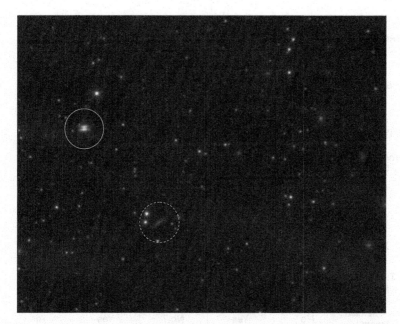

Figure 3.6. A lonely, massive egde-on disk in the ourskirts of the ρ Oph cloud (circle). The other disk-like object (dotted circle) is a distant galaxy (© ESO, VLT-ISAAC, Grosso et al., 2003).

obtain. Nowadays, telescopes are being used both on the ground and in space, covering almost all the electromagnetic spectrum, from mm wavelengths, to X-rays, across the IR and optical domains. Depending on the wavelength, it is possible to pierce the darkness of molecular clouds, and "see" inside them to watch the hidden birth of solar-like stars, most importantly in the IR to mm domains, and in the X-ray and gamma-ray ranges (e.g., Ryter, 1996).

In almost every case, one is able to distinguish three main components, which simultaneously evolve as star formation proceeds[5] (see, e.g., Shu et al., 1987; André and Montmerle, 1994; André et al., 2000, a summary and recent references are given in Feigelson and Montmerle, 1999, and Montmerle, 2005: Figure 3.7). At the so-called "protostellar stage", a vast, dense envelope (1,000–10,000 AU in radius) is detectable, and from the center emerges a "bipolar outflow". The envelope is so dense that its interior is invisible even at mm wavelengths; only its outer structure can be seen. It is now understood that the "seed" of a new star is formed from matter accreted from the envelope which "rains" on it under the pull of gravitation. The youngest observed protostars have an estimated age of $\sim 10^4$ years: this estimate is

---

[5] For a pioneering work, establishing (analytically!) the basic principles of early stellar evolution, see Hayashi (1966).

PROPERTIES	Infalling Protostar	Evolved Protostar	Classical T Tauri Star	Weak-lined T Tauri Star	Main Sequence Star
SKETCH	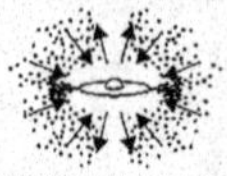	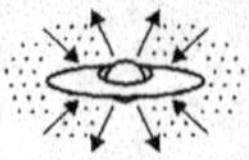	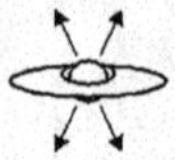	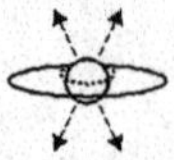	
AGE (YEARS)	10^4	10^5	$10^6 - 10^7$	$10^6 - 10^7$	$> 10^7$
mm/INFRARED CLASS	Class 0	Class I	Class II	Class III	(Class III)
DISK	Yes	Thick	Thick	Thin or Non-existent	Possible Planetary System
X-RAY	?	Yes	Strong	Strong	Weak
THERMAL RADIO	Yes	Yes	Yes	No	No
NON-THERMAL RADIO	No	Yes	No ?	Yes	Yes

Figure 3.7. A summary table of the various protostellar and stellar phases, with characteristic timescales and basic observational properties. (From Feigelson and Montmerle, 1999).

rather uncertain, but is consistent with the number deduced from the dynamical age of outflows (= size/velocity), and from their small number relative to their more evolved counterparts, like T Tauri stars (see below): indeed, one finds roughly 1 protostar for every 100 T Tauri stars, aged 1–10 Myr.

At an age $\sim 10^5$ years, the envelope is much less dense, since most of it has collapsed onto the disk. It becomes transparent at mm wavelengths, revealing a dense disk (500–1,000 AU in radius), from which the seed star continues to grow. The source of outflows become visible, in the form of highly collimated jets originating close to the central star, confirming earlier models in which molecular ourflows consist of cold cloud material entrained by the jet. (In fact, this jet-cloud interaction is believed by some authors to be the main agent to sustain the turbulent state of the cloud: this is the "feedback" mechanism mentioned above, see Matzner and McKee, 2000.) At this stage is revealed *the key three-component structure that governs the physics of star formation*: an outer envelope, an inner "accretion" disk, and matter ejected perpendicular to the disk (Figure 3.8). (The accretion disk probably exists from the start of collapse, but it cannot be detected because of the opacity of the envelope at the earliest stage.)

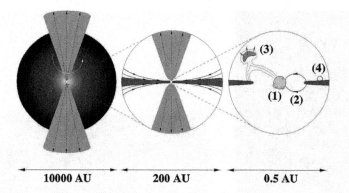

Figure 3.8. Sketch of the structure of a protostar, zooming on the star-dik interaction region, which is dominated by magnetic fields. This region is the seat of the "accretion–ejection" mechanism, by which the majority of the disk mass becomes accreted to form a star at the center, while the remainder is ejected. (From Feigelson and Montmerle, 1999).

As time passes, this three-component structure considerably evolves. The envelope is eventually exhausted after $\sim 10^6$ years, leaving a fully developed "real", luminous star, a massive circumstellar disk ($M_{disk} \sim 1\text{--}10\ M_{Jup}$), and a weak, optically visible bipolar jet. This is the start of the so-called "classical" T Tauri stage (after the name of the first-discovered star of this type, which, as it turns out, is quite atypical for its class; e.g., Bertout, 1989). Then, after a period which can be as long as 10^7 years, the disks of T Tauri stars (and jets) "disappear", or at least becomes undetectable, presumably via planet formation (see Section 3.2.1): this is the "weak" T Tauri stage. Why "weak"? Because the "classical" T Tauri stars have a very unusual optical spectrum, with very strong emission lines, in particular the Hα line of ionized hydrogen; in contrast, the "weak" T Tauri stars have "weak" emission lines, comparable to solar lines. A "weak" Hα line can be explained, in analogy with the Sun, by the presence of "active regions", that is, starspots scattered over the stellar surface, indicative, again as on the Sun, of locally strong magnetic fields. The strong Hα line of "classical" T Tauri stars cannot be explained by magnetic activity alone. Various arguments assign the strength of the Hα line in this case to matter falling onto the star, fed by the disk: this is the phenomenon known as "accretion" of matter (e.g., Bertout, 1989) – the very process by which the star grows to reach its final mass when the disk is exhausted, perhaps leaving young planetary bodies behind.

Let us now concentrate on young "classical" T Tauri stars, like HH30 in Taurus (Figure 3.9). All the observational evidence points to a causal relationship between the existence of jets and the presence of disks: it is clear, at least qualitatively, that the jet material is somehow coming from the disk, at the same time that accretion feeds the central star. This phenomenon, known as the *"accretion–ejection" phenomenon*, is absolutely central to our

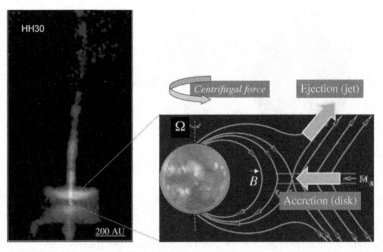

Figure 3.9. Left: HST image of the T Tauri star HH30, showing its edge-on disk and jet. The young Sun could have been such an object. *Right*: sketch of the theoretical magnetic structure used to model the accretion–ejection mechanism (the background drawing is taken from Ferreira et al., 2000). The "star" is an image of the magnetically active Sun, seen in X-rays by the Yohkoh satellite (see Section 3.2.1.2).

understanding of star formation and early evolution, and to a modern view of the solar nebula. Such a phenomenon is somewhat paradoxical: it implies that, for a star to form, it must lose mass! At least, a significant fraction of the mass ($> 10\%$ from observations, e.g., Muzerolle et al., 2001) must eventually be ejected. The point is that mass accumulation is not the only necessity to form a star: since (at we see at all scales in the universe, from clusters of galaxies to planets) rotation is always present, and seen in the form of circumstellar disks around forming stars, there must exist centrifugal forces that oppose gravity. For gravity to "win" and lead to star formation, *angular momentum* must be lost. Although the reasons for angular momentum loss in a forming star, and in particular the role of the circumstellar disk, are not entirely clear because of complex transport processes within them (see next subsection), it is well known that mass loss in the form of stellar winds (the solar wind is one example) is very efficient to spin down a star – provided the ejected matter remains coupled to the star. The only way to do it is to link the star and the wind by a *magnetic field*. Along this line of thought, in current models, accretion (mass gain), and ejection (mass and angular momentum loss), must be mediated by magnetic fields.

3.1.3. A STELLAR VIEW OF THE "PRIMITIVE SOLAR NEBULA"

To be more specific, let us now zoom again, this time on the region very close to the forming star, for instance a young T Tauri star, to which the primitive solar nebula must have been comparable in its first million years (Figure 3.9, left). A rich interplay between observations and theory, over the last decade, results in the following picture (e.g., Matsumoto et al., 2000; Shang et al., 2002; Ferreira and Casse, 2004, and refs therein; see Figure 3.9, right). Both the star and the disk are magnetized: (i) the star is surrounded by a "dipolar" magnetosphere that surrounds it like a tire, with a "closed", loop-like topology of the magnetic field; (ii) in contrast, the magnetic field lines connected to the disk are "open", above and below the disk. Then a special distance, called the "corotation radius" R_c, is naturally defined: this is the distance at which the "Keplerian" (i.e., orbital) velocity of a disk particle rotates at exactly the same speed as the star. (One could say, in analogy with the Earth's artificial satellites, that this is the "astrostationary" orbit.) At distances $r < R_c$ from the star, the intensity of the magnetic field is stronger than at R_c, and the magnetosphere rotates in a "rigid" fashion, the field lines being anchored on the stellar surface. At distances $r > R_c$, the magnetic field decreases rapidly and takes an open, spiral form as it becomes tied to the disk. The point at R_c thus has a very particular magnetic property: it is the border at which the magnetic field topology switches from closed (stellar component) to open (disk component). As such, it is also known as the "X-point" (or more exactly the "X-ring" in three dimensions in view of the assumed axial symmetry) because of its X-shaped magnetic configuration (Shu et al. 1997).

The existence of the X-point (in a 2-dimensional cut) holds the key to the majority of "accretion–ejection" theories. There are many discussions among theorists about its exact status. For instance, it is not clear why the magnetically defined X-point (at a distance R_x which depends only on the magnetic field intensity) should be exactly at the same location as the gravitationally defined corotation radius R_c (which depends only on the stellar mass and rotation velocity), in other words why should $R_x = R_c$. It is not clear either that the X-point should be that: a point (or more precisely an "X-ring", in three dimensions), since this would mean that at R_x there must be an infinite concentration of magnetic fields lines, which is physically impossible if matter is coupled to it (via some ionization, for instance due to X-rays), etc. But most theorists (and observers) agree, at least qualitatively, on the following general "accretion–ejection" picture, which will be sufficient for the purpose of this paper.

Because of its special gravitational and magnetic properties, seen in two dimensions, the X-point de facto behaves like a Lagrange point (Figure 3.9, right): (i) If a particle, initially located at this point, is pushed towards the interior ($r < R_x$), it will start falling freely on the star under the pull of gravity, along the corresponding "rigid" magnetic field line: this is called "magnetospheric accretion". (ii) Conversely, if a particle is pushed outwards, it will start following an open field line, and the centrifugal force will push it even further: this is how "centrifugal jets" are formed. Thus, the X-point is intrinsically unstable, analogous to the gravitational Lagrange point L1 between two celestial bodies (which here would be a three-dimensional magnetic "Lagrange ring" in a star-disk sytem), and flowing through it matter can either fall onto the star or be ejected. In practice, of course, matter does both: depending on the exact model, calculations predict that 1/10 to 1/3 of the disk matter should be ejected, the remainder falling on the star and thus feeding its growth (e.g., Johns-Krull and Gafford, 2002). What is important for our purpose is the location of the X-point: for a solar-mass star rotating relatively slowly as T Tauri stars are observed to rotate (period of a few days), R_x is on the order of several stellar radii. Since a typical T Tauri star has a radius $R_* \sim 3\ R_\odot$, this means $R_x \sim 0.1$ AU. In the well-studied T Tauri star AA Tau, for which the accretion disk is seen edge-on, it has been possible to reconstruct the magnetic structure of the inner hole, from a study of its active regions eclipsed by the disk (Bouvier et al., 1999, 2003). Although in this case the disk is found to be warped, which implies that the magnetic axis is tilted to the rotation axis, the size of the resulting magnetosphere is in good agreement with the theoretical picture (Figure 3.10), in particular

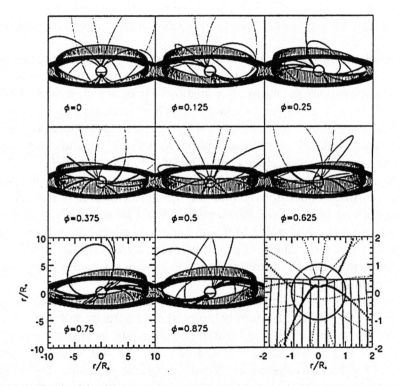

Figure 3.10. Sketch of the tilted magnetic structure inside the disk of the T Tauri star AA Tau. This sketch results from a study of the magnetic activity of the star, eclipsed by its warped accretion disk seen nearly edge-on (From Bouvier et al., 1999).

with the presence of an inner hole ~0.1 AU in radius. As a matter of fact, the magnetic field in the inner regions of several young stars has recently been directly detected, via the Zeeman effect (Donati et al., 2005). Compared to the present-day solar system, such a hole is well within the orbit of Mercury (0.4 AU). This is also the size of a pinhole compared to the disk sizes, which, as mentioned above, can be as large as 1,000 AU at an early stage.

Yet this "pinhole" region, which harbors the magnetic "central engine" for the accretion–ejection mechanism, may have played an important role for the early solar system. Indeed, because it is a region of tangled, unstable magnetic fields, and located close to the star, itself the seat of its own intense magnetic activity (as testified by the observed flaring X-ray emission, see Section 3.2.1.2), it is a place where any circumstellar material (gas, grain, possible small planetary bodies, etc.) will suffer a high dose of radiation, either in the form of hard photons (XUV activity), or energetic particles (accelerated in stellar flares like on the Sun, and which, while not directly observable, can be induced from X-rays). We shall return below (Section 3.2.1.2) to this important question, which connects the early evolution of circumstellar disks with the history of the young solar system.

To summarize, at an age of ~10^6 years, the young solar system – or the solar nebula – can be depicted as follows. At its center, a magnetized solar-like star has reached most of its final mass, ~1 $M_\odot$ This star is surrounded by a rigid magnetic "cavity" (its magnetosphere), about 0.1 AU in radius. At the "X-point", corotating with the star, lies the inner edge of the circumstellar disk, from which most (0.9–2/3) of the matter continues to fall on the star. The remainder is ejected on both sides perpendicular to the disk, in the form of a powerful, highly collimated jet, which evacuates most, if not all, its angular momentum. Beyond the X-point lies a dense circumstellar disk, either "small" (~50 AU, say) if it has been truncated in the vicinity of hot stars (as in a bright, UV-rich Orion-like environment), or "large", a few hundred AU in radius if it was born in a dark, quiet Ophiuchus-like environment. Within this disk, matter continues to flow from the outside to the inside, braked by viscosity, until the disk is exhausted – or because matter starts to assemble to form large grains and, ultimately, planetesimals and giant planets, as discussed in the next section below.

In this "dynamic" picture of the solar nebula, any "heavy" particle like a grain can have two fates when it arrives in the vicinity of the X-point: either it falls onto the star by accretion, or it is entrained outwards by "light" particles (the gas), but eventually falls far back onto the disk in a ballistic fashion because it is too heavy to be carried away along the open magnetic field lines by the centrifugal force. In this last case, it will have spent some time very close to the X-ring, and will have suffered there heavy irradiation by hard photons and energetic particles. These particles will then mix to the disk, holding specific "scars" from their passage near the X-ring. As explained in Section 3.2.2, this is how some models explain the mysterious presence of "extinct radioactivities" in meteorites.

3.2. The First 10 Million Years: the "Disk Era"

3.2.1. THE EVOLUTION OF CIRCUMSTELLAR DISKS AROUND YOUNG STARS AND IMPLICATIONS FOR THE EARLY SOLAR SYSTEM

3.2.1.1. *The path to planets: Astronomical timescale for the growth of dust grains*
JEAN-CHARLES AUGEREAU

Now that a dense circumstellar disk is installed around the central star, it must evolve: on the one hand, it continues (albeit at a lesser rate) to lose mass at its inner edge (by way of magnetic accretion), on the other hand, grains assemble via low relative velocity collisions to form larger, preplanetary bodies. But how long does it last? Infrared observations of T Tauri stars, which are sensitive to the presence of circumstellar material, show that disks disappear on widely different timescales. Figure 3.11 (Hillenbrand, 2006), which collects data from ~30 star-forming regions, shows that so-called "inner disks" (i.e., regions warm enough to radiate are near-IR wavelengths) are ubiquitous at young ages, and tend statistically to disappear after a few million years only. Quantitatively, the fraction is consistent with 100% at an age 1 Myr, and drops to less than 10% after 10 Myr, with some clusters containing no disk at all. Actually, this low fraction of "old" disks is a lower limit: mid-IR observations, which are sensitive to cooler disks, hence more distant from the central star, show that in some cases, like the nearby η Cha cluster, aged 9 Myr, the fraction of disks is closer to 40–60% (resp. Megeath et al., 2005; Lyo et al. 2003), suggesting that disks may live longer than previously thought. However, at this stage the disk mass is found to be too low to form even a Jupiter (10^{-3} M$_\odot$)—or perhaps they have already done so—so that the general conclusion is that *giant planets, if any, must have formed on timescales significantly shorter than 10 Myr.*

Therefore, the disk era is a critical period for planet formation, tightly constrained by astronomical observations. Submicron-sized dust particles composing young and massive disks constitute the raw material from which planets form. Tiny dust grains must coagulate to form large dust aggregates, pebbles, and then larger rocky bodies (planetesimals) before the dust disk becomes too tenuous. The formation of giant planets through the core-accretion scenario also requires the formation of planetary cores

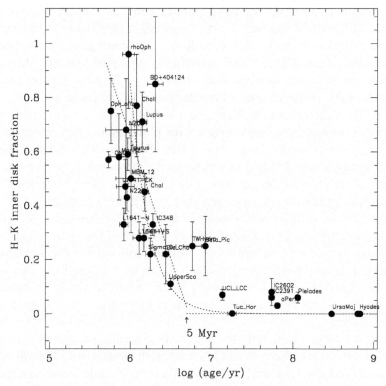

Figure 3.11. Disk fraction in young stellar clusters, as a function of their age. This fraction is consistent with 100% at young ages (less than 1 Myr), then declines over a timescale of a few Myr. After 10 Myr, with a few exceptions, all the disks around young stars have disappeared, presumably because of planet formation (Hillenbrand, 2006). This puts strong constraints on the formation of giant planets (which cannot be seen themselves).

before the disk has been mostly depleted of gas. A detailed investigation of how dust grains grow into large planetary embryos is presently one of the most important open questions, especially for the formation of the solar system: this is discussed in detail in Section 3.2.4. Here we give a broad outline of the results drawn from the study of circumstellar disks around young stars.

A key conclusion is that the planet formation process is observationally required to be both fast and common. The disappearance of circumstellar disks in less than 5–10 Myr is actually interpreted as a direct consequence of the formation of larger solid bodies decreasing the opacity and the dust emission (Haisch et al., 2001; Carpenter et al., 2005; Hillenbrand, 2006). Unless the circumsolar disk survived dissipation processes longer than usually observed for circumstellar disks, large solid bodies in the solar system should have then formed within less than a few million years, which is a major challenge for terrestrial planet formation theories.

In addition, the growth of solid particles in disks must be sufficiently generic for at least two reasons. First, a significant fraction of solar-like stars have been found during the last decade to host giant gaseous exoplanets[6]. The formation of our solar system may then not be peculiar, and the general conclusions derived from statistical analysis of star-forming regions may also apply to the early solar system. Nevertheless, it is not known whether these giant exoplanets formed through the accretion of gas onto a solid core or not. But there exists an independent evidence for planetesimal formation to occur routinely in disks around young stars. A large fraction of main sequence stars in the solar neighborhood are observed to be surrounded by tenuous disks composed of short-lived dust grains. The survival of these so-called "debris disks", which contain much less mass that young disks (typically 10^{-6} M$_\odot$), over hundreds of million years points indirectly towards the presence of reservoirs of meter-sized (or larger) bodies, similar to the Kuiper Belt in our solar system, which release grains by mutual collisions. (We see this phenomenon continuing in the present-day solar system in the form of zodiacal light.) The observation of debris disks is up to now one of the most convincing observational clues indicating that planetesimal formation in young disks is common.

Various processes contribute to the coagulation and growth of dust grains in disks of sufficiently high density. Current theoretical models of circumstellar disks (Section 3.2.4 and references therein) include Brownian motion, vertical settling, radial drift and turbulence. Interstellar like grains (smaller than about 0.1 micron) stay well mixed with the gas where the gas density is high enough. In that case, the low-velocity collisions between grains due to Brownian motion result in aggregates with fluffy structures. This process is particularly efficient in the disk mid-plane while collisions at the disk surface could be destructive. As aggregates grow and reach millimeter sizes, they settle to the disk mid-plane on short time-scales depending somewhat on the strength of the turbulence.

Can we test this model? The available tools to witness grain growth in disks remain limited. Observations can only probe the continuum, quasi-blackbody emission of grains with sizes of the order of the observing wavelength. Therefore, astronomers are basically limited to the direct detection of solid particles smaller than, at most, a few centimeters. Fortunately, disks become optically thin at millimeter wavelengths, which allows astronomers to probe the denser regions of disks where sand-like grains are expected to reside. At these wavelengths, the total disk flux directly relies on the dust opacity that is a function of the grain size, and it can be shown that grains several orders of magnitude larger than those found in the interstellar

---

[6] Visit *The Extrasolar Planets Encyclopaedia*: http://www.exoplanets.eu

medium are required to explain the (sub-)millimeter observations (Draine, 2006). At shorter wavelengths, the disk mid-plane is opaque and only the upper layers, mostly consisting of small grains according to the models, are accessible. But as the opacity of disks inversely depends on the wavelength, observations should probe deeper and deeper regions as the wavelengths increases (Duchêne et al., 2004). In practice, the interpretation of scattered light observations of disks at visible and near-infrared wavelengths is not straightforward as it strongly depends on the light scattering properties of the individual grains that are, unfortunately, hard to model properly. This type of study is also limited to a handful of objects, such as the circumbinary disk of GG Tau (Figure 3.12)

More statistically meaningful conclusions can be derived from the study of *dust mineralogy* through infrared spectroscopy. Especially, silicates that are composed of silicon, oxygen and, in most cases, a cation (often iron or magnesium), have characteristic solid-state vibrational bands that occur uniquely at infrared wavelengths. The emission spectra of silicates from optically thin disk surfaces display characteristic emission features around 10 microns due to the Si–O stretching mode, and around 20 microns due to the O–Si–O bending modes. Silicates are observed in the interstellar medium and in many solar system objects, including the Earth mantle where it appears

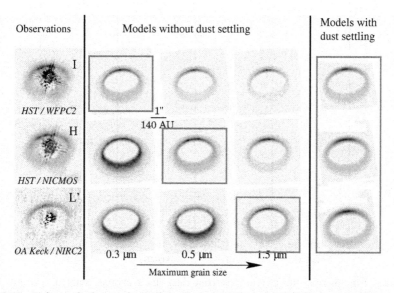

Figure 3.12. Left panel: observations of the dust ring about the GG Tau young binary system. In order to properly interpret the almost wavelength-independent appearance of the ring, dust settling toward the disk midplane must be taking into account in the models (*middle and right panels*). Such images may be unveiling the vertical stratified structure of the disk (Duchêne et al., 2004). The cold gas component of the GG Tau "ring world" has also been observed in the mm domain (Guilloteau et al., 1999).

mostly in the form of olivine (after its olive-green color). Their presence in disks around young solar-like stars is thus expected, by analogy with the solar system. But it is only recently that large fractions of T Tauri stars could be spectrally studied thanks to highly sensitive infrared space telescopes such as *Spitzer* (Figure 3.13). Silicates turn out to be ubiquitous in T Tauri disks in Myr-old forming regions. But, interestingly, the strength and shape of their emission features differ in many cases from those observed in the interstellar medium, indicating significant processing of silicates in young disks. As the silicate features strongly depend on grain size, the presence in disk atmospheres of dust particles several orders of magnitude larger than interstellar grains provides a natural explanation to the observations. Silicate emission features can thus be used as extremely valuable diagnostics of micron-sized solid particles in disks, as demonstrated for instance by Kessler-Silacci et al. (2006). Moreover, as the stars are fairly young (less than a few million years), this indicates fast grain growth in disks.

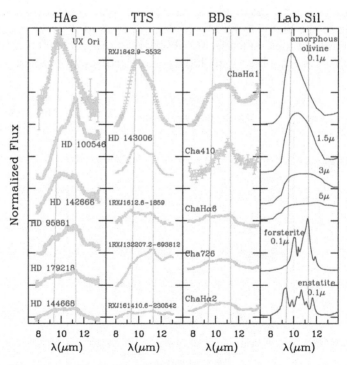

Figure 3.13. Ten micron silicate features from dust disks around stars of various masses (Natta et al. 2006). *From left to right*: Herbig Ae stars (HAe), a few times more massive than the Sun), T Tauri stars (TTS), and Brown Dwarfs (BDs). Examples of theoretical silicate emission features, based on their experimental optical properties, are dispayed in the extreme right panel (Lab.Sil.). By comparing the shape and strength of these profiles to the observations, one can infer the presence of large grains as well as the degree of crystallinity of the silicates.

The detection in circumstellar disks of crystalline silicates similar to those observed in solar-system cometary dust is an additional diagnostic of similar significant grain processing within the first 10 Myr. The silicates that are injected into the primordial stellar nebulae are in the form of amorphous silicates, as proven by the recurrent non-detections of crystalline silicates in the interstellar medium. The annealing (crystallization) of amorphous silicates can only happen in the inner regions of the disks where the temperature is sufficiently high. Also their incorporation in long-orbital period solar system objects points towards radial (and vertical) mixing of dust during the infancy of the solar system. Various mechanisms throughout the disk such as stellar/disk winds or radial and vertical mixing in (magneto-)turbulent disks can, in principle, transport dust grains with important implications on the disk chemistry and mineralogy. The observation of disks around young solar-like stars supports the radial mixing scenario since crystalline silicates are found at distances where the temperature is too low for *in situ* annealing. In contrast, it has been suggested that supernova shock waves in the outer solar nebula could have rapidly annealed the amorphous silicates *in situ* prior to their incorporation into comets, thereby eliminating the need for large-scale nebular transport processes (Harker and Desch, 2002; see also below, Section 3.2.2.3). But it is currently impossible, even for the solar system, to distinguish between the two proposed scenarios.

It is particularly interesting to note that the statistical analysis of large samples of T Tauri stars shows a poor correlation between the grain size in the upper layers of disks and the age of the star. The same conclusion can be drawn when the degree of crystallinity of the silicate grains is considered. Although star ages are difficult to estimate accurately,[7] these results tend to indicate that age is not the only parameter controlling disk evolution. The environment (ambient radiation field, bound companions, etc.) is likely to impact the disk evolution and hence the planet formation which could explain star-to-star variations at similar ages. Although the formation of meter-sized (or larger) rocks is a prerequisite to form terrestrial planets and planetary cores, one should keep in mind that this step is one of the less observationally nor theoretically constrained in the planet formation process. Exoplanet embryos are nevertheless expected to leave characteristic imprints in disks such as gaps or even to clean up the inner disk regions. The detection of gaps in young systems due to planet-disk coupling is still pending. In a few cases, the presence of depleted regions inside disks can be indirectly inferred from the lack of significant circumstellar emission at short IR wavelengths indicating a low amount of warm material close to the star. But such objects remain exceptional. Thus, at present, astronomers are only able to draft

---

[7] See Chapter 2

general trends without being able yet to further constrain the main steps that allow to go from sub-micron size grain to km-sized bodies, and this is a major problem in particular for theories of the formation of the solar system (Section 3.2.4).

3.2.1.2. *X-ray induced irradiation phenomena in circumstellar disks*
THIERRY MONTMERLE

A new factor probably plays an important role, for star formation as well as for planet formation. This is the ubiquitous X-ray emission from young stars, from the protostar stage (see, e.g., Feigelson and Montmerle, 1999; Micela and Favata, 2005, for reviews). X-rays are mainly detected in the form of powerful flares, very similar to those seen in the images of the Sun which the *Yohkoh* satellite sent us every day from 1991 to 2001,[8] but much more intense (10^3–10^5 times stronger than for the quiet Sun).

In the course of these flares, the stellar X-ray luminosity L_X amounts to 10^{-4} to 10^{-3} times the total luminosity of the young star (called the "bolometric" luminosity L_{bol}), reaching a few percent or more in exceptional cases. It can be shown that the "plasma" (a very hot, ionized gas, with a temperature of 10^7 to 10^8 K) that emits the X-rays is confined in very large magnetic loops (up to 2–3 R_*, i.e., about 10 $R_\odot$ for a T Tauri star). By analogy with the Sun, it is thought that the fast heating results from so-called "reconnection events", in which magnetic field lines of opposite polarities get in contact in a "short-circuit", and suddenly release the magnetic energy previously stored in the course of various motions, for instance when magnetic footpoints are dragged by convective cells (e.g., Hayashi et al. 2000). The plasma then cools radiatively typically in a few hours by the emission of X-ray photons.

Thanks to X-ray satellites (from the 90s), X-ray emission has been detected from hundreds of T Tauri stars, either concentrated in young stellar clusters like ρ Ophiuchi (e.g., Ozawa et al., 2005) or Orion (Getman et al., 2005), or else more widely dispersed as in the Taurus-Auriga clouds (e.g., Stelzer and Neuhäuser, 2001), and many other star-forming regions. For a given population of T Tauri stars (with and without disks), the X-ray detection rate is consistent with 100%. (Only the fainter stars are not all detected, for lack of sufficient sensitivity.) In other words, observations suggest that *all* young, solar-like stars, emit X-rays, at a level $L_X/L_{bol} \sim 10^{-4}$ to 10^{-3} with a fairly large dispersion (see, e.g., Ozawa et al., 2005), to be compared with $L_X/L_{bol} \sim 10^{-7}$ for the present-day Sun. This is for instance most vividly illustrated by the ~1,500 sources detected in the 10-day *Chandra* exposure of the Orion Nebula (Getman et al., 2005; Feigelson et al., 2005) (Figure 3.14).

---

[8] Visit http://www.solar.physics.montana.edu/YPOP/

Figure 3.14. The center of the Orion nebula cluster. *Left*: near-IR image (this is the same as Figure 3.3). *Right*: corresponding X-ray image by Chandra (Getman et al., 2005). Note the excellent identification between the IR and X-ray sources, demonstrating that all young stars emit X-rays at levels much higher than the Sun.

In addition to providing us information on the "magnetic state" of young stars (for instance relevant to ejection phenomena, or to the reconnection mechanisms, as discussed above), flare X-rays are important on another ground: they induce irradiation effects on the dense surrounding circumstellar and interstellar material: ionization and energetic particle interactions (this last point will be discussed in the next subsection; see Glassgold et al., 2000, 2005; Feigelson, 2005).

The first effect of X-ray irradiation is the ionization of the accretion disk. Direct evidence for irradiation has been found recently in the form of a fluorescence line of neutral iron at 6.4 keV in the spectra of nearly a dozen of T Tauri stars with disks, mainly in the ρ Oph and Orion clusters (Favata et al., 2005; Tsujimoto et al. 2005), and also by the presence of specific molecules in some disks (e.g., Greaves, 2005; Ilgner and Nelson, 2006, and references therein). Such a characteristic fluorescence line can be excited only in response to X-ray irradiation of the cold gas. On theoretical grounds, several authors have studied the case of an accretion disk irradiated by X-rays (e.g., Glassgold et al., 2000, 2005; Fromang et al., 2002; Matsumura and Pudritz, 2006). The results depends on details of the adopted disk model, but the main conclusion, sketched in Figure 3.15, is that the ionization of the accretion disk is dominated by X-rays, except in the densest equatorial regions where they cannot penetrate ("dead zone"), i.e., within a few AU of the central source. Everywhere else, the disk is partially ionized, and the ionization fraction $x_e = n_e/n_H$ is roughly comparable to that of the interstellar medium ($x_e \sim 10^{-7}$ or less).

The ionization state of the disk has another important consequence: the coupling of matter with magnetic fields. Indeed, even very weakly ionized

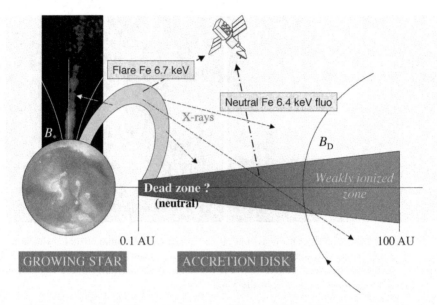

Figure 3.15. Sketch of the X-ray irradiation of circumstellar disks. Note in particular the emission of a neutral X-ray fluorescence line at 6.4 keV, detected in several systems. The other line, at 6.7 keV, is characteristic of a hot, X-ray emitting plasma, at temperatures of several 10^6 K. Note also the dense, neutral "dead zone", which, according to some authors, would be the most favorable to planet formation.

matter "sticks" to the magnetic field via collisions between neutral atoms and charged atoms (ions): this process, called ambipolar diffusion, has already been mentioned above in the context of star formation in molecular clouds (Section 3.1.1). When this process is coupled with the Keplerian motion of all atoms in a disk around the central object, it gives rise to a so-called "magnetorotational instability", discovered by S. Chandrasekhar and extensively studied by Balbus and Hawley (1991). This instability is invoked to explain the strong viscosity of accretion disks, and thus regulates accretion itself (e.g., Fromang et al., 2004). One interesting consequence is that such strong viscous coupling would not exist in the (neutral) dead zone, which would then undergo accretion only if some other, non-magnetic mechanism for an efficient viscosity were at work (which cannot be excluded). In fact, it has even been suggested that such a "protected", neutral region, of size ~20 AU, would be favorable for planet formation (Glassgold et al., 2000; Matsumura and Pudritz, 2006). Note, however, that the very existence of an extended dead zone has been challenged by recent numerical computations, which take into account turbulent motions within the disk, and which show that the neutral, dead zone volume tends to mix with the surrounding, ionized material: in the end, (weak) ionization probably dominates everywhere in the disk (Fleming and Stone, 2003), although this study does not take grains into account, which turn out to play an important role in the disk ionization (Ilgner and Nelson,

2006). Up to now, such a weak, but widespread ionization, which must also affect the growth of dust grains since they would be electrically charged, has not been taken into account in planet formation theories.

3.2.2. THE FIRST FEW MILLION YEARS AS RECORDED BY METEORITE DATA

MARC CHAUSSIDON, MATTHIEU GOUNELLE, THIERRY MONTMERLE

We now turn to the earliest stages of the solar system: the solar nebula, and the so-called *meteoritic record*. While most of the extraterrestrial flux to the Earth is in the form of micrometer size dust ($\approx$20,000 tons of the so-called micrometeorites par year), there are about 10 tons of meteorites of centimeter to meter size that fall on Earth every year. Most of these meteorites are likely to come from the asteroidal belt situated between Mars and Jupiter. Among meteorites, chondrites are primitive objects that have endured no planetary differenciation. This is attested by:

(i)   their mineralogical composition which reflects accretion of components formed at high temperature (the chondrules) and low temperature (the matrix), components that were never homogenized chemically and/or isotopically through melting and metamorphism,

(ii)   their bulk composition which is similar to the photospheric abundances of the elements in the Sun (Anders and Grevesse, 1989) and is thus considered as primitive, i.e. reflecting that of the forming solar system,

(iii) their age : they are the oldest rocks of the solar system, with a Rb/Sr age of $\approx$4.55 Gyr (Wasserburg, 1987). (This early pioneering result is now updated by more precise measurements: see below, Section 3.2.2.2.)

Other meteorites such as irons or some achondrites are the by-products of planetary differentiation : they have younger ages (by a few Myr) and have compositions reflecting the interplay of silicate–metal differentiation (core-mantle formation) and of silicate–silicate differentiation (crust-mantle formation). Thus chondrites, which are undifferentiated, can help us unravel the physico-chemical conditions in the solar nebula and the astrophysical context of the Sun's birth, while differentiated meteorites trace the formation and evolution of large planetary bodies.

3.2.2.1. *Chondrite components and physical conditions in the solar nebula*
MARC CHAUSSIDON, MATTHIEU GOUNELLE

Among chondrites, carbonaceous chondrites have a chemical composition most similar to that of the Sun, and are therefore believed to be our best proxy for the protosolar nebula (Brearley and Jones, 1998). The mineralogy, chemistry and isotopic composition of carbonaceous chondrites can help decipher the formation and history of the solar nebula. Carbonaceous chondrites are made of Calcium–Aluminium-rich Inclusions (CAIs), chondrules and matrix, chondrules representing by far the major component, 70–80% in volume (see recent reviews by Zanda, 2004 for chondrules and by MacPherson and Huss, 2003, for CAIs). (Figure 3.16)

Matrix is rich in volatile elements (e.g., H_2O, C,...) : it is a fine-grained component made of chondrule fragments and of minerals stable at low temperature. Matrix has endured extensive secondary processing in the parent-body evidenced by metamorphic transformations (solid state diffusion) and hydrothermalism due to fluid-rock interactions. Matrix is therefore not very useful for pinpointing the physical conditions in the solar nebula. At variance, CAIs and chondrules are *high temperature components* which were formed in the solar nebula and thus predate parent-body processes. CAIs are refractory components made of Al-, Ca-, Ti-rich silicates and oxides.

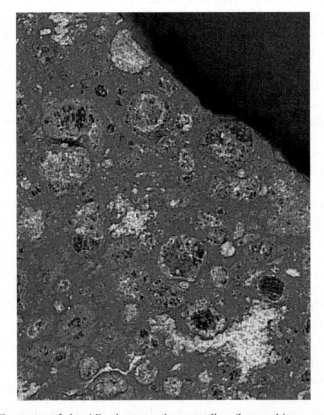

Figure 3.16. Fragment of the Allende meteorite, revealing (large white area in the bottom right-hand corner) the so-called "Calcium–Aluminium-rich Inclusions" (CAIs), in which evidence for short-lived, extinct radioactivies have been found.

Accordingly, they are generally considered from their mineralogy to be the first solids formed by condensation from the solar nebula gas at temperatures higher than 1800 K. Chondrules are spherical objects, made of iron–magnesium silicates (mostly olivine and pyroxene), metal and sulfides. Chondrules generally contain a large fraction of glassy mesostasis, implying that they were once molten and were subsequently quenched at a few to perhaps 1000 K/h (Hewins, 1997). Experimental studies show that CAIs cooled at $\approx$0.1 to $\approx$10 K/h (Stolper and Paque, 1986). Both CAIs and chondrules are thought to have formed in the solar nebula via complex high-temperature processes including condensation, evaporation, melting, etc.... Despite extensive studies, the exact mechanisms that led to the formation of CAIs and chondrules are still elusive, but it is clear that CAIs and chondrules formed at different times or locations. Chondrules probably formed at a higher pressure than CAIs and it is generally considered that chondrules formed later than CAIs, although this assumption cannot yet be demonstrated by direct dating of each component. In fact, the formation of chondrules and CAIs may overlap and there exists a variety of chondrules and CAIs which may have formed in different environments and may have complicated histories, involving precursors of variable composition and more or less extensive exchanges with the nebular gas (see Hewins et al., 2005, and references therein).

3.2.2.2. *Duration of the solar nebula*
MARC CHAUSSIDON, MATTHIEU GOUNELLE

Radioactive nuclide abundances in rocks are classically used to date rocks and to infer timescales for geological processes.[9] Among these, the U/Pb system which combines two parent/daughter couples (^{238}U decays to ^{206}Pb with a half life of 4.47 Gyr, and ^{235}U decays to ^{207}Pb with a half life of 0.7 Gyr), is the one which has provided the most accurate dating of meteoritic components. Accordingly, the CAIs appear to be the oldest component within chondrites, with a U/Pb age of 4567.2 $\pm$ 0.6 Myr (Manhès et al., 1988; Allègre et al., 1995; Amelin et al., 2002). Slightly different values, based on various methods, have appeared in the recent literature, giving an age of 4568.5 $\pm$ 0.4 Myr (Baker, 2005; Bouvier et al., 2005), but for the purpose of this article these values (which differ by ~1.3 $\pm$ 0.5 Myr) are consistent within the uncertainties. This age can be considered as giving the "time zero", t_0, for the start of the formation of the solar system; however this current experimental time resolution of ~1 Myr does not allow to build a precise chronology of the earliest processes which occurred in the solar accretion disk and which gave birth to the first solar system solids from the nebula. In fact, the

---

[9] See Chapter 2 on Chronometers (Section 2.2).

typical duration of these processes is most likely less than 1 Myr, so revealing them requires further improvements in the experimental methods.

Of particular interest for cosmochemistry are the so-called *extinct radioactive nuclides*, or short-lived radioactive nuclides, which have half-lives below a few Myr. The ones which have been identified in meteorites up to now are ^{7}Be ($T_{1/2} = 53$ days), ^{41}Ca ($T_{1/2} = 0.1$ Myr), ^{36}Cl ($T_{1/2} = 0.3$ Myr), ^{26}Al ($T_{1/2} = 0.74$ Myr), ^{10}Be ($T_{1/2} = 1.5$ Myr), ^{60}Fe ($T_{1/2} = 1.5$ Myr) and ^{53}Mn ($T_{1/2} = 3.7$ Myr). The presence of excesses of their radioactive daughter elements (e.g., positive correlation between ^{26}Mg/^{24}Mg and ^{27}Al/^{24}Mg ratios for ^{26}Al, see Figure 3.17) shows that these radioactive nuclides were present in various amounts both in CAIs and in chondrules when they formed. Because of their short half-lives, short-lived radioactivities can be used to constrain very tightly timescales (McKeegan and Davis, 2003 and refs therein): for instance the ^{26}Al/^{27}Al ratio decreases by a factor of two within 0.74 Myr. In this respect, ^{10}Be, ^{26}Al and ^{60}Fe are of special interest since they can (i) help constrain the chronology of the first million years of evolution of the solar system, and (ii) give nuclear clues to the astrophysical context of the Sun's birth.

Many measurements of the initial content of ^{26}Al in CAIs have led to the idea of a canonical ratio, ^{26}Al/^{27}Al $= 4.5 \times 10^{-5}$, that would have defined the starting time t_0 of the "protoplanetary" solar system (MacPherson et al., 1995)—which must not be confused with the starting time of the formation of the Sun as a star, which in this context *precedes* t_0 (see Section 2.5). CAIs are particularly well suited for the determination of the ^{26}Al/^{27}Al ratios, because since they are refractory objects they are enriched in Al relative to Mg. Less numerous measurements in chondrules, which are less enriched in Al relative to Mg and thus more difficult to analyze, suggest that chondrules formed with ^{26}Al/^{27}Al $< 1 \times 10^{-5}$ (Mostefaoui et al., 2002). Assuming an homogeneous distribution of ^{26}Al in the solar nebula, this yields a ~2–3 Myr age difference between CAIs and chondrules. This timescale has long been considered as the characteristic timescale of the solar nebula as determined by chondrite data. If true, such a long duration for the high temperature processes in the accretion disk implies that CAIs and chondrules must have been in some way "stored" in the nebula for 2–3 Myr before being accreted together to form the chondrites.

At this point it is important to stress that this so-called "chronological interpretation" of the ^{26}Al/^{27}Al variations relies entirely on the assumption mentioned above, which is very strong: the existence once in the early solar system of an *homogeneous* distribution of ^{26}Al (Gounelle and Russell, 2005). This assumption is far from being proven and very difficult to demonstrate in fact, simply because one would need very precise independent absolute ages for different objects to be able to test the homogeneity of their ^{26}Al/^{27}Al ratios. It could in fact well be that the difference in the initial content of ^{26}Al

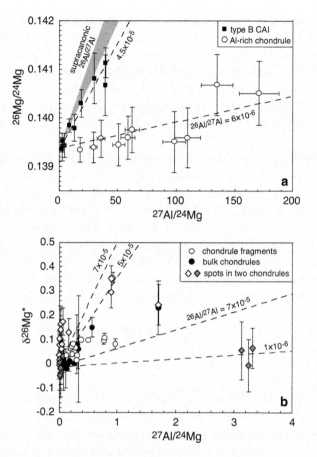

Figure 3.17. Variations in CAIs and chondrules of the Mg isotopic compositions (given as $^{26}Mg/^{24}Mg$ in the upper panel, or as the permil ^{26}Mg excesses noted $\delta^{26}Mg^*$ in the bottom panel) due to the radioactive decay of short-lived ^{26}Al (which decays to ^{26}Mg with a half-life of 0.7 Myr). These data show that ^{26}Al was present in the early solar system and that its abundance (given as $^{26}Al/^{27}Al$ ratios) can be used to constrain the chronology of the formation of CAIs and chondrules. A canonical $^{26}Al/^{27}Al$ ratio of 4.5×10^{-5} has been found by in situ ion microprobe analysis in most CAIs (data from Podosek et al., 1991, in the upper panel). Supracanonical $^{26}Al/^{27}Al$ ratios recently found in CAIs are shown by a grey field (data from Young et al., 2005 and Bizzarro et al., 2004) in the bottom panel. In this panel, the details of the distribution of ^{26}Mg excesses in chondrules (data from Galy et al., 2002; Bizzarro et al., 2004; Chaussidon et al., 2006) is not yet well understood : it can be interpreted as reflecting either the formation of some chondrules very early (i.e. at the same time than CAIs) or a late formation of chondrules in the nebula from a mixture of precursors including CAI material.

between chondrules and CAIs is due in part to spatial heterogeneities in the solar nebula. Predictions can be made on the existence of such an heterogeneity depending on the nucleosynthetic origin of ^{26}Al, either "last minute" injection from a nearby massive star or supernova, or production by irradiation processes around the early Sun (see next subsection). Note also that the picture has recently been complicated by results from the Oxford-UCLA laboratory (Galy et al., 2002; Young et al., 2005) who showed that the canonical ratio of CAIs might be due to a resetting event of an initially higher ratio of $\approx 7 \times 10^{-5}$, and that chondrules might have formed with an initial $^{26}Al/^{27}Al$ ratio significantly higher than 1×10^{-5}. The high $^{26}Al/^{27}Al$ ratios inferred for some chondrules can indicate that they formed very early (Bizzarro et al., 2004) or be simply due to the fact that they contain an inherited CAI component (Galy et al., 2002).

Recently, high precision data on Pb isotopes (U/Pb dating system) have been obtained on CAIs and chondrules (Amelin et al., 2002), providing absolute ages of the chondrites' components. CAIs have ages ranging from 4567.4 ± 1.1 Myr to 4567.17 ± 0.70 Myr, while chondrules have ages ranging from 4564.66 ± 0.3 Myr to 4566.7 ± 1 Myr. This confirms that the solar nebula could have lasted for several million years, and question the idea that *all* chondrules formed 2–3 Myr after CAIs. Most recent Pb data seem to indicate that a range of ages do exist for chondrules, from nearly as early as CAIs to a few Myr later.

3.2.2.3. *Extinct radioactivities and the astrophysical context of the birth of the Sun*
MARC CHAUSSIDON, THIERRY MONTMERLE

The presence of short-lived radioactivities in the solar nebula is intriguing. Because their half-life is shorter than the timescales needed to isolate the future solar system material from the interstellar medium, they need to have a "last minute origin", as opposed to the steady-state abundance of galactic cosmic-ray induced isotopes. They could have been injected in the nascent solar system by a nearby supernova, or made by irradiation of the nebular dust and/or gas by energetic solar protons, or both. The challenge of any mechanism is to explain most, if not all, of the extinct radioactivities at the same time.

The ubiquitous, intense and flare-like X-ray activity of young stars (Section 3.2.1.2) support the idea that irradiation could have been an important process in the early solar system (see previous sections; also Chaussidon and Gounelle, 2006, for a review of the traces of early solar system irradiation in meteoritic components, and Feigelson, 2005 for the relation with X-ray flares from young stars). Such an origin is supported by the presence of ^{10}Be ($T_{1/2} = 1.5$ Myr) and of ^7Be ($T_{1/2} = 53$ days) in Allende CAIs (McKeegan et al., 2000; Chaussidon et al., 2006), beryllium isotopes being made by flare-induced energetic particle irradiation (see Figure 3.18).

On the other hand, the presence of ^{60}Fe in sulfides and silicates from chondrites (Huss and Tachibana, 2004; Mostefaoui et al., 2004) implies the presence of a nearby supernova, since ^{60}Fe is too neutron-rich to have been made by in-flight spallation reactions (Lee et al., 1998), and is a signature of

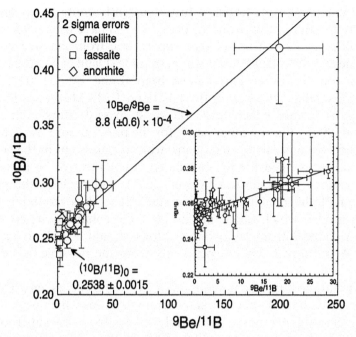

Figure 3.18. Variations in one Allende CAI of the $^{10}B/^{11}B$ ratios vs. the $^9Be/^{11}B$ ratio (data from Chaussidon et al., 2006). The isochrone-type correlation demonstrates that short-lived ^{10}Be (which decays to ^{10}B with $T_{1/2} = 1.5$ Myr) was present in the early solar system. The same CAI contains traces of the in situ decay of short-lived ^7Be ($T_{1/2} = 53$ days). Radioactive ^{10}Be and ^7Be can be produced in the early solar system by irradiation processes around the Young Sun (Chaussidon and Gounelle, 2006).

explosive nucleosynthesis in supernovae. The presence of a nearby supernova would imply in turn that the Sun was born in a crowded stellar environment, possibly similar to HII regions as observed in Orion (Hester and Desh, 2005, and above, Section 3.1.1). ^{26}Al also can be made in massive stars and supernova explosions (though other stellar processes, such as in novae and red giant stars, which are old, are also important sources, see Diehl et al. 2006), and it is thus tempting to explain the abundance of both isotopes in this way. As a matter of fact, one can find a set of parameters (time interval between the explosion and the birth of the solar system $\Delta t = 1$ Myr, dilution factor due to transport of supernova material to the presolar core $f = 10^{-4}$) that explains both abundances, and also other isotopes (Busso et al., 2003; Gallino et al., 2004). But this may not be completely realistic: as a rule, most star-forming regions are in some vicinity of OB associations in which massive stars evolve and explode sequentially. If they explode at a frequency shorter than ~1 Myr, the resulting ^{60}Fe and ^{26}Al nucleosynthesis will tend to be in a steady state locally (as long as the massive star formation episode lasts), rather than in separate bursts, subsequently decaying. For instance, in spite of the fact that the Orion Nebula Cluster currently contains no supernova, the 1.809 MeV gamma-rays from the decay of ^{26}Al have been detected by the GRO satellite, as a signature of past supernovae from massive stars excavating the nearby so-called Eridanus superbubble (Diehl et al., 2004). Present supernovae, on the other hand, have been found by the RHESSI and INTEGRAL satellites, by way of the 1.173 and 1.333 MeV emission of ^{60}Fe in the Cygnus region, which hosts the most massive stars in the Galaxy (Harris et al. 2005). On the other hand, for the presolar core the values of Δt and f are loosely constrained since some arbitrary mass cut in the supernova ejecta has to be invoked for the calculated abundances to be in agreement with observations in meteorites (Meyer, 2005), and neglects the preceding contribution of the precursor massive stars. It is thus troubling that the irradiation model, which takes into account the enhanced stellar energetic particle flux deduced from X-ray observations, should be able to account for the ^{26}Al/^{27}Al ratio in CAIs independently from the possible existence of a supernova.

At this point, it should be noted that, as early as 25 years ago, Montmerle (1979) identified about 30 massive star-forming regions which were, based on various observational criteria, tentatively associated with supernova (SN) remnants. These "special" regions, dubbed "SNOBs" (for "OB associations" or molecular clouds observed to be associated with supernovae) were at the time searched in connection with the identification of high-energy gamma-ray sources. For our purpose, this sample can be taken as examples of the reality of supernovae exploding in the close vicinity of young stars. It also shows that only a small fraction of all OB associations are in this situation at any given time. An illustration of the complexity of the problem is given, again, by the Orion star-forming region.

In 1895, E.E. Barnard discovered a faint, almost exactly (half-) circular ring in the outskirts of Orion, spanning several degrees in the sky. This spectacular structure, now known as Barnard's Loop, is shown in Figure 3.19. It is readily visible in the Hα line of ionized hydrogen. Its exact nature is still uncertain: proposed to be a supernova remnant by Öpik back in 1935, recent radio observations (Heiles et al., 2000) have shown that the emission is thermal, i.e., radiated only by an ionized gas, just like HII regions around young stars. The idea then is that this ring has been created by winds from the Belt stars (not the Trapezium; the Belt stars are the conspicuous stars visible with the naked eye on clear winter nights, see Figure 3.19). But it is also possible that the traditional radio signature of supernova remnants, i.e., the nonthermal emission from high-energy electrons accelerated at the shock front, is somehow buried in a more intense thermal emission – a classical dilemma in supernova remnant identifications. In either case, however, kinematics indicate that Barnard's Loop must have originated about 3×10^6 years ago close to the Orion nebula.

Figure 3.19. Barnard's Loop surrounding the Orion nebula, seen in Hα. It is unclear whether this extended structure is a the remnant of a supernova explosion or an ionized shell created by stellar winds from the Belt stars, but kinematical studies give it an age of 3×10^6 years. (Photograph by E. Mallart).

When considering the consequences on the possible irradiation of the "proplyds" of the Orion nebula (see Figures 3.3 and 3.4), one is therefore faced with two possibilities: (i) either Barnard's Loop is not the remnant of a supernova explosion having taken place 3×10^6 years ago, in which case *none* of the current young stars in Orion have been peppered with ^{60}Fe, (ii) or it is, then *only a fraction* of them have been. This can be seen by looking at the H–R diagram presented in Chapter 2 on chronometers (Figure 2.3). In this figure, we can easily see that the majority of young stars present in the Orion Nebula Cluster are in fact younger that 3×10^6 years, so that any ^{60}Fe spread by the explosion has disappeared, and these young stars cannot be contaminated by ^{60}Fe. Quantitatively, one can find after Figure 3.20 (taken from Palla and Stahler, 1999), that less than 40% of the older generation of young "suns" may have been effectively contaminated by the SN explosion, if

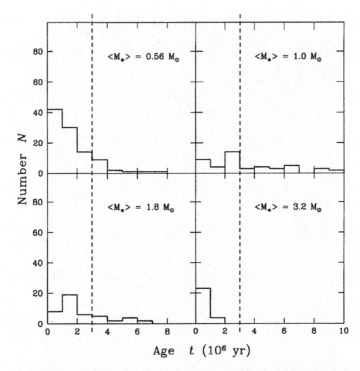

Figure 3.20. Age distribution in four mass ranges for stars in the Orion Nebula Cluster (Palla and Stahler, 1999). The dotted line indicates the estimated age of Barnard's Loop, 3 Myr. For stars of mass close to 1 $M_\odot$, less than 40% are older than 3 Myr and may have been contaminated by short-lived nucleosynthetic products of a supernova explosion, provided Barnard's Loop is a supernova remnant, which is still unclear.

there was such an explosion. The presence of ^{60}Fe in the early solar system would then not be the rule, even in an Orion-like birthplace.

In the end, and as discussed in the introductory sections, the birthplace of the Sun is still an unresolved question, although the birth of the Sun in a rich cluster seems to be favored by stellar statistics. To understand whether the Sun was born in a high-mass environment like Orion, or in a low-mass environment like ρ Ophiuchi or Taurus has however important implications not only for the astrophysical conditions for the Sun's birth itself, but also for the chronology of the early solar system. Indeed, depending on their origin, short-lived radionuclides were or were not homogeneously distributed in the solar nebula. Usually short-lived radionuclides are expected to be homogeneously distributed if coming from a supernova and heterogeneously distributed if originating from in situ irradiation by energetic particles. This comes from the fact that supernova material must be largely volatilized in the HII region and homogeneously mixed in the accreting disk where it is injected, though there is at present no definitive observation demonstrating whether a fraction of the supernova ejecta cannot be in the form of solid grains (Hester and Desch, 2005). At variance, irradiation models based on X-ray flare observations and the "X-ring" picture (Gounelle et al., 2001; 2004; see above, Section 1.3) predict possible variations in the production rate of short-lived radionuclides depending on parameters such as the fluence of the accelerated particles, their composition and the composition of the irradiated target. Irradiation, however, might also produce rather constant radionuclide abundances if characteristic time scales and compositions are considered for the irradiation. It is obvious that identifying the source of ^{26}Al and other short lived nuclides would provide a long-needed basis for the chronological use of these elements.

One could think of solving this problem by looking at young stars with circumstellar disks, in which particle irradiation is actually taking place as seen in X-rays, making use of the fact that, as mentioned above, ^{26}Al decays by the emission of 1.8 MeV gamma-rays, and is thus observable elsewhere than in the solar system by gamma-ray telescopes such as GRO. The calculation has been done by Montmerle (2002): taking into account that gamma-ray telescopes have a very wide field-of-view (several degrees in diameter), when pointed at a star-forming region they will integrate the flux of a whole star-forming region, i.e., several hundred young stars at the same time. As it turns out, even under the most optimistic assumptions the 1.8 MeV flux is undetectable, and dominated by the general ^{26}Al emission in the Galaxy, which is most conspicuous in massive star-forming regions because of successive supernova explosions (see Diehl et al. 2006).

In summary, while the presence of ^{60}Fe in CAIs shows that the forming solar system has been at least "polluted" by a nearby supernova, it is not clear whether this supernova has been responsible for the other extinct radioactivities, including ^{26}Al.

3.2.3. INTERMEZZO

THIERRY MONTMERLE

The study of circumstellar disks around solar-like stars, as described in Section 3.1.2, obviously has strong implications on our current views on the origin and formation of the solar system. We observe their global evolution (timescale for the disappearance of disks in stellar clusters), and we begin to understand their physicochemical evolution (growth and mineralogy of dust grains via IR spectroscopy) over a few million years, while large quantities of gas (detectable via mm observations) are still present. However, as already mentioned, once the grains reach sizes of a few mm, they are not observable any more. They start to be observable again much later (several tens of million years), but only as "debris" of collisions between macroscopic bodies, presumably planetesimals or asteroids, giving indirect evidence for ongoing planetary formation. Thus, there is a crucial observational time gap between the "primordial" disks (which would correspond to the solar nebula), and "second-generation" debris disks (in which planetary formation is well under way). Translated into sizes, this means that we have essentially no observational constraints on the transition between cm-sized grains (direct) and km-sized bodies (indirect).

In principle, the ~200 known exoplanetary systems should also give us clues about planetary formation in general, and the formation of the solar system in particular. However, at least as far as the formation of the solar system is concerned (which is our main concern here in the context of the origin of life as we know it), there are still many open problems. Indeed, while current observing techniques are sensitive mostly to super-giant exoplanets close to the central stars (the so-called "hot Jupiters", with masses between 5 and 10–15 Jupiter masses or more), and increasingly more to lower-mass exoplanets, exoplanetary systems with several planets, when they exist, do not resemble at all the solar system. The recently announced discovery of a distant 5-Earth-mass exoplanet orbiting a very low-mass, non-solar, star, by way of gravitational lensing towards the galactic center (Beaulieu et al. 2006), is not helpful either in this context, since nothing is known about the possible exoplanetary system it belongs to.

In the end, for most of the phase of planet formation we are essentially left with theory, backed by rare and difficult laboratory experiments. The goal is to put together mechanisms for grain agglomeration and destruction, and mutual dynamical interactions between bodies of various sizes and masses in the general gravitational field of the central star, and also in the presence of gas which affects the motion of small particles. In spite of the relative simplicity of the basic equations, this involves sophisticated numerical N-body simulations. The following section describes the basic theoretical concepts of planetary formation restricted to the solar system, with the goal of ultimately understanding the various steps that led to the formation of our planet, the Earth.

3.2.4. THE FIRST STAGES OF PLANETARY FORMATION IN THE SOLAR SYSTEM

ALESSANDRO MORBIDELLI

3.2.4.1. *From micron-size to kilometer-size bodies (1 Myr)*

In the proto-solar nebula, the most refractory materials condense first, gradually followed – while the local temperature drops – by more and more volatile elements. The dust grains thus formed float in the gas. Collisions stick the grains together, forming fractal aggregates, possibly helped by electrostatic and magnetic forces. Other collisions then rearrange the aggregates, and compact them. When the grains reach a size of about a centimeter, they begin to rapidly sediment onto the median plane of the disk in a time

$$T_{sed} \sim \Sigma/(\rho_p \Omega a) \sim (\rho v)/(\rho_p \Omega^2 a)$$

where Σ is the nebula surface density, ρ is its volume density, v is the r.m.s. thermal excitation velocity of gas molecules, ρ_p the volume density of the particles, a is the radius, and Ω is the local orbital speed of the gas (Goldreich and Ward, 1973; Weidenschilling, 1980). Assuming Hayashi's (1981) minimal nebula (minimal mass solar composition nebula, with surface density Σ proportional to $r^{-3/2}$, containing materials to create the planets as we know them), one gets

$$T_{sed} \sim 10^3/a \text{ years}$$

where a is given in cm. This timescale, however, is computed assuming a quiet, laminar nebula. If the nebula is strongly turbulent, or strongly perturbed from the outside or by the ejection of jets in the proximity of the star, the sedimentation can become much longer.

Once the dust has sedimented on the mid-plane of the nebula, the clock measuring the timescale of planet accretion starts effectively to tick. Thus, the time t_0 for planetary accretion is *not* the time t_0 usually used in stellar formation theory (start of the collapse of molecular cores) or the time t_0 of cosmochemists (the formation of the first CAIs, see above, Section 3.2.2.2). Linking the various times t_0 together is one of the major problems in establishing an absolute chronology for the formation of the solar system. In addition, notice that T_{sed} above depends on the heliocentric distance r. This means that time t_0 is different from place to place in the disk!

The growth from dust grains to kilometer-size planetesimals is still unexplained. There are two serious issues that remain unsolved. The first issue concerns the physics of collision between such bodies at speed of order 10 m/s (typical collision velocity for Keplerian orbits with eccentricity $e \sim 10^{-3}$), which is still poorly understood. For dust grains, current theories predict that collisions are disruptive at such speeds (Chokshi et al., 1993). However, recent laboratory experiments on collisions between micrometer size grains (Poppe and Blum, 1997; Poppe et al., 2000) give a critical velocity for accretion (velocity below which grains accrete, and above which they fragment) 10 times larger than predicted by the theory. The reason is probably that the fractal structure of the dust allows to absorb energy much better than envisioned *a priori*. This may help solving the fragmentation paradox for dust–dust collisions. When the agglomeration of dust builds larger bodies, though, the problem of collisional disruption becomes much more severe. Laboratory experiments cannot be done at this size range and one has to trust computer models. Specific simulations of this process with SPH (Smooth Particle Hydrodynamics) techniques have been done by Benz (2000) for basalt bodies (monolithic or rubble piles) of sizes in the range m to km. He found that low velocity collisions (5–40 m/s) are, for equal incoming kinetic energy per gram of target material, considerably more efficient in destroying and dispersing bodies than their high velocity counterparts. Furthermore, planetesimals modeled as rubble piles are found to be characterized by a disruption threshold about five times smaller than solid bodies. Thus, unless accretion can proceed avoiding collisions between bodies of similar masses, the relative weakness of bodies in meter to km size range creates a serious bottleneck for planetesimal growth. These apparently negative results, however, may once again depend on our poor understanding on the internal structure of these primitive small planetesimals. The simulations assume rocky objects, but it is still unclear how a puffy pile of dust becomes a solid rock. If the planetesimals were not yet 'solid rocks' maybe the impact energy could be dissipated more efficiently.

The second open problem on planetesimal growth concerns radial migration. Gas drag makes them fall onto the central star: gas being sustained by its own pressure, it behaves as if it felt a central star of lower mass, and therefore rotates more slowly than a purely Keplerian orbit at the same heliocentric distance. Solid particles tend to be on Keplerian orbits and therefore have a larger speed than the gas. So the gas exerts a force (drag) on the particles, the importance of which is given by the characteristic stopping time:

$$T_s = (m\Delta V)/(F_D),$$

where m is the mass of the grain, ΔV the difference between the particle velocity and the gas velocity, and F_D the gas drag. For small particles, the "Epstein" gas drag gives

$$T_s \sim \rho_p/\Sigma,$$

while for big particles, the "Stokes" drag gives

$$T_s \sim \rho_p a^2.$$

(Weideschilling, 1977). We then compare this time with the characteristic Keplerian time

$$T_K = 1/\Omega.$$

If $T_s \gg T_K$, the particle is almost decoupled from the gas. If $T_s \ll T_K$, it is strongly coupled to the gas, and it tends to move with the gas flow. The maximum effect occurs when $T_s \sim T_K$, which occurs for meter size particles. For the Hayashi minimal nebula, at about $r = 1$ AU from the Sun,

$\Delta V \sim 50$–55 m/s, and the particle's radial velocity induced by the gas drag is then of order 10–100 m/s (Weideschilling, 1977). So meter-sized particles should fall on the Sun in ~100–1,000 years, i.e., before they can grow massive enough to decouple from the gas.

One way out of this paradox is to have a density of solids larger than that of the gas. In this case, the growth timescale would be faster than the radial drift timescale. However, such a composition is not supported by observations nor by current theories.

Another possibility is the existence of vortices in the protoplanetary disk, due to the turbulent viscosity of the nebula (Tanga et al., 1996). In this model, 70–90% of the particles are trapped in anti-cyclonic vortices. Once trapped, the particles do not fall any more towards the Sun, but rather fall toward the center of the vortex. Such falling timescale varies from a few tens to a few thousand years, depending mainly on the size of the particle and on the heliocentric distance (which increases all dynamical timescales). Once at the center of the vortex, particles dynamics are stable over the lifetime of the vortex. Their relative velocities are reduced (particles tend to follow the gas stream lines, so they all tend to have the same velocity) and the local density is increased, enhancing the accretion process. Therefore, vortices would help the accretion of kilometer-size planetesimals in two ways: by stopping the drift of meter-sized bodies towards the Sun, and by speeding up the accretion process due to the accumulation of the bodies at the centers of the vortices.

Which of these two possible situations is the real one, profoundly affects the formation timescale. If there is no way to slow down the fall of growing planetesimals towards the Sun, then the formation of a multi-km object has to occur in about a few 1,000 years, probably by gravitational instability. If, on the contrary, the turbulence of the disks is an effective obstacle to the inwards drift, then the formation of planetesimals can take much longer. To add confusion (reality is never easy), it is likely that while the first planetesimals are building up, new dust is settling on the midplane or drift in from further heliocentric distances. So, different planetesimals can see different "times t_0". In other words, even if the formation of a planetesimal is locally very fast, the formation of a population of planetesimals can be ongoing for much longer.

3.2.4.2. *Formation of planetary embryos (1–10 Myr)*
One way or another, we now have a gas disk containing kilometer-size planetesimals. The dynamics of accretion starts to be dominated by the effect of the gravitational attraction among the planetesimals, which increases the collisional cross-sections. A *runaway growth* phase starts, during which the big bodies grow faster than the small ones, hence increasing their relative

difference in mass (Greenberg et al., 1978). This process can be summarized by the equation:

$$\frac{d}{dt}\left(\frac{M_1}{M_2}\right) = \frac{M_1}{M_2}\left(\frac{1}{M_1}\frac{dM_1}{dt} - \frac{1}{M_2}\frac{dM_2}{dt}\right) > 0,$$

where M_1 and M_2 are, respectively, the characteristic masses of the "big" and of the "small" bodies, and can be explained as follows.

Generally speaking, accretion is favored by a high collision rate, which occurs when the relative velocities are large, but also by large collisional cross-sections and gentle impacts, which occur when the relative velocities are low. Therefore the relative velocities between the different planetesimal populations govern the growth regime.

At the beginning of the runaway growth phase the large planetesimals represent only a small fraction of the total mass. Hence the dynamics is governed by the small bodies, in the sense that the relative velocities among the bodies is of order the escape velocity of the small bodies $V_{esc(2)}$. This velocity is independent of the mass M_1 of the big bodies and is smaller than the escape velocity of the large bodies $V_{esc(1)}$. For a given body, the collisional cross-section is enhanced with respect to the geometrical cross-section by the so-called *gravitational focusing* factor:

$$F_g = 1 + (V_{esc}^2/V_{rel}^2),$$

where V_{esc} is the body's escape velocity and V_{rel} is the relative velocity of the other particles in its environment. Because $V_{rel} \sim V_{esc(2)}$, the gravitational focusing factor of the small bodies $[V_{sec} = V_{vesc(2)}]$ is of order unity, while that of the large bodies $[V_{esc} = V_{esc(1)} \gg V_{esc(2)}]$ is much larger. In this situation one can show that mass growth of a big body is described by the equation

$$\frac{1}{M_1}\left(\frac{dM_1}{dt}\right) \sim M_1^{1/3}V_{rel}^{-2}$$

(Ida and Makino, 1993) Therefore, the relative growth rate is an increasing function of the body's mass, which is the condition for the runaway growth (Figure 3.21).

The runaway growth stops when the mass of the large bodies becomes important (Ida and Makino, 1993) and the latter start to govern the dynamics. The condition for this to occur is:

$$n_1 M_1^2 > n_2 M_2^2,$$

where n_1 (resp. n_2) is the number of big bodies (resp. small bodies). In this case, $V_{rel} \sim V_{esc(1)} \sim M_1^{1/3}$, and hence $(1/M_1)(dM_1 dt) \sim M_1^{1/3}$. The growing

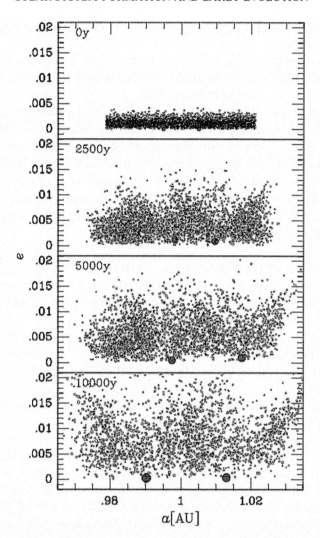

Figure 3.21. A simulation of the runaway growth process for planetary embryos. In a disk of equal mass planetesimals, two "seeds" (planetesimals of slightly larger size) are embedded. As time passes, the two seeds grow in mass much faster than the other planetesimals,, becoming planetary embryos (the size of each dot is proportional to its mass). While the growing planetary embryos keep quasi-circular orbits, the remaining planetesimals have their eccentricities (and inclinations) excited by the close encounters with the embryos. Notice also that the separation between the embryos slowly grows in time (i.e. passing from one panel to the subsequent one). From Kokubo and Ida (1998).

rate of the embryos gets slower and slower as the bodies grow, and the relative differences in mass among the embryos also slowly become smaller. In principle, one could expect the small bodies themselves to grow, narrowing their mass difference with the embryos. But in reality, the now large relative velocities prevent the small bodies to accrete with each other. The small bodies can only participate to the growth of the embryos: this phase is called *"oligarchic growth"*.

The runaway growth phase happens throughout the disk, with timescales that depend on the local dynamical time (Keplerian time) and on the local density of available solid material. This density will also determine the maximum size of the embryos and/or planets when the runaway growth ends (Lissauer, 1987). Assuming a reasonable surface density of solid materials, the runaway growth process forms planetary embryos of Lunar to Martian mass at 1 AU in 10^5–10^6 yr, separated by a few 10^{-2} AU. Beyond the so-called *snow line* at about 4 AU, where condensation of water ice occurred because of the low temperature, enhancing the surface density of solid material, runaway growth could produce embryos as large as several Earth mass in a few million years (Thommes et al., 2003).

3.2.4.3. *Formation of the giant planets (10 Myr)*

Observations and models of the interior of the giant planets give some important constraints on the composition and mass of the giant planets (see Guillot, 1999, and for a review, Guillot, 2005):

(i)   Jupiter has a mass of 314 M_E (Earth mass), and contains ~10–30 M_E of heavy elements;

(ii) Saturn has a mass of 94 M_E and contains ~10–20 M_E of heavy elements;

(iii) Uranus and Neptune have a mass of 14 and 17 M_E respectively, of which only ~1–2 M_E of hydrogen and helium.

To account for these constraints, the best current models for the formation of Jupiter and Saturn assume a three-stage formation (Pollack et al., 1996):

(1) the solid core accretes as explained in the previous section; beyond the snow line, the surface density of solid material is enhanced by a factor of several, due to the presence of ice grains. This allows embryos to grow to about 10 M_E on a timescale of a million years (Thommes et al., 2003).

(2) The accretion of the solid core slows down (see above), while a slow accretion of nebular gas begins, due to the gravity of the core. The gas accretion continues at a roughly constant rate over many million years, until a total mass of 20–30 M_E is reached;

(3) When the mass of the protoplanet reaches ~20–30 M_E the gas gravitationally collapses onto the planet. The mass of the planet grows exponentially and reaches hundreds of Earth masses in ~10,000 years: this is the "runaway" phase. (Figure 3.22)

The model explains well the properties of Jupiter and Saturn, in particular the existence of solid cores of about 5–15 Earth masses. There are however four main problems in the above scenario, which have not yet been solved.

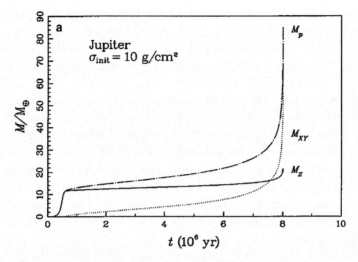

Figure 3.22. The growth of a Jupiter-mass planet. The solid curve gives the mass of metals as a function of time. The dotted curve gives the mass of the gas, and the dash-dotted curve the sum of the two, as a function of time. Notice that the growth of the solid core of the planet almost stalls after the first 0.5 Myr. During a ~7 Myr timespan, the planet slowly acquires an atmosphere, and, only when the total mass overcomes a critical threshold, a final exponential accretion of gas is possible. The timescale characterizing the slow accretion of the gas depends on the opacity of the atmosphere. From Pollack et al. (1996).

(I) When the planetary core reaches a mass of several Earth masses, its tidal interaction with the gas disk forces it to migrate very rapidly towards the Sun. (This is called "Type-I migration"). The estimated falling time is much shorter than the time required for the onset of the exponential accretion of the massive atmosphere. Thus, giant planets should not exist! Two ways out of this paradox have been proposed.

The first is that the gas disk was violently turbulent. In this case, the planetary cores would have suffered a random walk, rather than a monotonic infall towards the star (Nelson, 2005). Some would have collided with the star even faster than in the absence of turbulence, but others could be lucky enough to avoid collisions for a time long enough to start phase 3 above.

A second possibility (Masset et al., 2006) is that the gas disk surface density had a radial discontinuity. For instance, the inner part of the disk could be depleted by a factor of a few by the ejection of material in the polar jets, typical of young – magnetically active – stars (Section 3.1.3). If such a discontinuity exists, the planetary core would migrate towards the discontinuity, and stop there until the atmosphere is accreted.

(II) The second phase of giant planet accretion (the slow accretion of the atmosphere, prior to the onset of the runaway growth of the giant planet's mass) is also a problem, as it takes about 10–15 Myr, longer then the typical nebula dissipation time (Haisch et al., 2001; Hillenbrand 2006) (see above). To shorten the timescale of the second phase, two solutions have been proposed: to have an enveloppe of reduced opacity (Podolack 2003), or the migration of the planet, which continuously feeds the growth of its core (Alibert et al., 2005).

(III) The end of the exponential gas accretion is not yet fully understood. Most likely the growth of the giant planets is slowed down when a gap is opened in the gaseous disk, and is finally stopped when the nebula is dissipated by photo-evaporation from the central star. If this is true, then the full formation timescale of Jupiter and Saturn is of the order of the lifetime of the gas disk, namely of a few Myr. For Uranus and Neptune, it is generally assumed that the nebula disappeared before that the third phase of accretion could start. This would explain why these two planets accreted only a few Earth masses of gas.

(IV) The last problem is that, at the end of stage (III) above, the giant planets open a gap in the gas disk. Consequently they become locked in the radial evolution of the disk. As the disk's material tends to be accreted by the star, the giant planets have to migrate inwards. This migration, although slower than that discussed above for the cores, is nevertheless quite fast. (This is the so-called "Type-II" migration.) It is usually invoked to explain the existence of "hot Jupiters", massive extra-solar planets that orbit their star at distances smaller than the orbital radius of Mercury. But in our solar system this kind of migration evidently did not happen, or at least did not have a comparably large radial extent. Again, two solutions to this problem have been advanced.

The first possibility is that Jupiter and Saturn formed sufficiently late that the disk was already in the dissipation phase. Thus, the disk disappeared before it could significantly move the planets. In addition, this solution has the advantage of explaining why Jupiter and Saturn did not grow further in mass and why their massive atmospheres are enriched in heavy elements relative to the solar nebula composition (e.g., Guillot and Hueso, 2006).

A second possibility is that Jupiter and Saturn formed almost contemporaneously and on orbits that were close to each other. In this case, the gaps opened by the two planets in the disk would have overlapped (Masset and Snellgrove, 2001). This would have changed dramatically the migration evolution of the planets pair, possibly stopping or even reversing it. Of course the two solutions imply different formation timescales. If the planets stopped because the gas disappeared, the fact that Jupiter and Saturn did not migrate significantly implies that their formation timescale is of order of the nebula dissipation time (3–10 Myr; Hillenbrand 2006). In the opposite case, they might have formed even faster.

Finally, to be complete, a model of giant planet formation alternative to that of Pollack et al. has been proposed by Boss (see Boss, 2003 and references herein). In this model, the proto-planetary gas disk was massive enough to be unstable under its own gravity. In this situation, gaseous planets can form very rapidly, by a process that reproduces in miniature the one that led to the formation of the central star. There is no need of the presence of a massive solid core to trigger the capture of the giant planet's atmosphere. In this sense, Boss's model may bring a solution to the timescale problem related with the Pollack et al. model, discussed above. There is a still open debate in the literature on whether the clumps of gas observed in numerical simulations of gravitationally unstable disks are temporary features or would persist until the disk's dissipation. For the case of our solar system, the presence of massive cores inside all giant planets, and the limited amount of gas in Uranus and Neptune, make us think that the Pollack et al. model is more appropriate. It is possible, however, that some or several of the extra-solar planets observed so far formed through a gravitational instability mechanism.

3.3. The First 100 Million Years: the "Telluric Era"

3.3.1. FORMATION OF THE TERRESTRIAL PLANETS AND PRIMORDIAL SCULPTING OF THE ASTEROID BELT

ALESSANDRO MORBIDELLI

After a few 10^5 years of runaway growth, the embryos in the terrestrial planet region and in the asteroid belt region have Lunar to Martian masses. They govern the local dynamics and start perturbing each other. The system becomes unstable, and the embryos' orbits begin to intersect (Chambers and Wetherill, 1998). Because of mutual close encounters, the embryos' dynamical excitation (increase of eccentricity and inclination) moderately increases, and accretional collisions among embryos start to occur. The situation drastically changes when Jupiter and Saturn acquire their current masses. These two planets strongly perturb the dynamical evolution of the embryos in the asteroid belt region between ~2 and 5 AU. The embryos acquire a strong dynamical excitation, begin to cross each other, and cross rather frequently the orbits of the embryos in the terrestrial planets region. The collision rate increases. Despite the high relative velocity, these collisions lead to accretion because of the large mass of the embryos.

The typical result of this highly chaotic phase – simulated with several numerical N-body integrations – is the elimination of all the embryos originally situated in the asteroid belt and the formation a small number of terrestrial planets on stable orbits in the 0.5–2 AU region in a timescale ~100 Myr (Figure 3.23).

This scenario has several strong points:

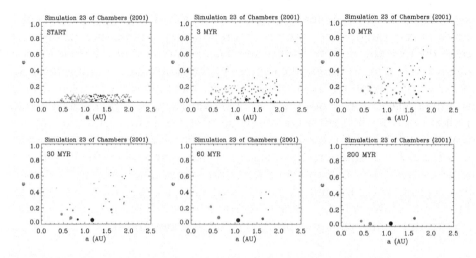

Figure 3.23. The growth of terrestrial planets from a disk of planetary embryos. Each panel shows the semi-major axis and eccentricity of the bodies in the system, the size of each dot being proportional to the mass. The color initially reflects the starting position of each embryo. When two (or more) embryos collide, the formed object assumes the color corresponding to the embryo population that has mostly contributed to its total mass. A system of four terrestrial planets, closely resembling our solar system, is formed in 200 Myr. From Chambers (2001).

(i)   Planets are formed on well separated and stable orbits only inside ~2 AU. Their number typically ranges from 2 to 4, depending on the simulations, and their masses are in the range Mars mass – Earth mass (Chambers and Wetherill, 1998; Agnor et al., 1999).

(ii)   Quasi-tangent collisions of Mars-mass embryos onto the proto-planets are quite frequent (Agnor et al., 1999). These collisions are expected to generate a disk of ejecta around the proto-planets (Canup and Asphaug, 2001), from which a satellite is likely to accrete (Canup and Esposito, 1996). This is the standard, generally accepted, scenario for the formation of the Moon (see below, Section 3.2)

(iii) The accretion timescale of the terrestrial planets is ~100 Myr. This is compatible with several constraints on the chronology of accretion coming from geochemistry (Allègre et al. 1995). On the other hand, Hf–W chronology seems to indicate that the formation of the Earth's core occurred within the first 40 Myr (Yin et al., 2002; Kleine et al., 2002). This might suggest that the Earth accretion was faster than it appears in the simulations. However, if the cores of the embryos are not mixed with the mantles during the collisions – as indicated by SPH simulations (Canup and Asphaug, 2001) – this timescale would measure the mean differentiation age of the embryos that participated to the formation of the Earth, and not the time required for our planet to accrete most of its mass.

(iv)   All the embryos located beyond 2 AU are eliminated in 2/3 of the simulations (Chambers and Wetherill, 2001). They either are dynamically ejected from the solar system, or collide with the Sun, or are accreted by the forming terrestrial planets.

(v)   In the same time, the small planetesimals are subject to the combined perturbations of the giant planets and of the embryos (Petit et al., 2001). The dynamical excitation increasing very rapidly (timescale 1–2 Myr), most of the small planetesimals are eliminated in a few million years by either the ejection from the solar system, or the collision with the Sun or with a growing planet. In the asteroid belt (2–4 AU range), this leads to a remaining population of small bodies (the asteroids) on stable orbits with quite large eccentricities and inclinations, which contains only a very small fraction of the total mass initially in the region. This scenario explains well the current mass deficit of the asteroid belt, the eccentricity and semi-major axis distribution of the largest asteroids and other more subtle properties of the asteroid belt population, such as the partial mixing of taxonomic types.

However, this scenario of terrestrial planet formation suffers from some weaknesses:

(i)   The final orbits of the planets formed in the simulations are typically too eccentric and/or inclined with respect to the real ones. This could be due to the fact that the current simulations neglect the so-called phenomenon of *dynamical friction*, namely the effect of a large population of small bodies, carrying cumulatively a mass comparable to that of the proto-planets. Dynamical friction should damp the eccentricities and inclinations of the most massive bodies.

(ii)   Obliquities of the terrestrial planets should have random values. However in reality, only one planet has a retrograde spin (Venus). Moreover all planetary obliquities are compatible with an initial 0-degrees obliquity, modified by the subsequent evolution in the framework of the current architecture of the planetary system (Laskar and Robutel, 1993).

(iii)   The planet formed in the simulations approximately at the location of Mars is typically too massive.

3.3.2. THE FORMATION OF THE MOON

ALESSANDRO MORBIDELLI

As explained above, the currently accepted model for the Lunar formation is that of a giant impact occurring during the formation of the Earth. The current view of terrestrial planet formation implies several giant impacts, and impacts with an angular momentum similar of that of the Earth–Moon system are not rare, particularly during the end of the accretion process (Agnor et al., 1999). Simulations of a Moon-forming impact have been done since 1986, using SPH simulations. The most advanced, recent high-resolution simulations have been done by Canup and Asphaug (2001) and Canup (2004a). A very detailed review on the Moon formation can be found in Canup (2004b).

In the SPH simulations, three critaria has been used to judge the degree of success: the formation of a circumplanetary disk with about a Lunar mass outside of the Roche radius of the Earth, a mass of iron in the protolunar disk that is about 10% of the total mass (the fraction present inside the Moon), and a total angular momentum of the Earth-disk system of order of that of the current Earth–Moon system. In essence, in case of an "early" formation of the Moon, when the proto-Earth was only 60% of the current Earth mass, the impactor needs to be of about 30% of the total mass (proto-Earth + impactor). If the impact is late, the impactor can be of about 10% of the total mass, namely of order of the mass of Mars. The authors privilege the "late impact" scenario, because otherwise the Moon would have accreted too many siderophile elements after its formation, assuming that it accreted about 10% of the mass that is required to collide with the Earth to complete the Earth's formation and a chondritic compostion of such material.

The differentiation of the Moon can be dated using the Hf–W chronometer, and turns out to have occurred at about 40 Myr after CAI formation (Yin et al., 2002; Kleine et al., 2002). Thus, if the Moon-forming impact was the last one (or close to the last one), the Earth formed (i.e., received its last giant impact) in a similar timescale. If on the contrary, the Moon-forming impact was an early one, the formation of the Earth might have taken longer. The Hf–W chronometer seems to indicate an age of 40 My also for the differentiation of the Earth, but in this case the interpretation is less straightforward because, in case there is little equilibration between the core and the Mantle during the giant impacts, the overall mechanical accretion process of the Earth could have taken significantly longer (Sasaki and Abe, 2004).

In the SPH simulation of the Moon-forming impact, about 80% of the material that ends in the proto-lunar disk comes from the impactor, rather than from the proto-Earth. The Moon and the Earth have a very similar composition under many aspects (for instance the oxygen isotopic composition is identical). The logical interpretation is then that the proto-Lunar impactor and the proto-Earth had very similar compositions (and identical oxygen isotope composition). This however seems in conflict with the results of N-body simulations of the accretion of terrestrial planets. These simulations show that a planet forms by accreting material in a stochastic way from a wide variety of heliocentric distances. Given that the oxygen isotope compostion of Mars and of all meteorite classes differ from each other, it then seems unlikely that the proto-Earth and the lunar impactor could have exactly the same resulting composition. A solution of this apparent paradox has been proposed by Pahlevan and Stevenson (2005). During the formation of the proto-lunar disk there might have been enough isotopic exchange with the proto-Earth to equilibrate the two distributions. Another possibility is that the fraction of the proto-Lunar disk of terrestrial origin was in reality much larger than estimated in the SPH simulations. In fact these simulations assume a non-spinning Earth, so that the totality of the angular momentum of the Earth-Moon system is carried by the impactor. This has the consequence of having the impactor on a quasi-tangent trajectory relaitve to the proto-Earth, which maximizes the amount of impactor material that goes in geocentric orbit. If the Earth was spinning fast at the time of the Moon formation, then the impactor might have had a more head-on trajectory and less of its material would have contaminated the proto-Lunar disk. SPH simulations with a fast rotating Earth have never been done up to now.

The accretion of the Moon from a protolunar disk has been simulated by Ida et al. (1997) and Kokubo et al. (2000). The Moon forms very quickly, in about 1 year, at a distance from the Earth of about 1.2 to 1.3 Roche radii. In many cases, the Moon is accompanied by many moonlets, left-over of the accretion process in the disk. More rarely, two Moons of similar mass are formed. The subsequent evolution of these systems, simulated in Canup et al. (1999), typically leads to an end-state with a single Moon on a stable orbit.

3.3.3. TOWARDS 1 GYR: THE EARLY EVOLUTION OF THE EARTH

BERNARD MARTY

3.3.3.1. *Formation and closure of the Earth's atmosphere*
The formation of the terrestrial atmosphere is a suite of complex processes which are not yet fully understood. Indeed its abundance and isotopic compositions differ from those of the solar nebula and of other known cosmochemical components like comets or primitive meteorites. The general consensus is that our atmosphere was formed by contributions of several volatile-rich components and that it was subject to several episodes of loss to space throughout its early history. Some of these loss events were non-fractionating, that is, elements and isotopes were lost in the same proportion. This might have been the case of large impacts. Because the isotopic composition of some of the atmospheric elements such as noble gases differ from known end-member compositions, one has to consider also that

isotopic fractionation took place, and that there are several escape processes for the atmosphere that have been demonstrated to isotopically fractionate noble gases. Among these, there are the *thermal loss*, in which the velocity of a given isotope is higher than that of another given isotope, allowing the former to escape at a larger rate, and the *pick-up ion loss* in which atoms at the top of the atmosphere are ionized during charge transfer from solar wind ions, light isotopes being statistically more prone for such ionization. Hence it is convenient to define an epoch at which the atmosphere became closed to further loss into space.

During at least the first tens of Myr of its existence, while the Sun was still on its way to the main sequence, the Earth was subject to large-scale igneous (volcanic) events, during which the proto-mantle exchanged volatile elements with surface reservoirs. It is likely that a primitive atmosphere existed at this time, but any chemical record of it has been erased owing to the high thermal state of the Earth evidenced by core formation and magma ocean episodes. Records of atmospheric processes at this time cannot be found directly at the Earth's surface at Present. The record of extinct radioactivity systems in which parents differ from daughters by their respective volatilities give strong clues on the timing of terrestrial differentiation and the early cycle of volatile elements. Noble gases are chemically inert and their isotopic composition can only be modified by kinetic fractionation, or mixing with nucleosynthetic components, or through nuclear reactions including extinct radioactivity decays.

Xenon, the heaviest stable noble gas, is of particular interest because some of its isotopes are the radioactivity products of three different decay systems covering contrasted time intervals. Iodine is a volatile element for which one isotope, ^{129}I decays with $T_{1/2} = 15.7$ Myr to ^{129}Xe. During terrestrial magma ocean episodes, xenon, which has presumably a higher volatility than iodine, was degassed preferentially to it. The amount of ^{129}Xe in the atmosphere, in excess of the non-radiogenic xenon composition, corresponds to a "closure" interval of about 100 Myr (Allègre et al., 1995). Put in other words, only a tiny fraction of radiogenic ^{129}Xe has been retained in the atmosphere, showing evidence that the atmosphere was open to loss in space for at least several tens of Myr. The terrestrial mantle has kept even less ^{129}Xe. Heavy xenon isotopes e.g., ^{136}Xe, are produced by the spontaneous fission of ^{244}Pu ($T_{1/2} = 82$ Myr) so that the combination of both chronometers allows one to date the closure of the terrestrial mantle at 60–70 Myr (Kunz et al., 1998). This age could be interpreted as averaging the period during which the magma ocean episodes declined enough to quantitatively retain is volatile elements. Notably, heavy Xe isotopes are also produced by the spontaneous fission of ^{238}U which decays with a half-life of 4.45 Gyr, so that it is possible to compute closure ages based on ^{244}Pu–^{238}U–^{136}Xe. Results indicate a closure age of about 400–600 Myr

(Yokochi and Marty, 2005), which could correspond to a significant decline in mantle geodynamics. Xe isotope variations are therefore consistent with large-scale differentiation within the first 100 Myr and also with prolonged mantle activity at rates much higher than at Present during the Hadean. A comparable scenario was derived from other radioactive tracers, notably the ^{146}Sm–^{142}Nd and ^{147}Sm–^{143}Nd systems, and this view is fully consistent with models linking the thermal state of the Earth with the rate of mantle convection.

3.3.3.2. Geological evidence: Core differentiation, magnetic field

The time it took to build the Earth and the other terrestrial planets has been investigated from two different approaches, theoretical modeling on one hand (see previous sections), and absolute chronology on the other hand. In short, numerical simulations indicate that the growth of terrestrial planets was a geologically fast process. In the turbulent nascent solar system, dust accreted into small bodies, for which gravity became the main coalescent agent, within 10^4–10^5 years. Models predict that bodies with sizes up to that of Mars accreted within a few Myr. Independent evidence for rapid growth stems from coupled variations of ^{142}Nd, ^{182}W, ^{129}Xe and $^{131-136}$Xe produced by extinct radioactivities of ^{144}Sm ($T_{1/2} = 106$ Myr), ^{182}Hf (9 Myr), ^{129}I (15.7 Myr) and ^{244}Pu (82 Myr) observed in Martian meteorites (Halliday et al., 2001; Marty and Marti, 2002). These geochemical tracers attest that Martian differentiation including core formation, crustal development and mantle degassing took place within ≤ 15 Myr, which is remarkably short, and contemporaneous to the differentiation of parent bodies of meteorites. Furthermore, they also indicate that these heterogeneities were not re-homogenized later on as is the case of the Earth, and therefore attest for a low mantle convection rate on Mars, if any.

As dicussed previously, it took about 100 Myr to complete the Earth, considerably longer than the time of Mars-sized object formation (Wetherhill, 1980). This longer time interval is the result of the decreasing probability for collisions to occur in an increasingly depleted population of growing bodies. This modeling approach does not allow one to get more detailed chronological definition within this time frame, which needs to be constrained independently. The tungsten isotopic composition of the Earth compared to that of the nascent solar system as recorded in primitive meteorites supports large-scale differentiation within a few tens of million years. ^{182}Hf decays into ^{182}W with a half-life $T_{1/2} = 9$ Myr. During metal–silicate differentiation, tungsten is siderophile, that is, concentrates preferentially in metal phase whereas hafnium prefers the silicates (lithophile). If, in a differentiated body, this event took place when ^{182}Hf was still present, the isotopic composition of tungsten is different from the solar one as recorded for example in undifferentiated meteorites.

In pioneering attempts to use this chronometer, no difference in the tungsten isotopic composition was found between carbonaceous chondrites (the most primitive meteorites found so far) and terrestrial silicates, leading the authors to conclude that the last global Hf–W differentiation of the Earth happened after all [182]W decayed, in practice ≥ 60 Myr after the start of solar system condensation t_0 (Lee and Halliday, 1995). More recently, differences between chondrites and the Earth have been found (Kleine et al., 2002; Yin et al., 2002), implying a mean metal–silicate differentiation of 30 Myr for the Earth if terrestrial differentiation was a single and global event, with a possible range of 11–50 Myr for more realistic accreting conditions. A collision between a Mars-sized object and a growing proto-Earth is consistent with the unique Earth–Moon angular momentum. Numerical models for such a giant impact indicate that the proto-Earth was severely disrupted while material from the proto-mantle and the impactor spiraled at high temperature and formed the Moon (see above, Section 3.3.2). It is therefore logical, even if it has not yet been demonstrated, to ascribe to this collision the last global differentiation between metal and silicate recorded in the Hf–W system. Recent tungsten data for lunar basalts indicate a [182]W anomaly for the lunar mantle, interpreted as record for differentiation at 45 ± 4 Myr (Kleine et al., 2005). These authors proposed that this age represented the end of magma ocean episodes on the Moon.

The earliest record of a geomagnetic field dates back to the Archean (Hale and Dunlop, 1984; Yoshihara and Hamano, 2004), some 1 Gyr after the formation period. The terrestrial magnetic field has a major role in preventing solar wind ions to interact extensively with the top of the atmosphere, and create isotope fractionation of atmospheric elements by pick-up charge exchange. It also tends to preserve the surface of the Earth from cosmic-ray bombardment that are lethal for the development of organic chemistry. However, there is no evidence for the existence of magnetic field induced by the geodynamo during the first Gyr.

References

Abergel, A. et al.: 1996, *Astron. Astrophys.* **315**, L329–L332.

Adams, F. C. and Myers, P. C.: 2001, *Astrophys. J.* **553**, 744–753.

Adams, F. C., Hollenbach, D., Laughlin, G. and Gorti, U.: 2004, *Astrophys. J.* **611**, 360–379.

Adams, F. C., Proszkow, E. M., Fatuzzo, M., and Myers, P. C.: 2006, *Astrophys. J.* **641**, 504–525.

Agnor, C. B., Canup, R. M. and Levison, H. F.: 1999, *Icarus* **142**, 219–237.

Alibert, Y., Mordasini, C., Benz, W. and Winisdoerffer, C.: 2005, *Astorn. Astrophys.* **434**, 343–353.

Allègre, C. J., Manhes, G. and Gopel, C.: 1995, *Geochim. Cosmochim. Acta.* **59**, 1445–1456.
Amelin, Y., Krot, A. N., Hutcheon, I. D. and Ulyanov, A. A.: 2002, *Science* **297**, 1678–1683.
Anders, E. and Grevesse, N.: 1989, *Geochim. Cosmochim. Acta* **53**, 197–214.
André, P. and Montmerle, T.: 1994, *Astrophys. J.* **420**, 837–862.
André, P., Ward-Thompson, D. and Barsony, M.: 2000, in V. Mannings, A. P. Boss and S. S. Russell (eds.), Protostars and Planets IV, University of Arizona Press, Tucson, pp. 59–97.
Baker, J., Bizzarro, M., Wittig, N., Connelly, J. and Haack, H.: 2005, *Nature* **436**, 1127–1131.
Balbus, S. A. and Hawley, J. F.: 1991, *Astrophys. J.* **376**, 214–233.
Bate, M. R. and Bonnell, I. A.: 2004, in H. J. G. L. M. Lamers, L. J. Smith, and A. Nota (eds.), *The Formation and Evolution of Massive Young Star Clusters. ASP Conf. Ser.*, 322, 289.
Beaulieu, J.-Ph. et al.: 2006, *Nature* **439**, 437–440.
Benz, W.: 2000, *Sp. Sci. Rev.* **92**, 279–294.
Bertout, C.: 1989, *Ann. Rev. Astron. Astrophys.* **27**, 351–395.
Bizzarro, M., Baker, J. A. and Haack, H.: 2004, *Nature* **431**, 275–278.
Boss, A. P.: 2003, *Astrophys. J.* **599**, 577–581.
Bouvier, A., Blichert-Toft, J., Vervoort, J. D., McClelland, W. and Albarède, F.: 2005, *Geochim. Cosmochim. Acta Suppl.* **69**(Suppl. 1), A384.
Bouvier, J. et al.: 1999, *Astron. Astrophys.* **349**, 619–635.
Bouvier, J. et al.: 2003, *Astron. Astrophys.* **409**, 169–192.
Brearley, A. and Jones, R. H.: 1998, in J. J. Papike (ed.), Planetary materials. *Rev. Mineral.* **36**, 3/1–3/398.
Busso, M., Gallino, R. and Wasserburg, G. J.: 2003, *Pub. Astr. Soc. Austr.* **20**, 356–370.
Canup, R. M.: 2004a, *Icarus* **168**, 433–456.
Canup, R. M.: 2004b, *Ann. Rev. Astron. Astrophys.* **42**, 441–475.
Canup, R. M. and Asphaug, E.: 2001, *Nature* **412**, 708–712.
Canup, R. M., Levison, H. F. and Stewart, G. R.: 1999, *Astr. J.* **117**, 603–620.
Canup, R. M. and Esposito, L. W.: 1996, *Icarus* **119**, 427–446.
Carpenter, J. M., Wolf, S., Schreyer, K., Launhardt, R. and Henning, T.: 2005, *Astr. J.* **129**, 1049–1062.
Chambers, J. E. and Wetherill, G. W.: 1998, *Icarus* **136**, 304–327.
Chambers, J. E. and Wetherill, G. W.: 2001, *Met. Plan. Sci.* **36**, 381–399.
Chambers, J. E.: 2001, *Icarus* **152**, 205–224.
Chaussidon, M. and Gounelle, M.: 2006, in D. Lauretta and L. Leshin (eds.), *Meteorites & Early Solar System II.* Tucson, University of Arizona Press, pp. 323–340.
Chaussidon, M., Libourel, G. and Krot, A. N.: 2006, *Lunar Planet. Sci.* **36**, 1335.
Chaussidon, M., Robert, F. and McKeegan, K. D.: 2006, *Geochim. Cosmochim. Acta* **70**, 224–245.
Chokshi, A., Tielens, A. G. G. M. and Hollenbach, D.: 1993, *Astrophys. J.* **407**, 806–819.
Clark, P. C., Bonnell, I. A., Zinnecker, H. and Bate, M. R.: 2005, *Mon. Not. Roy. Astr. Soc.* **359**, 809–818.
Crutcher R. M.: 2005, in *Magnetic fields in the Universe: From Laboratory and Stars to Primordial Structures, AIP Conf. Proc.*, vol. 784, pp. 129–139.
Diehl, R., Halloin, H., Kretschmer, K., Lichti, G. G., Schönfelder, V., Strong, A. W., von Kienlin, A., Wang, W., Jean, P., Knödlseder, J., et al.: 2006, *Nature* **439**, 45–47.
Diehl, R., Cerviño, M., Hartmann, D. H. and Kretschmar, K.: 2004, *New Astr. Rev.* **48**, 81–86.
Donati, J.-F., Paletou, F., Bouvier, J. and Ferreira, J.: 2005, *Nature* **438**, 466–469.
Draine, B. T.: 2006, *Astrophys. J.* **636**, 1114–1120.
Duchêne, G., McCabe, C., Ghez, A. M. and Macintosh, B. A.: 2004, *Astrophys. J.* **606**, 969–982.
Ehrenfreund, P. and Charnley, S. B.: 2000, *Ann. Rev. Astr. Ap.* **38**, 427–483.

Falgarone, E., Hily-Blant, P., Pety, J. and Pineau Des Forêts, G.: 2005 in *Magnetic fields in the Universe: From Laboratory and Stars to Primordial Structures, AIP Conf. Proc.*, vol. 784, pp. 299–307.

Favata, F., Micela, G., Silva, B., Sciortino, S. and Tsujimoto, M.: 2005, *Astron. Astrophys.* **433**, 1047–1054.

Feigelson, E. D.: 2005, *Met. Planet. Sci.* **40**(Suppl.), 5339.

Feigelson, E. D. and Montmerle, T.: 1999, *Ann. Rev. Astron. Astrophys.* **37**, 363–408.

Feigelson, E. D., Getman, K., Townsley, L., Garmire, G., Preibisch, T., Grosso, N., Montmerle, T., Muench, A. and McCaughrean, M.: 2005, *Astrophys. J. Suppl. Ser.* **160**, 379–389.

Ferreira, J., Pelletier, G. and Appl, S.: 2000, *Mon. Not. R. Astron. Soc.* **312**, 387–397.

Ferreira, J. and Casse, F.: 2004, *Astrophys. J.* **601**, L139–L142.

Fleming, T. and Stone, J. M.: 2003, *Astrophys. J.* **585**, 908–920.

Fromang, S., Balbus, S. A., Terquem, C. and De Villiers, J.-P.: 2004, *Astrophys. J.* **616**, 364–375.

Fromang, S., Terquem, C. and Balbus, S. A.: 2002, *Mon. Not. R. Astron. Soc.* **329**, 18–28.

Gallino, R., Busso, M., Wasserburg, G. J. and Straniero, O.: 2004, *New Astr. Rev.* **48**, 133–138.

Galy, A., Young, E. D., Ash, R. D. and O'Nions, R. K.: 2002, *Science* **290**, 1751–1753.

Getman, K. V., Feigelson, E. D., Grosso, N., McCaughrean, M. J., Micela, G., Broos, P., Garmire, G. and Townsley, L.: 2005, *Astrophys. J. Suppl. Ser.* **160**, 353–378.

Glassgold, A. E., Feigelson, E. D. and Montmerle, T.: 2000, in V. Mannings, A. P. Boss and S. S. Russell (eds.), Protostars and Planets IV, University of Arizona Press, Tucson, pp. 429–455.

Glassgold, A. E., Feigelson, E. D., Montmerle, T. and Wolk, S.: 2005, in A. N. Krot, E. R. D. Scott and B. Repurth (eds.), *Chondrites and the Protoplanetary Disk. ASP Conf. Ser.*, vol. 341, pp. 165–182.

Goldreich, P. and Ward, W. R.: 1973, *Astrophys. J.* **183**, 1051–1061.

Goodwin, S. P., Whitworth, A. P. and Ward-Thompson, D.: 2004, *Astron. Astrophys.* **423**, 169–182.

Gounelle, M. and Russell, S. S.: 2005, *Geochim. Cosmochim. Acta* **69**, 3129–3144.

Gounelle, M., Shu, F. H., Shang, H., Glassgold, A. E., Rehm, K. E. and Lee, T.: 2004, *Lunar Planet. Sci. Conf.* **35**, 1829.

Gounelle, M., Shu, F. H., Shang, H., Glassgold, A. E., Rehm, K. E. and Lee, T.: 2001, *Astrophys. J.* **548**, 1051–1070.

Greaves, J. S.: 2005, *Mon. Not. R. Astron. Soc. Letters* **364**, L47–L50.

Greenberg, R., Wacker, J. F., Hartmann, W. K. and Chapman, C. R.: 1978, *Icarus* **35**, 1–26.

Grosso, N., Alves, J., Wood, K., Neuhäuser, R., Montmerle, T. and Bjorkman, J. E.: 2003, *Astrophys. J.* **586**, 296–305.

Guillot, T.: 1999, *Science* **286**, 72–77.

Guillot, T.: 2005, *Ann. Rev. Earth Plan. Sci.* **33**, 493–530.

Guillot, T. and Hueso, R.: 2006, *Mon. Not. R. Astron. Soc.* **367**, L47–L51.

Guilloteau, S., Dutrey, A. and Simon, M.: 1999, *Astr. Astrophys.* **348**, 570–578.

Haisch, K. E., Lada, E. A. and Lada, C. J.: 2001, *Astrophys. J. Lett.* **553**, L153–L156.

Hale, C. J. and Dunlop, D.: 1984, *Geophys. Res. Lett.* **11**, 97–100.

Halliday, A. N., Wänke, H., Birck, J. L. and Clayton, R. N.: 2001, *Space Sci. Rev.* **96**, 1–34.

Harker, D. E. and Desch, S. J.: 2002, *Astrophys. J. Lett.* **565**, L109–L112.

Harris, M. J., Knödlseder, J., Jean, P., Cisana, E., Diehl, R., Lichti, G. G., Roques, J.-P., Schanne, S. and Weidenspointner, G.: 2005, *Astron. Astrophys.* **433**, L49–L52.

Hayashi, C.: 1966, *Ann. Rev. Astron. Astrophys.* **4**, 171–192.

Hayashi, C.: 1981, *Prog. Theor. Phys.* **70**(Suppl.), 35–53.
Hayashi, M., Shibata, K. and Matsumoto, R.: 2000, *Adv. Sp. Res.* **26**, 567–570.
Heiles, C., Haffner, L. M., Reynolds, R. J. and Tufte, S. L.: 2000, *Astrophys. J.* **536**, 335–346.
Hester, J. J. and Desch, S.: 2005, in A. N. Krot and E. R. D. Scott and B. Reipurth (eds.), *Chondrites and the Protoplanetary Disk. ASP Conf series*, vol. 341, pp. 107–130.
Hewins, R. H.: 1997, *Ann. Rev. Earth Planet. Sci.* **25**, 61–83.
Hewins, R. H., Connolly, H. C. Jr., Lofgren, G. E. and Libourel, G.: 2005, in A. N. Krot, E. R. D. Scot and B. Reipurth (eds.), *Chondrites and the Protoplanetary Disk. ASP Conf. Series*, vol. 341, pp. 286–316.
Hillenbrand, L. A.: 1997, *Astron. J.* **113**, 1733–1768.
Hillenbrand, L. A.: 2006, in Livio M. (ed.), *A Decade of Discovery: Planets Around Other Stars. STScI Symposium Series 19*, in press.
Huss, G. R. and Tachibana, S.: 2004, *Lunar Planet. Sci. Conf.* **35**, 1811.
Ida, S., Canup, R. M. and Stewart, G. R.: 1997, *Nature* **389**, 353–357.
Ida, S. and Makino, J.: 1993, *Icarus* **106**, 210–227.
Ilgner, M. and Nelson, R. P.: 2006, *Astron. Astrophys.* **445**, 205–222.
Johns-Krull, C. M. and Gafford, A. D.: 2002, *Astrophys. J.* **573**, 685–698.
Kessler-Silacci, J. E., Augereau, J.-C. and Dullemond, C. P. et al.: 2006, *Astrophys. J.* **639**, 275–291.
Kleine, T., Metzger, K. and Palme, H.: 2005, *Lunar Planet. Sci.* **36**, 1940, CD-ROM.
Kleine, T., Munker, C., Mezfer, K. and Palme, H.: 2002, *Nature* **418**, 952–955.
Kokubo, E. and Ida, S.: 1998, *Icarus* **131**, 171–178.
Kokubo, E., Ida, S. and Makino, J.: 2000, *Icarus* **148**, 419–436.
Kroupa, P.: 2002, *Science* **295**, 82–91.
Krumholz, M. R., McKee, C. F. and Klein, R. I.: 2005, *Nature* **438**, 332–334.
Kunz, J., Staudacher, T. and Allègre, C. J.: 1998, *Science* **280**(5365), 877–880.
Laskar, J. and Robutel, P.: 1993, *Nature* **361**, 608–612.
Lee, D. C. and Halliday, A.: 1995, *Nature* **378**, 771–774.
Lee, T., Shu, F. H., Shang, H., Glassgold, A. E. and Rhem, K. E.: 1998, *Astrophys. J.* **506**, 898–912.
Lissauer, J.: 1987, *Icarus* **69**, 249–265.
Lyo, A.-Ran, Lawson, W. A., Mamajek, E. E., Feigelson, E. D., Sung, Eon-Chang and Crause, L. A.: 2003, *Mon. Not. R. Astron. Soc.* **338**, 616–622.
MacPherson, G. J., Davis, A. M. and Zinner, E. K.: 1995, *Meteoritics* **30**, 365–386.
MacPherson, G. J. and Huss, G. R.: 2003, *Geochim. Cosmochim. Acta* **67**, 3165–3179.
Matsumura, S. and Pudritz, R. E.: 2006, *Mon. Not. R. Astron. Soc.* **365**, 572–584.
Matsumoto, R., Machida, M., Hayashi, M. and Shibata, K.: 2000, *Progr. Th. Phys.* **138**(Suppl.), 632–637.
Matzner, C. D. and McKee, C. F.: 2000, *Astrophys. J.* **545**, 364–378.
Megeath, S. T., Hartmann, L., Luhman, K. L. and Fazio, G. G.: 2005, *Astrophys. J. Lett.* **634**, L113–L116.
Manhès G., Göpel C. and Allègre C. J.: 1988, Comptes Rendus de l'ATP Planétologie, 323–327.
Marty, B. and Marti, K.: 2002, *Earth Plan. Sci. Letts.* **196**(3–4), 251–263.
Masset, F., Morbidelli, A., Crida, A. and Ferreira, J.: 2006, *Astrophys. J.* **642**, 478–487.
Masset, F. and Snellgrove, M.: 2001, *Mon. Not. R. Astron. Soc.* **320**, L55–L59.
McKeegan, K. D. and Davis, A. M.: 2003, in H. Holl and K. Turekian (eds.), Treatise on Geochemistry, Elsevier-Pergamon, Oxford, pp. 431–460.
McKeegan, K. D., Chaussidon, M. and Robert, F.: 2000, *Science* **289**, 1334–1337.

Meyer, B. S.: 2005, in A. N. Krot, E. R. D. Scott and B. Reipurth (eds.), *Chondrites and the Protoplanetary Disk*, ASP Conf series, vol. 341, pp. 515–526.

Micela, G. and Favata, F.: 2005, *Sp. Sci. Rev.* **108**, 577–708.

Montmerle, T.: 1979, *Astrophys. J.* **231**, 95–110.

Montmerle, T.: 2002, *New Astr. Rev.* **46**, 573–583.

Montmerle, T.: 2005, in M. Gargaud, B. Barbier, H. Martin and J. Reisse (eds.), Lectures in Astrobiology, Springer, Heidelberg, pp. 27–59.

Morbidelli, A. and Levison, H. F.: 2004, *Astron. J.* **128**, 2564–2576.

Mostefaoui, S., Kita, N. T., Togashi, S., Tachibana, S., Nagahara, H. and Morishita, Y.: 2002, *Meteoritics Planet. Sci.* **37**, 421–438.

Mostefaoui, S., Lugmair, G. W., Hoppe, P. and El Goresy, A.: 2004, *New Astron. Rev.* **48**, 155–159.

Motte, F., André, P. and Neri, R.: 1998, *Astron. Astrophys.* **336**, 150–172.

Muzerolle, J., Calvet, N. and Hartmann, L.: 2001, *Astrophys. J.* **550**, 944–961.

Natta, A., Testi, L., Calvet, N., Henning, Th., Waters, R. and Wilner, D.: 2006, in: B. Reipurth (ed.), *Protostars and Planets V*. Tucson, University of Arizona Press, in press (eprint arXiv:astro-ph/0602041).

Nelson, R. P.: 2005, *Astron. Astrophys.* **443**, 1067–1085.

Ozawa, H., Grosso, N. and Montmerle, T.: 2005, *Astron. Astrophys.* **429**, 963–975.

Padoan, P., Jimenez, R., Juvela, M. and Nordlund, Å.: 2004, *Astrophys. J.* **604**, L49–L52.

Pahlevan, K. and Stevenson, D. J.: 2005, *Lunar Planet. Sci. Conf.* **36**, 1505.

Palla, F. and Stahler, S. W.: 1999, *Astrophys. J.* **525**, 772–783.

Petit, J.-M., Morbidelli, A. and Chambers, J.: 2001, *Icarus* **153**, 338–347.

Pety, J. and Falgarone, E.: 2003, *Astron. Astrophys.* **412**, 417–430.

Podolak, M.: 2003, *Icarus* **165**, 428–437.

Podosek, F. A., Zinner, E. K., MacPherson, G. J., Lundberg, L. L., Brannon, J. C. and Fahey, A. J.: 1991, *Geochim. Cosmochim. Acta* **55**, 1083–1110.

Pollack, J. B., Hubickyj, O., Bodenheimer, P., Lissauer, J. J, Podolak, M. and Greenzweig, Y.: 1996, *Icarus* **124**, 62–85.

Poppe, T. and Blum, J.: 1997, *Adv. Space. Res.* **20**, 1595–1604.

Poppe, T., Blum, J. and Henning, T.: 2000, *Astophys. J.* **533**, 454–471.

Reipurth, B. and Bally, J.: 2001, *Ann. Rev. Astron. Astrophys.* **39**, 403–455.

Ryter, C.: 1996, *Astr. Sp. Sci.* **236**, 285–291.

Sasaki, T. and Abe, Y.: 2004, *Lunar Planet. Sci. Conf.* **35**, 1505.

Shang, H., Glassgold, A. E., Shu, F. H. and Lizano, S.: 2002, *Astrophys. J.* **564**, 853–876.

Shu, F. H., Adams, F. C. and Lizano, S.: 1987, *Ann. Rev. Astron. Astrophys.* **25**, 23–81.

Shu, F. H., Shang, H., Glassgold, A. E. and Lee, T.: 1997, *Science* **277**, 1475–1479.

Stelzer, B. and Neuhäuser, R.: 2001, *Astron. Astrophys.* **377**, 538–556.

Stolper, E. and Paque, J. M.: 1986, *Geochim. Cosmochim. Acta* **50**, 1785–1806.

Tanga, P., Babiano, A., Dubrulle, B. and Provenzale, A.: 1996, *Icarus* **121**, 158–177.

Thommes, E. W., Duncan, M. J. and Levison, H. F.: 2003, *Icarus* **161**, 431–455.

Throop, H. B. and Bally, J.: 2005, *Astrophys. J.* **623**, L149–L152.

Tsujimoto, M., Feigelson, E. D., Grosso, N., Micela, G., Tsuboi, Y., Favata, F., Shang, H. and Kastner, J. H.: 2005, *Astrophys. J. Suppl. Ser.* **160**, 503–510.

Wasserburg, G. J.: 1987, *Earth Plan. Sci. Lett.* **86**, 129–173.

Weidenschilling, S. J.: 1977, *Mon. Not. R. Astron. Soc.* **180**, 57–70.

Weidenschilling, S. J.: 1980, *Icarus* **44**, 172–189.

Wetherhill, G. W.: 1980, *Ann. Rev. Astron. Astrophys.* **18**, 77.

Wetherhill, G. W.: 1990, *Ann. Rev. of Earth Plan. Sciences* **18**, 205–256.

Yin, Q., Jacobsen, S. B., Yamashita, K., Telouk, P., Blichert-Toft, J. and Albarède, F.: 2002, *Nature* **418**, 949–952.

Yokochi, R. and Marty, B.: 2005, *Earth Plan. Sci. Letts* **238**, 17–30.

Young, E. D., Simon, J. I., Galy, A., Russell, S. S., Tonui, E. and Lovera, O.: 2005, *Science* **308**, 223–227.

Yoshihara, A. and Hamano, Y.: 2004, *Precambr. Res.* **131**, 111–142.

Zanda, B.: 2004, *Earth Planet. Sci. Lett.* **224**, 1–17.

Earth, Moon, and Planets (2006) 98: 97–151
DOI 10.1007/s11038-006-9088-4

4. Building of a Habitable Planet

HERVÉ MARTIN
Laboratoire Magmas et Volcans, Université Blaise Pascal, Clermont-Ferrand, France
(E-mail: martin@opgc.univ-bpclermont.fr)

FRANCIS ALBARÈDE
Ecole Normale Supérieure, Lyon, France
(E-mail: albarede@ens-lyon.fr)

PHILIPPE CLAEYS
DGLG-WE, Vrije Universiteit Brussel, Brussels, Belgium
(E-mail: phclaeys@vub.ac.be)

MURIEL GARGAUD
Observatoire Aquitain des Sciences de l'Univers, Université Bordeaux 1, Bordeaux, France
(E-mail: gargaud@obs.u-bordeaux1.fr)

BERNARD MARTY
Ecole Nationale Supérieure de Géologie, Nancy, France
(E-mail: bmarty@crpg.cnrs-nancy.fr)

ALESSANDRO MORBIDELLI
Observatoire de la Côte d'Azur, Nice, France
(E-mail: morby@obs-nice.fr)

DANIELE L. PINTI
GEOTOP-UQAM-McGill, Université du Québec à Montréal, Montréal, Qc, Canada
(E-mail: pinti.daniele@uqam.ca)

(Received 1 February 2006; Accepted 4 April 2006)

Abstract. Except the old Jack Hills zircon crystals, it does not exit direct record of the first 500 Ma of the Earth history. Consequently, the succession of events that took place during this period is only indirectly known through geochemistry, comparison with other telluric planets, and numerical modelling. Just after planetary accretion several episodes were necessary in order to make life apparition and development possible and to make the Earth surface habitable. Among these stages are: the core differentiation, the formation of a magma ocean, the apparition of the first atmosphere, oceans and continents as well as the development of magnetic field and of plate tectonics. In the same time, Earth has been subject to extra-terrestrial events such as the Late Heavy Bombardment (LHB) between 3.95 and 3.8 Ga. Since 4.4–4.3 Ga, the conditions for pre-biotic chemistry and appearance of life were already met (liquid water, continental crust, no strong meteoritic bombardment, etc...). This does not mean that life existed as early, but this demonstrates that all necessary conditions assumed for life development were already present on Earth.

Keywords: Hadean, Archaean, continental growth, atmosphere and ocean formation, Late Heavy Bombardment

After planetary accretion several stages were necessary in order to make life apparition and development possible, in other words to make the Earth surface habitable. Among these stages are: the differentiation of Earth surface, the apparition of the first atmosphere, oceans and continents as well as the development of magnetic field and of plate tectonics.

4.1. Terrestrial Differentiation

FRANCIS ALBARÈDE

The Earth's primordial mineralogical composition and structure as well as the petrological nature of its surface are totally unknown. Plate tectonics, mountain building and erosion have long ago destroyed any geological record of this period. As developed in part 4.3, the oldest terrestrial material is a mineral, a zircon from Jack Hills (Australia), dated at 4.4 Ga, the oldest rock we can hold in our hand is a 4.1 Ga old gneiss from Acasta (Canada) and the oldest terrestrial swath of continental crust large enough for a field geologist to pace is 3.8 Ga old and located at Isua (Greenland), (Figure 4.1 and part 4.3). In order to understand when and how quickly the modern dichotomy between continental and oceanic crusts has emerged and whether it is relevant to the dynamics of the early Earth and the origin of life, we need to understand the structure of the planet left to us at the end of planetary accretion. To assist us in this task, we can rely on the oldest terrestrial

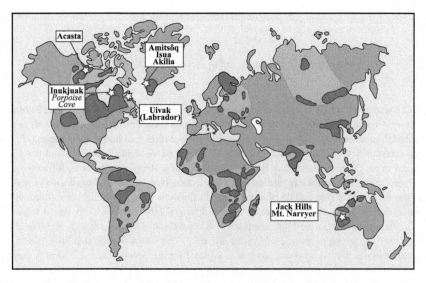

Figure 4.1. Simplified map showing the emplacement of the localities where very old continental crust has been dated. Exposed Archaean terranes are in red, and areas underlain by Archaean rocks are in orange.

material, but as explained in part 4.3, the record until 700 Ma after the Earth's formation comes out rather blurred. We can also proceed by comparison with other planets which have not apparently been as completely resurfaced as the Earth. First and foremost, intense cratering attests to the antiquity of the surface of both Mars and the Moon, while under similar standard, Venus's surface is young. Radiometric ages of Martian and Lunar samples are occasionally old, which supports cratering evidence, similarly chronological data on extinct radioactivities also supported the early Earth differentiation hypothesis. The parent body of some basaltic meteorites (Eucrites Parent Body = EPB) is also a particularly old object that can provide valuable information on the early inner solar system.

4.1.1. BULK COMPOSITION OF THE EARTH ($\sim$4.568 GA[1])

In order to understand how the Earth formed, it is first necessary to estimate 'what' our planet is made of. Our current knowledge of the Earth's composition is based on a number of observations and constrained by a number of deductions from our understanding of solar system formation processes:

- The composition of the most primitive carbonaceous chondrites (CI) is (except for gas, mainly H and He) identical to that of the solar photosphere. Any other type of meteorite has lost a substantial fraction of volatile elements (i.e. H_2O, K, Cl, S) due to the radiative activity of the young Sun, (most likely during its T-Tauri phase), and during the metamorphic alteration of the meteorite parent body.
- Planetary accretion is a quick process: the giant planets (Jupiter–type) are essentially gaseous and must form before the end of the T-Tauri phase (few Ma) which blew off all the nebular gas. Modern meteorite chronology using U-Pb and extinct radioactivity supports this scenario.
- Exchange of material across the solar system is very active, notably fed by the destabilization of planetary embryo orbits by Jupiter.
- Primordial material cruising around the Earth has a very limited life expectancy in orbit (<100 Ma).

These simple guidelines lead to the conclusion that, if the Earth accreted mostly from local material, our planet must be strongly depleted in volatile elements and consequently its interior must be particularly dry. Such depletion can be illustrated using alkali element ratio: in CI carbonaceous chondrites, K/U = 60,000 (refractory U is used as a reference), compared with 12,000 in the Earth, and 3000 in the Moon (Taylor, 2001). On the other hand, since a long time, it has been noticed that the modern $^{87}Sr/^{86}Sr$ of the Earth ($\sim$0.7045) is much too unradiogenic with respect to $^{87}Rb/^{86}Sr$ in

---

[1] The age of Earth formation (t_0) has been chosen as 4568 ± 1 Ma (Amelin et al., 2002; Bouvier et al., 2005).

chondrites; which points to Earth depletion in Rb (Gast, 1960). The Earth mainly formed from dry, volatile-element depleted indigenous material which is no longer available for analysis. Undepleted 'wet' material (ice-covered asteroids, comets) occasionally thrown off out of the outer solar system by Jupiter has been added providing the 'veneer' required by the geochemistry of platinum group elements and adding at the same time the water necessary to form the terrestrial ocean. Consequently, Earth probably formed by accumulation of planetary embryos of different sizes; the ultimate collision being the Moon-forming event resulting of the impact of the Mars-size embryo with the Earth.

In addition, oxygen isotopic ratios indicate that Earth's brew was a special one (Clayton, 1993). From one object to another, isotopic abundances can vary for three reasons:

(1) Thermodynamic fractionation, due to a temperature-dependent population of vibrational modes, in a common stock and which shows a smooth dependence with mass; the consistency of $\delta^{17}O$–$\delta^{18}O$ values is the base for assigning different meteorites to a single planetary body, notably Mars (SNC) and the eucrite–howardite–diogenite parent body;

(2) The survival of poorly mixed different nucleosynthetic components, as in refractory inclusions;

(3) Radiogenic ingrowths in systems with different parent/daughter ratios.

Both cases (1) and (2) are very clearly shown in a $\delta^{17}O$–$\delta^{18}O$ plot (Figure 4.2). There, the Earth and the Moon lie on a same fractionation line with the predicted slope of 0.5. This coincidence is one of the strongest arguments in favour of the two bodies having accreted on adjacent orbits. Only few CI carbonaceous chondrites and the enstatite ordinary chondrites lie on the terrestrial fractionation line, which gave rise to the conjecture that the Earth may be made of enstatite chondrites (Javoy, 1995). The $\Delta^{17}O$ elevation of the points representing the different planetary objects above the Earth–Moon fractionation line is a measure of their nucleosynthetic ^{16}O excess and indicates by how much their constituting material differed from Earth: oxygen being the most abundant element in the rocky planets, this difference should not be taken lightly.

4.1.2. ENERGETIC OF PLANETARY DIFFERENTIATION PROCESSES

Upon contraction of the solar nebula and collapse of the disk onto its equatorial plane, dust and particles collide with each other. Shock waves emitted by the contracting Sun are now thought to be largely responsible for the melting of chondrules (Hood and Horanyi, 1993). The presence of the

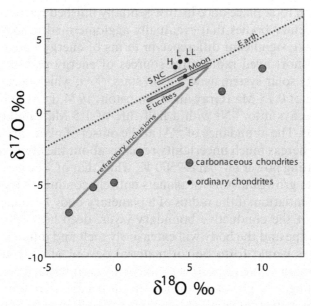

Figure 4.2. Plot of $\delta^{17}O$ vs. $\delta^{18}O$ in different planetary materials. These numbers represent excesses or deficits of ^{17}O and ^{18}O with respect to ^{16}O in a reference material, which is seawater. Thermodynamic isotopic fractionation is mass-dependent and moves the points representing actual samples along lines with a slope equal to $(17-16)/(18-16) = 0.5$. The most visible of these lines is the Earth-Moon common fractionation trend. $\delta^{17}O$ represents the elevation of a point above this line. Different planetary objects (planets or groups of meteorites) have different $\delta^{17}O$ values and therefore were created from material with a very different history in the solar nebula or even prior to the isolation of the nebula. SNC = Martian meteorites.

decay products of the short-lived ^{60}Fe in meteorites attests to the explosion of supernova(e) in the neighbourhood of the solar nebula (Wasserburg and Busso 1998). However, the energy emitted by all these sources has been largely re-irradiated into space and consequently it did not efficiently contribute to planetary differentiation.

Gravitational energy appears to be the main source of planetary energy for Mars-size objects and larger: the temperature increment associated with gravitational energy released by material falling at the surface of a planet with radius R, mean density ρ, and specific heat C_p is $(3/5)G\pi\rho\ R^2/C_p$ in which G is the universal constant of gravitation. This increment is ~1600 K for the Moon, 7600 K for Mars and 38,000 K for the Earth. Most of this energy is, however, irradiated back into space by the molten material and impact vapour and for most of the growth history of the planet, much smaller numbers should be considered. Another related source of energy is core formation, which, for the Earth, induced a temperature increase of worth more than 1500 K, and that remains largely buried into the planet. Whether core segregation took place once in an essentially finished Earth or took place in several planetary bodies that eventually agglomerated to make the Earth should not make significant difference in terms of energy budget.

Two main short-lived radioactive sources of energy played a prominent role in the early solar system as well as in stars: ^{26}Al, which decays into ^{26}Mg with a half-life of 0.75 Ma (Gray and Compston, 1974; Lee et al., 1976), and ^{60}Fe, which decays into ^{60}Ni with a half- life of 1.5 Ma (Shukolyukov and Lugmair, 1993). The abundance of ^{26}Al at the outset of planetary formation is well-known, whereas much uncertainty remains about the abundance of ^{60}Fe: the overall heating power of ^{26}Al is 9500 K, while that of ^{60}Fe could be as high as 6000 K. The growth history of planets entirely conditions its thermal history and differentiation: if the radius of a planetary body becomes larger than the thickness of the conductive boundary layer, deep heat will only slowly diffuse and escape and the body will extensively melt and differentiate. Slowly growing objects would form out of material devoid of ^{26}Al and ^{60}Fe and indefinitely remain a pile of porous rubble with no internal structure.

The early stages of planetary formation must have therefore seen a relatively large population of small (1–100 km) to medium-sized (100–1000 km) objects orbiting the Sun and colliding with each other. Some of these planetary objects, even the smallest, were molten and differentiated into a metallic core, an ultramafic mantle and a basaltic crust. We can think of the EPB as one of these. Others were only warm porous objects. The later stages differ from one planet to the other. The largest planets (Earth, Venus and Mars) were certainly large enough for the release of accretion energy to induce wholesale melting of their upper 500–2000 km giving rise to a magma ocean.

4.1.3. THE CHRONOMETERS OF ACCRETION AND DIFFERENTIATION

The last decade watched the emergence of powerful dating techniques (see part 2.2). In addition to substantial improvements on Secondary Ionization Mass Spectrometry (SIMS) and traditional Thermal Ionization Mass Spectrometry (TIMS), the advent of Multiple Collector Inductively Coupled Plasma Mass Spectrometry (MC ICP MS) produced most significant advances. For 'absolute' U-Pb dating, it is now possible to obtain U-Pb ages with a precision of a fraction of a million years. Similar precisions hold for extinct radioactivities such as ^{26}Al–^{26}Mg, ^{60}Fe–^{60}Ni, and ^{182}Hf–^{182}W.

In fact measuring 'absolute' ages in a rock or in a mineral consist in reality, in obtaining a cooling age. For instance in the U-Pb system the measured date is the time at which a particular phase stops exchanging Pb with surrounding minerals. Lead exchange stops at ∼800 °C in pyroxene and at ∼500 °C in phosphate (Cherniak et al., 1991; Cherniak, 2001). As discussed in part 4.1.4., planetary bodies went through a stage of metamorphic heating or even wholesale melting which may have lasted several million years. External (whole-rock) Pb–Pb isochrons potentially date the original U/Pb fractionation stage, such as the segregation of metal (core) from the silicate minerals and melts (mantle and crust) or the differentiation of different objects in the planet (mantle, crust). This technique however relies on a stringent assumption not necessarily met in natural objects: all the samples dated must have formed at precisely the same time and they had the same Pb isotope composition at that time, which is only likely to happen for samples with strong genetic relationships.

In contrast, extinct radioactivities date the extraction of material from the nebula (see part 2.2). Observing in a particular mineral an excess or a deficit with respect to the chondritic reference of a radiogenic nuclide, (i.e. ^{26}Al $\rightarrow$ ^{26}Mg), signals that the host rock picked up Al from the solar nebula at the time ^{26}Al still existed. The excess may vanish during metamorphic perturbation or melting, but it cannot be created after ^{26}Al has disappeared.

Because of the distinctive geochemical behaviour of the parent and daughter isotopes, different extinct radioactivities reflect different planetary events:

- The fastest accretion chronometers ^{26}Al–^{26}Mg ($T_{1/2} = 0.75$ Ma) and ^{60}Fe–^{60}Ni ($T_{1/2} = 1.5$ Ma) date extraction of the planetary material from the nebula.
- Preferential uptake of W and Ag by metal makes the ^{182}Hf–^{182}W ($T_{1/2} = 8.9$ Ma) and ^{107}Pd–^{107}Ag ($T_{1/2} = 6.5$ Ma) pairs ideal to date planetary core formation.
- Fractionation of the lithophile (silicate-loving) elements makes the ^{146}Sm–^{142}Nd chronometer ($T_{1/2} = 103$ Ma) suitable for the first 100's of Ma of planetary evolution, in particular with respect to the existence of a magma ocean (see below).

4.1.4. PETROLOGY OF PLANETARY DIFFERENTIATION

The Earth, Moon, Mars, Venus are all differentiated planets, which indicates that they went through a molten stage with ensuing core-mantle mantle-crust segregation. Planetary differentiation is the outcome of phase separation, metal-silicate first, then mineral-liquid silicates. Regardless of the composition of the molten planets, the mineral phases in equilibrium with most melts are fortunately few: olivine, three pyroxenes, an Al-bearing phase, and ilmenite. At higher pressures typical of the terrestrial mid-mantle, other phases such as perovskite may be present but their incidence on planetary differentiation is unclear.

Plagioclase stability is a parameter that plays a prominent role in planetary differentiation (Figure 4.3). Among Al-bearing phases stable in mantle

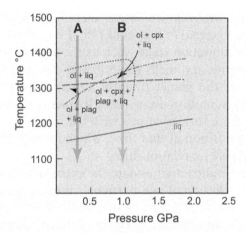

Figure 4.3. Solubility of minerals in a basaltic melt as a function of temperature and pressure. Pressure is proportional to depth, but also to the acceleration of the planet gravity (g). Cooling at low pressure (arrow A), on Vesta or the Moon, for example, leads to the formation of large volumes of buoyant plagioclase (plagioclase crystallizes before pyroxene) whereas cooling at higher pressure (arrow B), as in the Earth, where pyroxene crystallizes before plagioclase favours crystallization of dense olivine and pyroxene.

conditions, plagioclase is the only one stable at the lower pressure and consequently at shallowest depth. Plagioclase has three important properties related to its loose structure: (a) contrary to any other mantle phase, it is less dense that the parent melt where it crystallizes and therefore it floats to the surface instead of sinking; (b) it is fairly compressible and for that reason becomes more soluble in melt as pressure increases; (c) it more likely precipitates in dry melts rather than in hydrous melts. On a planet, plagioclase stability is first modulated by gravity: pressure is proportional to gravity acceleration g and $g = (4/3) G \pi \rho R$: in the Earth, plagioclase production in dry melts extends down to 30 km whereas in the Moon plagioclase is stable up to 180 km. In addition, by cutting Si–O–Si bonds to produce Si–OH groups, water breaks down the plagioclase component in melts. Hence, a planet with a strong gravity field (Earth) retained water whereas weak gravity field planets (Moon) are almost dry. Consequently, a molten planet with a deep plagioclase production zone such as the Moon, will produce a thick lithosphere of buoyant plagioclase-bearing rocks (gabbros, anorthosites), which will not be possible on bigger planets (Earth).

A different consequence of a strong gravity and of a wet surface is the high-temperature reaction of water with hot mantle and melts. The most abundant mineral dominating the shallow mantle and precipitating from magmatic liquids is olivine. It reacts with water to give serpentine at 500 °C, talc and other hydrous minerals at higher temperature. The density of all these minerals is remarkably low and tends to form a wet lid on top of the hot mantle (Boyet et al., 2003). This process also produces enormous quantities of hydrogen by damp oxidation of the ferrous component of olivine (fayalite + water → magnetite + silica + hydrogen) (Barnes and O'Neil, 1969; Sleep et al., 2004).

4.1.5. PLANETARY MAGMA OCEANS AND THE SURFACE OF THE PLANETS AT THE END OF ACCRETION (~4.56 TO ~4.45 GA)

The chronological framework emerging from recent data is tightening up. The first planetary objects probably formed shortly before 4.568 Ga (Amelin et al., 2002): the time from the condensation of the first refractory grains to the accretion of ~100 km size objects did not last more than a few Ma (Bizzarro et al., 2004). The smallest objects, such as the parent body of H and L ordinary chondrites, took 5–15 Ma to cool below the phosphate closure temperature (~500 °C) of Pb (Göpel et al., 1994). From ^{182}Hf–^{182}W evidence, terrestrial core formation was nearly complete within 30 Ma after solar system isolation (Kleine et al., 2002; Yin et al., 2002b).

By then, most of the terrestrial upper mantle was probably molten as a result of impacts on a high-gravity planet. The concept of a widespread magma ocean, however, goes back to the Moon and the Apollo missions (Wood et al., 1970). The first geochemical observations on lunar basalts indicated that they formed from a mantle source with an apparent deficit in two elements concentrated in plagioclase, Eu and Sr. Such a deficit could not result from solid-state processes and therefore it was argued that the lunar mantle went through a stage of wholesale melting during which enormous amounts of buoyant plagioclase-rich cumulates were floated towards the surface. Apollo 14 and subsequent missions brought back samples of anorthosite from the highlands which were identified with these buoyant cumulates. Since then, similar observations were made in eucrites. Other arguments, notably the ubiquity of ilmenite in soil minerals, have been successfully used to strengthen the hypothesis of a lunar magma ocean.

For decades, the existence of a terrestrial magma ocean met a strong resistance. The main reason was that the mineral assemblage that would crystallize in such a gigantic pool of magma would be less rich in plagioclase than the lunar highlands and would mostly contain minerals whose density is greater than magma density; therefore it would sink as it forms, letting heat escape so fast that the magma ocean would freeze on time scales of 10^4–10^6 years. This interpretation, however, ignores the existence of Earth's hydrosphere, a term which loosely bounds liquid water, vapour and gas. It is a modern concept based on orbital dynamics and on D/H ratios across the solar system that water was added to the surface from the outer solar system (Morbidelli et al., 2000). The interior of the Earth was nearly dry. Reaction between any form of water with rocks of mantle or even basaltic composition forms buoyant hydrous minerals (i.e. serpentine). These rocks, transformed into serpentinite, form rafts covering the magma surface and drastically reducing heat radiation into space. In 2003, the case for a terrestrial magma ocean was bolstered by the discovery of ^{142}Nd anomalies in 3.8 Ga old basalts at Isua (Greenland) (Boyet et al., 2003; Caro et al., 2003). These anomalies indicate that, 3.8 Ga ago, the mantle source of Isua basalts was still preserving the record of a major Sm/Nd fractionation event that took place within the first 100 Ma of the Earth's history. As for the Moon, such a fractionation can only be achieved by melt/mineral segregation: the current interpretation of these hard-won results is therefore that the ^{142}Nd excess records the wholesale melting and crystallization of the terrestrial upper mantle. The case of the magma ocean was further strengthened by Boyet and Carlson (2005) who identified a difference of ^{142}Nd abundances between the Earth's mantle and crust on the one hand, and chondrites on the other hand; this finding not only demonstrates that the terrestrial mantle was largely molten $\sim$30 Ma after the formation of the solar system, but also that this event was accompanied by the irreversible segregation at the base of the mantle of some material with properties distinct from those of the rest of the mantle. It is remarkable that ^{142}Nd exists in lunar rocks (Nyquist et al., 1995) but they are more subdued than those found for the Isua mantle source, which suggests that, as expected from experimental data, Sm/Nd fractionation was reduced by the saturation of the magma with plagioclase and probably that the magma ocean lasted longer in the Moon than in the Earth.

4.1.6. THE EARLY CRUST: TOWARD PLATE TECTONICS AND CONTINENTS (4.5–4.4 GA)

The nature of the earliest Earth surface is still in the eye of the beholder. Boyet and Carlson (2005) argue that the difference of ^{142}Nd abundances between the Earth's mantle and chondrites correspond to the foundering of the primordial crust at the core-mantle boundary where it may have formed the D'' seismic layer. By 4.5–4.4 Ga, plate tectonics was probably nowhere to be seen and continental crust in the modern sense could not form. In order to create plates, a relatively rigid mantle is needed, which, upon upwelling creates oceanic crust and its underlying lithosphere. Progressive cooling of the lithosphere creates negative buoyancy, which is resisted by plate bending and mantle viscous drag (Conrad and Hager, 1999). Even if we forget about the details of crust formation, dewatering and /or melting of the sinking plates seem to be required to form a SiO_2-rich crust. None of these characteristics are likely to exist on top of a cooling magma ocean overlain by a lithospheric lid rich in hydrous Mg-bearing minerals. Stronger constraints on the evolution of the terrestrial magma ocean are a prerequisite to the understanding of the initiation of the plate tectonics regime.

To wrap up with some speculations with what the earliest Earth's surface may have looked like, it was probably a mixed bag of basaltic to komatiitic flows erupted over or intruded into hydrous mafic and ultramafic rocks (serpentinite and amphibolite). The topographic lows were covered by whichever liquid water could condense from the thick dense atmosphere, in which nitrogen, ammonia, methane, and water gas dominated. Ammonia and methane were permanently destroyed by solar irradiation of the upper atmosphere but also permanently produced by widespread serpentinization of freshly erupted magma. Wet rocks flow more easily than their dry equivalent: arguably, by softening the underlying mantle, continuous foundering of lithosphere and its hydrous surface rocks, was instrumental into the transition to plate tectonics. Wet melts are not prone to plagioclase crystallization: occasional patches of more differentiated rocks (granites and granodiorite) would occur produced by melting of amphibolites sinking locally in the underlying mantle. They may represent parent rocks for the rare 4.4 Ga old Jack Hill zircons (Wilde et al., 2001): such rocks certainly herald the upcoming plate tectonic regime. The precise timing of the transition is not well understood but substantial progress is being made. The unique 4.4 Ga old zircon from Jack Hill has typical mantle ^{18}O abundances and may have formed from mantle-derived magmas (similar zircons are present in Icelandic rhyolites and even in lunar soils). In contrast, most of the $\delta^{18}O$ values of the 'younger' 4.3 Ga-old crystals are indistinguishable from those of the modern continents and bear the imprint of a source affected by low-temperature geological processes (Cavosie et al., 2005). In order to resolve this conundrum, Harrison et al. (2005) measured the Hf isotope compositions of the oldest Jack Hill zircons; they found evidence of Lu/Hf fractionation typical of continental crust extraction and concluded that the earliest crust formation events took place between 4.3 and 4.4 Ga ago. Although the transition between the magma ocean stage and plate tectonics may have lasted tens or even hundreds of million years, the oldest continental crust therefore formed in less than 150–200 Ma after the accretion of the planet.

4.2. Late Contributions

4.2.1. ATMOSPHERE

BERNARD MARTY

4.2.1.1. *The origin of terrestrial volatiles*
The Earth is highly depleted in volatile elements when compared to bodies located further away from the Sun, like giant planets or comets. Even

asteroids located 2–3 AU from the Sun (the Earth is by definition located at 1 AU from the Sun) and which are sampled by meteorites falling onto the Earth are richer in volatile elements like water, carbon, nitrogen, halogens and noble gases. The scarcity of volatile elements may result from the combination of several factors. The classical condensation model for the formation of the solar system stipulates that the existence of a temperature gradient in the protosolar nebula would have prohibited condensation of volatile elements in the inner solar system. Volatile elements such as rare gases, carbon and nitrogen found in carbonaceous chondrites are mostly hosted by carbonaceous phases which are prone to destruction during high temperature processes (<1300 K in laboratory experiments under low oxygen fugacity (fO_2) and <800 K in oxidizing condition). A fraction of volatile elements found in primitive meteorites is implanted from the solar corpuscular irradiation, presumably at an early stage of solar activity. These surface-sited components are also released at relatively low temperature. This view is consistent with the apparent concentration gradient of moderately volatile elements (e.g., alkali elements) among the Earth, Mars, Vesta, ordinary chondrites and carbonaceous chondrites (see part 4.1.1). However, recent models for the formation of the solar system propose that terrestrial planets grew up from multiple collisions with smaller bodies, the planetesimals. These later, originated from areas much wider than those specific of the different terrestrial planets (Morbidelli, 2002). Thus volatile elements including water could have been carried by impacting bodies originating from area "wetter" than the 1 AU one (Morbidelli et al., 2000) like the outer asteroid belt, and the Kuiper belt. In such a case, the volatile element gradient could be somewhat coincidental. The Earth is highly volatile-depleted, which could be the result of the giant lunar impact (see below), Mars could represent an Earth that did not experience such a giant impact, ordinary chondrites could represent non differentiated bodies whereas carbonaceous chondrites could originate from region beyond giant planets like comets. This scenario is not without problems. The most important is the similar oxygen isotopic compositions of both the Moon and the Earth, which can hardly result from mixing by any collisional process. If we take into account the diversity of O isotopic signatures among planetary bodies, it seems difficult to achieve such a similarity for two bodies formed by a random mix of smaller bodies from different regions of the mid plane.

The energy released by the Moon-forming impact might have degassed vigorously the proto-mantle. Indeed, lunar basalts are desperately dry and do not contain fluid inclusions even in their most refractory minerals, and the potassium content of the Moon is only half that of the Earth, which is itself about half of the chondritic abundance. This event took place within a few tens of Ma after the solar system started to form. Hence, the fate of terrestrial volatiles is linked to this destruction, which probably resulted in the blow off of any primitive atmosphere and in degassing of the proto-mantle. However, this degassing might not have been perfect as in the case of the Moon. Some primordial volatiles (i.e. volatile elements that were never at the Earth's surface) remain in the mantle, such as the rare isotope ^{3}He (Clarke et al., 1969; Mamyrin et al., 1969), and a neon component different from atmospheric neon and which isotopically resembles to solar neon (Sarda et al., 1988; Marty et al., 1989). The trapping of volatile elements in planetary bodies might have occurred when gas-rich dust coalesced and when it was buried by newly accreting material. Alternatively, these elements could have been "ingassed" from a massive atmosphere into a melted proto-mantle (Mizuno et al., 1980). A third, and not exclusive, possibility is that volatile elements were contributed at a late stage of terrestrial accretion to an essentially volatile-free proto-Earth after the lunar cataclysm. Each type of model has its advantage and inconvenient; in the following we present arguments mainly based on isotopic signatures, which allow further insight into both the source of terrestrial volatiles and the timing of their trapping.

Volatile elements in the solar system present a large variety of isotopic and elemental abundances that probably reflects processing in the solar nebula of an original gas, which is best preserved in the Sun, to some extent in the atmosphere of giant planets, and possibly in some cometary families. For hydrogen and nitrogen, isotopic variations are so large (a factor 7 for D/H and a factor 3 for ^{15}N/^{14}N) that mass-dependent fractionation seems difficult to advocate as the main source of isotopic variations. Nucleosynthetic anomalies with much larger effects are found in specific micro-phases of primitive meteorites but they cannot account for the mentioned large-scale isotopic variations in the solar system. Thus, D/H and N isotope variations measured among different reservoirs of the solar system are attributed by default to mixing with exotic components isotopically fractionated at very low temperature during e.g., ion-molecule exchanges in the interstellar medium.

The diversity of isotopic components for some of the key volatile elements presents the advantage of allowing insight into the origin of volatile elements trapped in planetary bodies and their atmospheres. Important advances in our understanding of the origin of atmospheric volatiles have stem from the ability to identify volatile elements in the terrestrial system in term of solar versus planetary origin. A pre-requisite for this approach is a good knowledge of the isotopic composition of volatile elements in the gas from which the solar system formed. The best analogue to this is probably the Sun itself, but our knowledge in this field is still very limited. The isotopic composition of light noble gases (He, Ne and Ar) has been measured in lunar soils implanted by the solar corpuscular irradiation (the low energy solar wind – $\sim$1 keV/amu – and more energetic emissions such as solar energetic particle – SEP, up to 1 MeV/amu) and in aluminium foils exposed by Apollo astronauts. These experiments have shown that the solar ^{20}Ne/^{22}Ne ratio is

30% lower than the atmospheric ^{20}Ne/^{22}Ne ratio. The isotopic composition of other volatile elements is not well known and measuring the isotopic composition of the solar wind was the goal of the Genesis mission which collected the solar wind for 27 months and returned (brutally) irradiated targets to the Earth on Sept. 8th, 2004 (to date, the samples have not yet been analyzed). The pre-deuterium burning hydrogen isotopic composition of the solar nebula is known from the D/H and ^{3}He/^{4}He analyses of the atmosphere of Jupiter by the Galileo project (Mahaffy et al., 1998). Studies show that the solar nebula is depleted by 87% in D relative to terrestrial oceans. Recently, depth profiling analysis of lunar soil grains by ion probe together with single grain laser analysis of lunar soils by static mass spectrometry allowed Hashizume et al. (2000) to propose that solar nitrogen is depleted by at least 24% in ^{15}N relative to terrestrial atmospheric N, a result subsequently confirmed by Galileo data analysis (Owen et al., 2001). These experiments corroborate that planetary bodies like the Earth and primitive meteorites (see below) are enriched in heavy, and rare, isotopes of H, N and Ne relative to the proto-solar nebula, thus making these elements important tracers of provenance for these elements.

A remarkable advance in our knowledge was the discovery of a mantle neon end-member having an isotopic composition similar to that inferred for the solar nebula (Sarda et al., 1988; Marty et al., 1989; Honda et al., 1991). Solar gas could have been trapped in accreting dust and buried during terrestrial accretion (Trieloff et al., 2000). However, the efficiency of this process appears too small (by several orders of magnitude) to supply the terrestrial inventory (Podosek and Cassen, 1994). Alternatively, nebular gas could have been trapped in the terrestrial mantle by dissolution of a massive, hydrogen-rich, solar-like atmosphere into a melted proto-mantle (Mizuno et al., 1980). The atmospheric pressure necessary to account for the noble gas inventory of the mantle is reasonable (few bars up to 100 bars, depending on models – Sasaki and Nakazawa, 1984; Porcelli and Pepin, 2000), and total or partial melting of the proto-mantle, under a massive atmosphere during accretion appears as realistic. In such a case, the limitation of the process is the lifetime of the solar nebula, less than 10 Ma at best (Podosek and Cassen, 1994), which imposes the Earth to have grown at a size sufficient enough to retain a massive hydrogen-rich atmosphere during this time interval. It also requires the proto-mantle to keep at least partially dissolved solar gases during the lunar cataclysm, despite extensive degassing of lunar material. Dissolution of solar noble gases in the core (Tolstikhin and Marty, 1998; Porcelli and Halliday, 2001) might have offered such an harvest and may also account for the association of solar-like neon with mantle plumes, which source is considered as very deep-seated, at the core-mantle boundary.

Recently, the isotopic composition of neon in the convective mantle which is the source of mid-ocean ridge basalts has been found to be similar to that of gas-rich meteorites, suggesting a mantle-scaled Ne isotope heterogeneity (Ballentine et al., 2005). It also indicates that a major fraction of terrestrial volatiles were contributed by chondritic matter, possibly after the Moon-forming event. Dust irradiated by the Sun in the early solar system and snowing gently at the Earth's surface could have been the vector of such contribution, but in this case, a process able to subduct this dust into the mantle without degassing it has to be found. Cold subduction in the Hadean could have done the job (Tolstikhin and Hofmann, 2005), but the occurrence for such a process is highly speculative.

Other sources than the solar nebula, are also required in order to match the isotopic composition of hydrogen and nitrogen in both terrestrial atmosphere and oceans. The D/H ratio of the oceans water (150×10^{-6}) coincides with the average value of primitive meteorites and is 7 times the nebular ratio (Robert et al., 2000). In addition, the atmosphere is enriched in ^{15}N relative to what has been claimed to be the signature of the proto-solar nebula (Hashizume et al., 2000). Such isotopic ratios are inconsistent with direct derivation from the solar nebula, but are consistent with a chondritic origin (Dauphas et al., 2000). Thus, in addition to a solar nebula contribution whose remnant is found in the deep mantle, another component likely carried by planetary bodies such as meteorites has contributed to the mantle inventory of volatile elements. It seems logical to consider that this contribution took place after the Moon-forming event because, as argued above, such event resulted in a drastic degassing of the mantle, only its deepest regions or the core having kept a record of the solar nebula.

4.2.1.2. *Early evolution of the atmosphere*

Atmospheric erosion due to large impacts could have played a significant role in reducing the atmospheric reservoir. Indeed, the speed of ground on the other side of the Earth due to a major collision might have exceeded the liberation speed of 11 km/s for the Earth and therefore triggered major atmosphere (and ocean) loss. However, the efficiency of this process in the Hadean is subject to discussion as it is unclear if the contribution of volatile elements by the impactors exceeded or not the loss of these elements. A potentially efficient process consists in gravitational escape of light elements, enhanced by the photo-dissociation of water by UV from the young sun, into molecular hydrogen and atomic H, and escape of the latter (Hunten et al., 1987). The so-called hydrodynamic escape process is able to drag atoms and molecules heavier than H, so that the escape of volatile elements from the atmosphere will last during the enhanced UV flux period, which could have lasted up to 150 Ma in the case of weak T-Tauri phase (Pepin, 1991). Atmospheric escape takes place when the velocity of gaseous atoms exceeds the liberation velocity for the planet, and is therefore favoured by high atmospheric temperature. Here too, giant collisions might have played a role by increasing the latter.

The study of extinct radioactivities provides evidence that the atmosphere was open to loss to space during the Hadean. The amount of radiogenic ^{129}Xe generated by decay of ^{129}I and of $^{131-136}$Xe from the fission of ^{244}Pu only represents a fraction of the total Xe isotopes generated by the decays of these elements in bulk Earth. Degassing of the mantle during the Hadean can perfectly account for the depletion of this reservoir in Xe isotopes. However, the atmosphere must also have lost most of radiogenic and fissiogenic Xe isotopes. Taking into account the half-life of ^{244}Pu (82 Ma), such a lost must have lasted for at least 200 Ma, possibly up to 600 Ma (Tolstikhin and Marty, 1998). Some of the atmospheric noble gases are severely isotopically mass fractionated compared to potential precursors (solar or asteroidal compo-nents), which suggests independently escape from the atmosphere that favoured light isotopes relative to heavy ones. As discussed above, atmospheric neon is depleted by 30% in the light ^{20}Ne isotope relative to ^{22}Ne when compared to mantle neon, a difference that might have arisen from mass fractionation during atmospheric loss. However, several meteoritic Ne data encompass the atmospheric isotopic ratio, leaving open the possibility that air Ne was contributed by asteroidal bodies (Marty, 1989). Xenon, which is the heaviest noble gas, is severely fractionated by 3% per amu relative to solar or asteroidal precursors, which is fully consistent with isotope fractionation during hydrodynamic escape (Sasaki and Nakasawa, 1988). However, the problem with this element is that it is also elementally depleted relative to lighter noble gases, e.g., the atmospheric Xe/Kr ratio is 5 times lower than the chondritic Kr/Xe ratio, exactly the opposite of what would be expected from Xe isotopes. This long-standing problem, known as the xenon paradox, indicates that the early evolution of the atmosphere was probably very complex and involved several episodes of exchanges between the mantle, the atmosphere, the outer space and some of the precursors (Pepin, 1991; Tolstikhin and Marty, 1998). In this respect, comets that may be depleted in Xe might have contributed lighter volatile elements whereas a residue of isotopically fractionated Xe attested previous episodes of atmospheric loss (Dauphas, 2003).

The role of comets in supplying volatile elements to the early Earth has been often advocated. Indeed, it seems that, when compared to inner regions of the solar system, the Kuiper belt has a mass deficit, which is interpreted as reflecting a depletion of bodies due to gravitational perturbations by giant planets, or to the existence of a yet undiscovered Mars-sized planet that later on disaggregated, or to external factors like the transit of a nearby star. Small bodies caught in resonance from nearby giant planets are ejected and part of them falls towards the inner solar system. As a result, one would expect the inner solar system to be battered by cometary objects ejected from outer regions of the solar system and providing volatile elements to terrestrial planets. Cometary data prone to investigate quantitatively this possibility are scarce, but some tentative inferences can be made in the case of noble gases, water and nitrogen. First, the hydrogen isotopic ratio measured by remote sensing in a few comets indicates that these objects are richer in D than both the oceans and chondrites, with a D/H ratio about 2 times that of the former. Thus, the contribution of cometary material seems to be very limited. A budget for argon, hydrogen and platinum group elements (which are believed to have been added by asteroidal bodies after the major metal-silicate differentiation), as well as the lunar cratering record led Dauphas et al. (2000) to conclude that the fraction of cometary material in the late contributing veneer (here the amount of undifferentiated material that felt onto Earth after metal-silicate differentiation) was lower than 1%. A similar conclusion is derived from N isotope systematics, indeed, cometary N seems to be depleted in ^{15}N (Jewitt et al., 1997) as the Sun is. However, recent data for cometary CN show that some of the N compounds are enriched in ^{15}N (Arpigny et al., 2003), and no definite conclusion can be proposed for this element.

4.2.1.3. *A tentative scenario*

The formation of the atmosphere was a multi-stage process during which several cosmochemical reservoirs contributed volatile elements. There is strong evidence that some of them derived from the solar nebula, possibly through dissolution of a massive solar-like atmosphere. Accumulation of such a massive atmosphere occurred when the nebular gas was present, that is, ≤10 Ma after start of solar system condensation (ASSC) at most. Therefore, planetary bodies large enough to retain a hydrogen-rich, probably transient, atmosphere existed already within this timeframe. This is in full agreement with time ranges provided by extinct radioactivity products measured in differentiated meteorites (particularly in Martian meteorites which show evidence for metal-silicate-melt differentiation within the first 9 Ma; Halliday et al., 2001) on one hand, and with planetary build up modelling showing the existence of Mars-sized bodies and of giant planets within few Ma ASSC.

The occurrence of a massive hydrogen-rich atmosphere had several important consequences for the thermal and chemical state and evolution of the proto-Earth. During the magma ocean stage (part 4.1.5.) solar-like gases could dissolve into magmas. The low oxygen fugacity (fO_2) imposed by hydrogen might have played a major role in metal reduction that led to the formation of the core, in an active magma ocean where the oxygen fugacity would be buffered by atmospheric hydrogen. The remnant of the solar component might have been kept in the deep mantle or in the core where it might have survived the last giant impact forming the Moon. The ^{182}Hf–^{182}W extinct radioactivity system indicates that this event took place at about 30 Ma after Earth accretion, within a possible range of 11–50 Ma (Kleine et al., 2002; Yin et al., 2002a). A magma ocean episode on the Moon might have lasted 45 Ma, and on Earth, it might have been much longer from obvious thermal consideration. The $^{129}I-^{244}Pu-^{129}Xe-^{136}Xe$ systems have recorded a mantle closure time at about 60–70 Ma ASSC (Yokochi et al., 2005) which might correspond to a strong decline in large-scale magma ocean episodes.

Presumably, later on, the proto-mantle was contributed by meteoritic Ne, water, nitrogen and other volatile elements (C, other noble gases). Compared to sun, planetary material (e.g., meteorites, asteroids) is highly depleted in noble gases (relative to H, C, N etc.), such that contribution of this material would dominate the budget of major volatiles without overprinting noble gas characteristics inherited from previous processes. Thus, isotopic fraction-ation related to atmospheric escape is recorded in noble gases whereas the major volatile element isotopic ratios (e.g., D/H, $^{13}C/^{12}C$, $^{15}N/^{14}N$) have kept a less disturbed record of their chondritic origin.

The study of extinct radioactivities involving volatile elements like Xe demonstrates that the early atmosphere was open to loss in space during long time intervals, possibly covering the whole Hadean. The Hadean mantle was also very active, convecting at a rate one order of magnitude or more higher than at Present, as indicated by the extensive loss of fissiogenic Xe from ^{244}Pu (half-life of 82 Ma) which occurred well before the loss of iso-topes produced by long-lived radioactivities like ^{40}Ar or ^{4}He (Yokochi et al., 2005).

The amount of nitrogen and water that was supplied by "late" contri-bution (i.e. after the last global metal-silicate segregation) of asteroidal material, as estimated from the platinum group element (PGE) budget of the mantle, is consistent,(within a 2–3 times range) with the terrestrial inventory of oceans and atmosphere (Dauphas et al., 2000; Dauphas and Marty, 2002). Part of water and nitrogen might have been cycled through time between the mantle and the surface, so that the match between the PGE budget and the surface inventory is fairly good (Figure 4.4). This implies that the main volatile elements that are present in the atmosphere and in the oceans could have been carried by "late" veneer contributions having taken place pre-sumably after the Moon-forming event. It is unclear if these contributions were the product of big impacts or were delivered by snowing dust. The latter could have been efficient in providing part of (Marty et al., 2005), or even most of, volatile elements present in the atmosphere and the oceans (Maurette, 1998).

The amount of CO_2 assumed to be contributed by this veneer is com-parable to the quantity of carbonates trapped in sediments. The latter is probably a lower limit of the carbon inventory since today, due to recycling; the Earth mantle contains a significant fraction of carbon. The calculated equivalent partial pressure of CO_2 is about 100 atmospheres, which is comparable to that observed today in Venus atmosphere. Such CO_2 partial

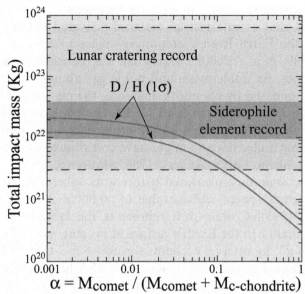

$$\alpha = M_{comet} / (M_{comet} + M_{c\text{-chondrite}})$$

Figure 4.4. The total mass of impacting bodies on Earth after the Lunar cataclysm as a function of the fraction of cometary material in the impacting bodies (cometary plus asteroidal). The lines represent solutions constrained by the D/H ratios of the different endmembers. The orange area represents the mass flux constrained by the Lunar cratering record, and the light orange area is the mass fraction constrained by the highly siderophile element (HSE) content of the terrestrial mantle. The latter is not equilibrated with the terrestrial core; it is attributed to addition of chondritic material after core formation. The D/H ratio of comets 3 times larger than that of chondrites imposes that the cometary fraction cannot by higher than 10% at best to be compatible with the mass flux recorded on the Lunar surface, and about ∼1% to be consistent with the mass flux constrained by the HSE record. Taking into account the amount of noble gases present in the atmosphere will reduce the cometary contribution to about 0.1% or less (Dauphas et al., 2000; Dauphas and Marty, 2002).

pressure would lead to a greenhouse surface temperature of several hundreds of °C. This scenario implies that a very efficient process of condensation of the oceans and of carbonation of atmospheric CO_2 took place in the early Hadean such that the Earth reached temperatures below the water boiling temperature at about 4.4 Ga. Indeed, oxygen isotope data of some 4.4 Ga old zircons point to water–rock interaction below 100 °C (see part 4.3.4.).

4.2.1.4. Contributions from accretion stages to present
Independent lines of evidence such as the prominence of carbonaceous chondrites over ordinary chondrites in the asteroid belt (Gradie et al., 1989), the occurrence of carbonaceous chondrite debris in the lunar regolith (Keays et al., 1970) and in meteorites (Zolensky et al., 1996; Gounelle et al., 2002),

remnants of a cometary shower in terrestrial sediments (Farley et al., 1998), strongly suggest that extraterrestrial material delivered volatile elements to the surface of the Earth from accretionary stages to Present (e.g. Chyba, 1990; Morbidelli et al., 2000). Mass balance considerations based on hydrogen isotopes, on noble metal and noble gas abundances indicate that this flux was dominated by chondritic material, the mass fraction of comets being $\sim10^{-3}$ or less (Dauphas et al., 2000; Dauphas and Marty, 2002). Carbonaceous chondrites, comets and interplanetary dust particles (IDPs) are rich in organic molecules and might have contributed directly to the pre-biotic inventory of the Earth (Anders, 1989; Maurette, 1998).

The flux of cosmic dust measured before atmospheric entry in the near-Earth interplanetary space yields a value of 30,000 ± 20,000 tons/yr (Love and Brownlee, 1993). Cosmic dust represents the largest contribution of extraterrestrial matter to the Earth's surface at present, as the meteorite mass flux over the 10 g^{-1} kg interval accounts for only $\sim$3–7 tons/yr (Bland et al., 1996). IDPs collected in the stratosphere by NASA have typical sizes below 50 μm. They probably sample a mixed population of debris originating from the asteroidal belt and from the Edgeworth–Kuiper belt (e.g. Brownlee, 1985; Brownlee et al., 1993), with possible contribution from silicates and organics of interstellar origin (Messenger, 2000; Aléon et al., 2001; Messenger et al., 2002, 2003; Aleon et al., 2003).

Objects in the size range 25–200 μm (exceptionally up to 400 μm and larger), labelled micrometeorites (MMs), dominate the mass flux of cosmic dust onto Earth (e.g. Love and Brownlee, 1993). Micrometeorites share mineralogical and chemical similarities with CM and/or CR chondrites (Kurat et al., 1994; Engrand and Maurette, 1998), and D/H ratios of MMs point to an extraterrestrial, likely to be asteroidal (CM-CI-type), origin of trapped water (Engrand et al., 1999).

In the following, we make an estimate of the nitrogen contribution by cosmic dust to the atmosphere through time, following a recent study (Marty et al., 2005). The case of nitrogen perfectly illustrates the fate of other at-mophile elements; it can be extended to that of water as far as orders of magnitude are concerned. An estimate of the nitrogen delivery flux can be assessed by assuming a N content of $\sim$1000 ppm characteristic of carbona-ceous chondrites and in particular of CM chondrites. This value may be a lower limit since some of the micrometeorites might contain a larger initial N content as suggested by the sometimes high trapped noble gas content (see preceding section), although the frequency of such particles is unknown. IDPs can contain N at the percent level (Aleon et al., 2003). However the IDP mass flux is $\sim$0.01 times that of micrometeorites, (Love and Brownlee, 1993), so that their nitrogen contribution appears limited, perhaps at the same level as the one from MMs. Comets contain about 10,000–40,000 ppm N (see Dauphas and Marty, 2002, and references herein) and might have

contributed significantly to cosmic dust. However, mass balance considerations involving volatile and siderophile elements together with D/H ratios of
the terrestrial oceans suggest that the bulk cometary contribution to the
terrestrial inventory of volatile elements was very limited, of the order of 10^{-3}
relative to the total mass of contributors (Dauphas et al., 2000; Dauphas and
Marty, 2002). It must be noted that the nitrogen content of cosmic dust
necessary to account for N isotope variations in lunar soils has been estimated at ~1000 ppm on average over most of the lunar history (Hashizume
et al., 2002).

The cosmic dust flux might not have varied dramatically since 3.8 Ga ago,
except for a significant increase in the last 0.5 Ga (Grieve and Shoemaker,
1994; Culler et al., 2000; Hashizume et al., 2002). A near-constant planetary
contribution within a factor of 2 since 3.8 Ga ago is also consistent with the
cratering record at the lunar surface (Hartmann et al., 2000). Thus, we assume, as done previously (Chyba and Sagan, 1992), that the cosmic dust flux
remained constant since 3.8 Ga at a rate similar to the present-day one.
Another important question is the flux ratio between dust and larger objects.
The present-day mass ratio between dust and meteorites is of the order of
10^2–10^3. Although highly imprecise, the mass contribution due to large objects, which is dominated by rare events of km (or more) sized objects, might
have been comparable to the cosmic dust flux, averaged over the last 3 Ga
(Kyte and Wasson, 1986; Anders, 1989; Trull, 1994). A constant cosmic dust
flux similar to the present-day one (~30,000 tons/yr) integrated over 3.8 Ga
could have supplied ~4×10^{15} mol N_2, to the atmosphere, which makes only
~3×10^{-5} times the atmospheric N_2 amount (1.38×10^{20} mol N_2, (Ozima
and Podosek, 2002) (Figure 4.5). Such a limited contribution is unlikely to
have had discernable isotopic effects on atmospheric nitrogen. This view is
consistent with available data for Archaean and Proterozoic sediments
indicating that the atmospheric N isotope composition has been near-constant since 3.5 Ga (Sano and Pillinger, 1990; Pinti, 2002).

Some argue a steep decline of the cratering rate between 4.5 and 3.8 Ga,
there is compelling evidence that a spike of bombardment took place 4.0–
3.8 Ga ago and that in the time interval 4.4–4.0 Ga ago the impacting flux
was not dramatically high (Hartmann et al., 2000, part 4.5). The total mass
of impactors during the last spike of bombardment around 3.9 Ga is ~$6 \times$
10^{21} g (Hartmann et al., 2000). The content of siderophile elements in the
ancient highlands suggests that the amount of interplanetary mass accumulated by the Moon in the 4.4–4.0 Ga period is about the same as that
required in order to form the 3.9 Ga basins. Consequently, the post 4.4 Ga
contribution to the lunar surface might have been ~1.2×10^{22} g (Hartmann
et al., 2000). This contribution included both bolides (evidenced by remnants
of the lunar cataclysm) and cosmic dust, all of them being integrated in the
siderophile element record. A lower limit for the Earth's efficiency over Moon

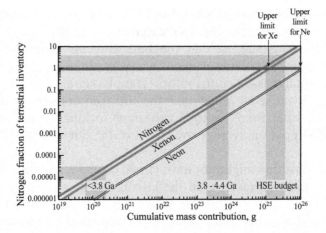

Figure 4.5. ET nitrogen contribution to the terrestrial N inventory (atmosphere, crust and mantle, as a function of the cumulative mass contribution to the Earth. A mean N content of 1000 ppm is assumed for contributing material. A *y*-value of 1 (horizontal grey bar) means that the total terrestrial budget can be supplied by contributing material. Cosmic dust contribution from the Archaean (<3.8 Ga) to Present is likely to be negligible for the nitrogen budget. Contributions during the Hadean (3.8–4.4 Ga ago) could have been significant (up to 10%). The highly siderophile element (HSE) budget is based on the amount of these elements in the mantle that did not equilibrate with the core and presumably were supplied after terrestrial differentiation (data from Chyba, 1991). The lower limit of this contribution allows to supply the total amount of nitrogen present in the atmosphere, the crust and the mantle, showing that the case of nitrogen parallels that of HSE. The xenon and neon potential contributions are computed assuming a chondritic composition (Mazor et al., 1970) and put constraints on the total delivery of nitrogen. Assuming mixing between a pre-existing, isotopically fractionated atmosphere and a chondritic-like delivery, the amounts of Xe and Ne that can be brought cannot exceed those of the pre-existing atmosphere because their isotopic compositions differ markedly. Xenon is particularly sensitive to this effect and confirms that the total amount of nitrogen that was delivered after terrestrial differentiation is compatible with the lower limit of the HSE budget.

to collect cosmic dust is given by the ratio between the surfaces of the two bodies (13). Gravitational focusing of cosmic dust (Kortenkamp et al., 2001) possibly increases the collection efficiency of the Earth by a factor of ~3 relative to the Moon (Hashizume et al., 2002). Therefore, the range of extraterrestrial material collected by Earth between 4.4 and 3.8 Ga might have been ~ (2.4–7.2) × 10^{23} g. This represents a contribution of (0.07–2.1) × 10^{19} mol N_2, that is, 6–18% of present-day atmospheric nitrogen. Nitrogen has been actively exchanged between the surface and the mantle (Marty and Dauphas, 2003), so that it is more appropriate to consider the total nitrogen budget of the Earth, which is 2.8 ± 1.0 × 10^{20} mol N_2 (Marty, 1995; Marty and Dauphas, 2003). Thus, the total post-4.4 Ga contribution of nitrogen to Earth might have been ~10% of total terrestrial N (Figure 4.5). An upper limit of the nitrogen contribution by extraterrestrial matter to Earth can be set from the highly siderophile element (HSE) budget of our planet which might have been delivered after metal/silicate differentiation towards the end of terrestrial accretion (e.g. Chyba, 1991, and references herein). If nitrogen is a siderophile element as recently argued (Hashizume et al., 1997), the amount now measured in both mantle and atmosphere might have been delivered after core formation during "late veneer" events. The amount of HSE present in the mantle corresponds to the contribution of 1×10^{25}–4×10^{25} g of chondritic material (Chyba, 1991), thus to a delivery of $(3-14) \times 10^{20}$ mol N_2. This is clearly in the range of the nitrogen budget of the Earth (3×10^{20} mol N_2), thus suggesting that cosmic dust, which dominates now the extraterrestrial flux of matter to Earth, could have supplied nitrogen present in the atmosphere, but only during the late building stages of our planet. A summary of nitrogen delivery through time is given in (Figure 4.4).

4.2.2. OCEAN

DANIELE L. PINTI

The establishment of a precise chronology of the formation of the terrestrial oceans depends on the ability to answer two fundamental questions: (1) how and when did water arrive on the Earth? (2) when did internal heat flow from the accreting Earth decreased to a critical threshold allowing water to be liquid on its surface?

The origin of water on Earth has been at the centre of much debate in the last decades, yet few models have been formulated on the origin of the oceans (Abe, 1993; Sleep et al., 2001; Pinti, 2005) because of the evident lack of geological record. Actually, the scientific community agrees with the hypothesis that water was brought very early in the history of the solar system, possibly during the last phases of the Earth accretion, which implies that the oceanic water inventory was available since the beginning (e.g. Morbidelli et al., 2000). Theories on an "expanding ocean", with water delivered all along the Earth history seem to be implausible and not supported by the geological record (de Ronde et al., 1997; Harrison, 1999), although several authors still claim its existence to account for climate and sea-level variations during eons (Deming, 1999). These variations are likely the result of plate tectonics, continental drifting and the opening of new oceanic basins, rather than variations of the volume of the oceans.

However, contrasting opinions subsist on the extraterrestrial carrier that brought water on Earth, splitting in two the scientific community. Some authors propose a dominant cometary origin (Frank et al., 1986; Delsemme, 1999), which has found much credit among the biologists because comets contain plenty of organic molecules that could be considered as the primordial "bricks" of life. The other part of the community, supported by strong isotopic evidence and dynamical models of the solar system evolution (Dauphas et al., 2000; Morbidelli et al., 2000; Robert, 2001; Dauphas and Marty, 2002; Dauphas, 2003; Robert, 2003), proposes that a few asteroidal embryos of chondritic composition are the water delivers.

4.2.2.1. *Water accretion to Earth (4.56–4.49 Ga = t_0 + 11–70 Ma)*
The prevailing opinion on the origin of the oceans is that the water comes from the outer solar system, brought by comets or asteroids colliding the freshly formed Earth, the usually called "late veneer scenario" (e.g. Javoy, 1998; Javoy, 2005 and references therein) (Figure 4.6). Isotopic and mass balance calculations, based respectively on the hydrogen isotopic ratio of water (D/H), the amount of water in different extraterrestrial reservoirs (e.g., meteorites, comets) and numerical simulations of extraterrestrial fluxes suggests that delivery of water by comets is less than 10% of the total (Dauphas et al., 2000; Robert, 2001). Compilation of the isotopic ratio of hydrogen

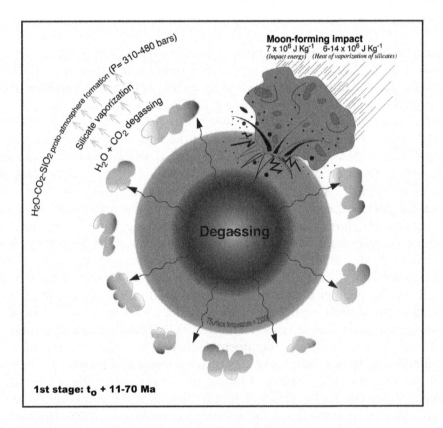

Figure 4.6. Sketch representing the main phases of the formation of the terrestrial oceans (t_0 + 11–70 Ma).

(D/H) measured in different extraterrestrial reservoirs and compared with that of the modern seawater (D/H = 155.7×10^{-6} or $\delta D = 0$ by definition) suggests that few embryos of asteroids (with a D/H = $149 \pm 6 \times 10^{-6}$, Morbidelli et al., 2000; Robert, 2003) or alternatively consistent fluxes of large IDPs (AMM, Antarctic Micrometeorites with a D/H = $154 \pm 16 \times 10^{-6}$, Maurette et al., 2000) could be the carrier of water on Earth. The chemical and isotopic composition of both these two reservoirs is close to that of the carbonaceous chondrites (Robert, 2003), the latter ranging from that of Orgueil (CI; fully hydrated) to that of Murchinson (CM; containing about 40% of hydrated minerals). Independent estimates based on the excess of siderophile elements of the platinum group in the mantle suggest that contribution from comets could be less than 10^{-3} by mass of the total (Dauphas and Marty, 2002). Numerical simulations indicate that impact probability of comets is too low to supply all of the terrestrial water (Zahnle and Dones, 1998; Morbidelli et al., 2000). Very recently, new simulations have been carried out in order to explain the cataclysmic Late Heavy Bombardment (LHB) period that lasted between 10 and 150 Ma, at around 4.0–3.8 Ga ago (Gomes et al., 2005, see part 4.5). In these simulations, the rapid migration of the giant planets destabilized the planetesimal disk thus resulting in a sudden massive delivery of these planetesimals to the inner solar system. During this burst, the total amount of cometary material delivered to the Earth is 1.8×10^{23} g, which corresponds to about 6% of the current ocean mass.

   The hypothesis of Maurette et al. (2000) of adding water to Earth's oceans by accreting cosmic dust (AMM) is interesting but it fails against the unknown fluxes of extraterrestrial particles hitting the early Earth and the capacity of AMM to furnish the total Earth's water inventory. This latter includes both the mass of the oceans (1.37×10^{24} g; 98% of the total surface water inventory) and an unknown amount of water that could reside in the Earth's mantle, as suggested by its high degree of oxidation. Water masses up to 50 times the amount in the modern ocean have been postulated for the primitive mantle (Abe et al., 2000), while high-pressure experiments in hydrous mineral phases suggest that Earth's lower mantle may actually store about five times more water than the oceans (Murakami et al., 2002). The minimum estimation for the whole mantle is of 1.93×10^{24} g, only 1.4 times the present-day oceanic inventory (Dreibus and Wanke, 1989). Accepted values of the volume of water for the most "wet" mantle scenarios are from 5 to 10 times that of the modern oceans.

   Best estimates of the present-day accretion rate of cosmic dust measured before atmospheric entry yields a value of 30,000 ± 20,000 tons/yr (see part 4.2.1.) while initially Love and Brownlee (1993) gave a slightly higher value of 40,000 ± 20,000 tons/yr. On Earth surface, the flux is much less and estimates obtained from abundance of cosmic dust from Antarctic cores yields values from 5300 ± 3,100–16,000 ± 9300 tons/yr with a global average of

14,000 tons/yr (Yada et al., 2004). The amount of water in the AMM has been measured to range from 2 to 8 wt.% (Engrand et al., 1999). Assuming a cosmic dust accretion rate ranging from 2200 to 60,000 tons/yr, integrated to the whole Earth history (4.5 Ga), the total amount of water would be between 1000 and 100,000 times less than the present-day oceanic inventory.

Cosmic dust accretion rate has been relatively constant in the last 3.8 Ga within 2 times the present flux (Hartmann et al., 2000). However, the fluxes in the period between 4.5 and 3.8 Ga are unknown and highly debated. Marty et al. (2005) propose that the total amount of collected cosmic dust might have been 2.4–7.2×10^{23} g, between 4.4 and 3.8 Ga. In term of added water, this represents a contribution of 0.3–4% of the present-day oceanic inventory. Integrating 2 times the present flux over the remaining 3.8 Ga, does not significantly change the figure. Theoretical extrapolations of Hartmann et al. (2000) indicated a declining flux since the start of the Earth accretion, with an initial flux of planetesimal impactors at t_0, on the order of 2×10^9 the present flux. Assuming a ratio of 100–1000 between meteorites and the cosmic dust, as at the present-day, the initial flux of cosmic dust could have been as high as $10^6 \times$ the present flux (Maurette, 2001). Assuming a present flux ranging from 20,000 to 60,000 tons/yr, a water content of 2–8 wt%, a dust/planetary ratio of 100–1000 and integrating accretion rates from Hartmann (Hartmann et al., 2000) (Figure 4.7), we can calculate the "age" of the oceans (i.e. the time needed to add water to the Earth). Assuming maximum rates of cosmic dust accretion and water contents, our computations give us a minimum time of 70 Ma to collect the present ocean inventory (1.37×10^{24} g; Figure 4.7) since the time t_0 of the Earth formation. The total maximum amount of water that could be brought to Earth in 4568 ± 1 Ma by cosmic dust however, is only 116% of the present oceanic inventory (1.59×10^{24} g; Figure 4.7), which means that other sources are needed to account for the whole Earth water content (hydrosphere + mantle = 3.3–15.1×10^{24} g). If we assume minimum fluxes of AMM hitting the Earth, after 70 Ma from the solar system formation, only 0.8% of the total oceanic inventory will be delivered. This latter figure strongly supports the existence of alternative carriers of terrestrial water. The above calculation does not take into account the peak of the so-called LHB, which lasted from between 10 and 150 Ma, at 4.0–3.8 Ga (Gomes et al., 2005, see part 4.5). During this spike, the amount of asteroidal material that struck the Moon is calculated to be 3–8×10^{21} g (Hartmann et al., 2000; Gomes et al., 2005), which corresponds to an amount on Earth of 0.6–1.7×10^{23} g. Assuming again an asteroid/cosmic dust ratio of 100–1000, the total amount of water added during the spike varies from 1.3 to 130×10^{18} g, which can be considered as negligible.

Based on dynamical models of primordial evolution of the solar system, Morbidelli et al. (2000) calculated the orbits and timescales of all possible

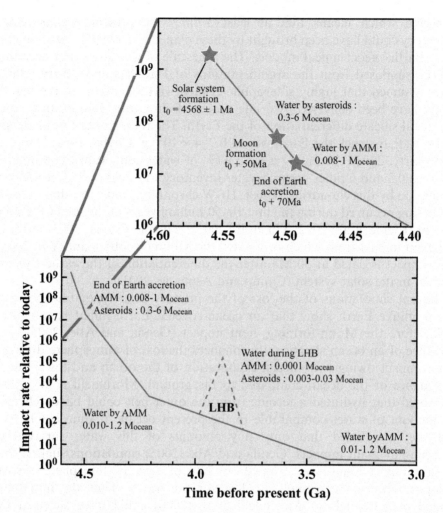

Figure 4.7. Empirical curve of impact declining flux during Hadean, as reported by Hartmann et al. (2000), modified. The cumulated amount of water transported by a minimum and maximum flux of cosmic dust composed principally of AMM of chondritic composition (Engrand et al., 1999) is calculated for each segment of the curve. The total amount of water delivered by asteroids and AMM during the LHB have been also reported. Small figure top-right is a zoom on the first period of Earth's accretion. It is possible to note that most of the oceanic and mantle water has been likely transported within the first 100 Ma of the Earth history.

water carriers in the primordial solar system. They concluded that towards the end of the solar system formation (several tenths of millions years), the Earth began to accrete few planetary embryos originally formed in the outer asteroid belt, heavily hydrated with a water content of about 10 wt.% deduced from average contents of hydrated carbonaceous chondrites (Boato, 1954; Robert and Epstein, 1982; Kerridge, 1985; Robert, 2003). A 10% of the Earth mass (6×10^{27} g) could derive from this late accretion of primitive material, which means that at least 5 times the present oceanic water inventory could have been brought by these planetary embryos, satisfying the "wet Earth" geochemical models. The timescale of this late water accretion can be supposed from the argumentations of Dauphas and Marty (2002). They assumed that highly siderophile element (HFSE) excess in the mantle might have been delivered by a late veneer of chondritic composition, after the metal/silicate differentiation of the Earth. The total amount of planetary embryos that struck the Earth was of $0.7–4 \times 10^{25}$ g (Chyba, 1991; Dauphas and Marty, 2002). Assuming a 6–22 wt% of water content of these objects, between 0.3 and 6 times oceanic water inventory was delivered after the core formation by this chondritic veneer. Hf-W chronology indicates that the core formation occurred during the first 10–50 million years of the life of the solar system (Kleine et al., 2002; Yin et al., 2002a; Yin and Ozima, 2003), prior of the giant impact from which originated the Moon (Jacobsen and Yin, 2003). This impact is dated at 50 Ma after the differentiation of the earliest plane-tesimal in the solar system (Canup and Asphaug, 2001).

Recent simulations of the loss of the primary solar-type atmosphere of the primitive Earth show that an ocean-covered Earth might have existed well before the Moon-forming giant impact (Genda and Abe, 2005). The presence of an ocean significantly enhances the loss of atmosphere during a giant impact owing two effects: evaporation of the ocean and lower shock impedance of the ocean compared to the ground. Morbidelli et al. (2000) proposed that hydrated asteroids from the outer belt could have delivered an amount of water comparable to the present ocean, at time $t_0 + 10$ Ma, but they concluded that only tiny amounts of this water survived to stripping by giant impacts. Genda and Abe (2005) simulations suggest that a giant impact could have stripped out 70% of a proto-atmosphere of solar composition, with essentially no loss for the ocean. Thus the initial hy-drated planetesimals of Morbidelli et al. (2000) could have survived the giant impact period and participated to the oceanic water budget. It is interesting to note that for all these models, the time of water accretion is within the time of the Earth formation. Both hydrated planetary embryos and AMM could have sustained the water flux on Earth between $t_0 + 10$ and 70 Ma.

4.2.2.2. Ocean formation (4.49–4.39 Ga = t_0 + 70–165 Ma)

If water arrived very early on the accreting Earth, it does not mean neces-sarily that a stable and cold ocean existed in a very early period of its history. Earth became potentially habitable once it developed a solid proto-crust to separate the hot interior of the planet from a cooler surface environment featuring liquid water. Giant impacts such as the Moon-forming one has likely molten the Earth surface and vaporized proto-oceans to form a thick

CO_2–H_2O proto-atmosphere (Pinti, 2005), yet this stage lasted likely less than 1 Ma (Abe, 1993). A molten surface can be sustained only by the blanketing effect of a massive proto-atmosphere. Yet, a hot runaway water greenhouse maintained by interior temperatures could have existed only for a short geological time (Sleep et al., 2001) (Figure 4.8). The Earth's interior would cool down, in the presence of a runaway water greenhouse, of 700 K in only 1.8 Ma (Sleep et al., 2001). During this very short period, after the Moon-forming impact, molten surface might have facilitated the exchange of volatiles with the Earth interior and part of the accreted water would be partitioned to the mantle, leaving at the surface the amount needed to form the present oceanic inventory.

The second stage of formation of the ocean is the production of a tiny basaltic proto-crust covering the Earth surface. This could have took place when the interior heat flow, produced by the gravitational energy flux released by accretion of the Earth, decreased to a threshold value of 150–160 W m^{-2} (Abe, 1993; Sleep et al., 2001). This could have been reached

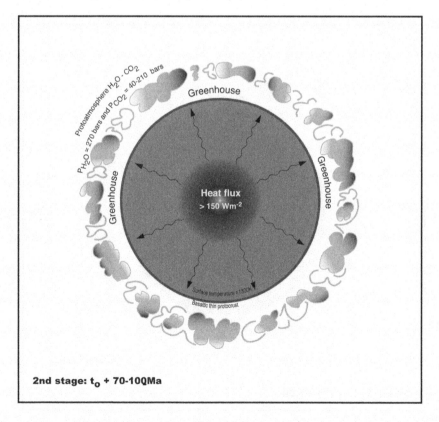

Figure 4.8. Sketch representing the main phases of the formation of the terrestrial oceans (t_0 + 70–100 Ma).

at the end of the Earth accretion, when the Earth radius was 97% of the present size (Abe, 1993).

The ending stage of the Earth accretion, at $t_0 + 50$–70Ma, was characterized by a mechanically fragile proto-crust that was partially molten and resurfaced by meteoritic impacts and extensive volcanism. Assuming an atmospheric pressure at the surface of several hundred bars (270 bars of H_2O and $= 40$ bars of CO_2, Pinti, 2005), water starts to condense and precipitates when the surface temperatures were below 600 K. Condensation of the oceans could have been a very rapid process, lasted few thousand years (Abe, 1993) (Figure 4.9). Although formed, the oceans were not yet habitable. Temperatures were too high for the survival of any form of life, and large-scale hydrothermal alteration of the proto-crust produced very saline water (halite-dominated brines, Sleep et al., 2001), where few forms of life could have survived.

A stable oceanic habitat took place when surface heat flow was less than 1 W m^{-2}, and best estimates following the model of Abe (1993)

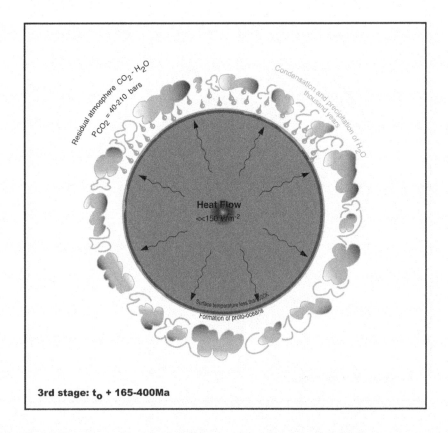

Figure 4.9. Sketch representing the main phases of the formation of the terrestrial oceans ($t_0 + 165$–400 Ma).

indicate that such a situation was reached at $t_0 + 400$ Ma (Figure 4.10). Independent estimates can be obtained by assuming that the temperature of the ocean was controlled by the greenhouse effect of a CO_2 residual atmosphere, after the condensation of water in the oceans (Sleep et al., 2001). The maximum temperature acceptable for living organism is ≤110 °C (thermophiles, Lopez-Garcia, 2005). This can be taken as one of the minimum conditions acceptable for a habitable ocean. This temperature is reached at equilibrium with a residual atmosphere of 25 bars of CO_2 (Sleep et al., 2001). The initial atmosphere might have contained from 40 to 210 bars of CO_2 (Pinti, 2005). The most effective way to remove CO_2 from the primitive atmosphere is weathering of silicates in the proto-crust (during Hadean, weathering could have been a fast process, at rates of 10,000 times higher than the present ones, Abe, 1993), formation of carbonates and recycling of produced $CaCO_3$ to the mantle. Zhang and Zindler (1993) proposed a maximum initial recycling rate of CO_2 in the mantle of 1.5×10^{14} mol/yr. At this rate, 15–185 bars of CO_2 could have

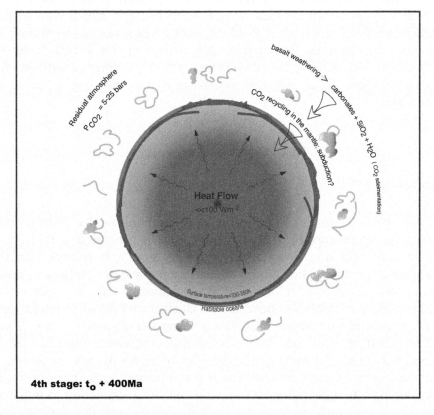

Figure 4.10. Sketch representing the main phases of the formation of the terrestrial oceans ($t_0 + 400$ Ma).

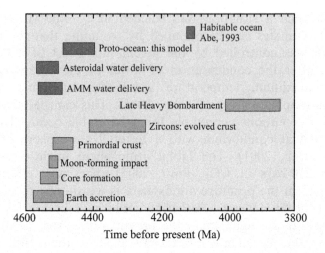

Figure 4.11. Chronology of the main phases of the Earth formation and differentiation, together with the phases of extraterrestrial water delivery and ocean formation.

been removed from the atmosphere in a span of time ranging from 19 to 140 Ma. The upper limit is close to a $t_0 + 165$ Ma age, which corresponds to the minimum age of an evolved continental proto-crust containing liquid water, as suggested by the Jack Hills detrital zircon W74/2–36 Pb-Pb age (4404 ± 4 Ma, Wilde et al., 2001) (Figure 4.11). Clement conditions for the establishment of a habitable ocean were thus possibly in place at $t_0 + 165$ Ma.

4.3. The First Continents

HERVÉ MARTIN

On Earth, plate tectonics is a powerful set of mechanisms that continuously rebuilds the planet surface; indeed oceanic crust is generated in mid ocean ridge systems and due to its relatively high density, it returns into the mantle through subduction zones, the average time for such a cycle is 60 Ma and no ocean older than 200 Ma is known today. In other words oceanic crust is not able to constitute an efficient archive of Earth history. On the opposite, continental crust has a lower density such that it is not easily recycled into the mantle. Consequently, continental crust is able to undergo and to record most of the episodes of Earth history. In addition, continental crust average composition is granodioritic, such that, contrarily to oceanic basalts, minerals such as zircon can easily crystallize in it. These minerals are extremely important in geochronology and this for two reasons: (1) zircon is a silicate of zirconium ($ZrSiO_4$) where sometimes Th and

U can replace Zr. Due to the radioactivity of these elements zircon is "easy" to date. (2) zircon is extremely resistant to weathering and metamorphism, consequently it can resist to mechanisms as alteration, erosion and even sometimes to partial melting. This resistance allows it to efficiently record ages of old events.

4.3.1. THE 3.8–3.9 GA ARCHAEAN CRUST

4.3.1.1. *Amitsôq TTG (Greenland)*
Since the field work of McGregor (1968; 1973) it has been recognized that Greenland Archaean terrains were among the oldest in the world. The oldest components of this craton were called the Amitsôq gneisses, which were first dated by Rb–Sr whole rock isochron method at 3.98 ± 0.17 Ga (Black et al., 1971) and 3.74 ± 0.1 Ga (Moorbath et al., 1972). More recent works showed that the so-called Amitsôq gneisses were not homogeneous, but on the contrary very diverse, and in order to account for this heterogeneity, Nutman et al. (2004) proposed to refer these formations as "Itsaq Gneiss Complex"; both denominations are used in geological literature. Zircon dating allowed to more precisely determine the ages of the genesis of these gneisses which range between 3.88 and 3.60 Ga. A statistical analyse performed on ca. 2000 zircons from this area pointed to the existence of 3 main episodes of continental crust genesis at Amitsôq: at ~3.80 Ga; ~3.7 Ga and ~3.65 Ga. Even if rarer, ages older than 3.85 Ga were also measured. A zircon extracted from a gneiss sample, tonalitic in composition, gives the oldest reliable age obtained in this area: 3.872 ± 0.010 Ga (Nutman et al., 1996). This age is interpreted as being that of crystallisation and emplacement of the tonalitic magma. The same authors (Nutman et al., 2000) also obtained an age of 3.883 ± 0.009 Ga on a zircon core, whereas the main part of the mineral was dated at 3.861 ± 0.022 Ga; here too these ages are supposed to give the time of crystallization of the parental magma.

Amitsôq gneisses outcrop over about 3000 km^2; they are well exposed and generally well preserved of weathering (Figure 4.12). They derived from magmatic rocks Tonalitic, Trondhjemitic and Granodioritic (TTG) in composition (Nutman and Bridgwater, 1986; Nutman et al., 1996). TTG is the widespread composition of the Archaean crust between 3.8 and 2.5 Ga, crust which is considered as generated by hydrous basalt melting, possibly in subduction like environment (Martin et al., 2005). The ages obtained on the Amitsôq gneisses indicate that as early as 3.87 Ga ago, these mechanisms were already active and efficient, able to generate the huge volumes of continental crust observed at Amitsôq.

Figure 4.12. 3.87 Ga old Amitsôq gneisses (Greenland) that emplaced as TTG (Tonalite, Trondhjemite, Granodiorite) magmas. On this picture they are cut by a pegmatitic dyke related to later granitic intrusion. (Photo G. Gruau)

4.3.1.2. *Isua and Akilia greenstone belt (Greenland)*

Supracrustal rocks (i.e. volcanic and sedimentary) also outcrop as lenses into the Amitsôq gneisses. Some lenses as at Isua can be big (~35 km long), but they are generally less than 1km long (Akilia Island for instance). Contrarily to TTG that emplaced deep into the crust, supracrustal deposited or erupted at the surface of the Earth (Figure 4.13). At Isua, zircons gave an age of 3.812 ± 0.014 Ga (Baadsgaard et al., 1984). In Akilia Island, the Amitsôq tonalite dated at 3.872 ± 0.010 Ga by Nutman et al. (Nutman et al., 1996) is intrusive (dyke) into a banded iron formation (BIF) thus demonstrating that this sedimentary rock is older than 3.872 Ga.

Figure 4.13. Isua (Greenland) gneisses, these rocks emplaced as sediments more than 3.87 Ga ago; they represent the oldest known terrestrial sediment (Photo G. Gruau).

The Isua and Akilia supracrustal rocks are extremely important, first of course because they represent the oldest huge volumes of rocks so far recognized, but also because:

- They contain sediments, thus demonstrating the existence of a hydrosphere (ocean) as old as 3.87 Ga.
- The origin of sediments has been subject to active discussion, some authors consider them as mostly of pure chemical origin, or even due to transformation (metasomatism) of pre-existing magmatic rocks (Rose et al., 1996; Fedo, 2000; Fedo et al., 2001; Myers, 2001; Fedo and Whitehouse, 2002; Bolhar et al., 2004). However, recent work (Bolhar et al., 2005) clearly demonstrates that at least some parts of Isua supracrustals correspond to true clastic sediments of mixed mafic and felsic provenance. The weathering, erosion and transport of the clasts are strong arguments in favour of emerged continents as early as 3.87 Ga.
- Both Isua and Akilia sediments contain carbonaceous inclusions, (generally into apatite crystals) whose $\delta^{13}C$ average isotopic constitution of -30 to -35 could be interpreted as biological signature (Mojzsis et al., 1996). More recently, U-rich sediments from the same locality were interpreted as indicators of oxidized ocean water resulting of oxygenic photosynthesis (Rosing and Frei, 2004). However the reliability of these biological signatures is still subject to a very active controversy and the so-called bacteria *Isuasphaera Isua* as well as *Appelella ferrifera* are now considered as artefacts (Appel et al., 2002; Westall and Folk, 2003).
- Since at least 30 years, evidence of horizontal tectonics have been recognized in Amitsôq area (Bridgwater et al., 1974) and more recently it has been proposed that the early Archaean crustal accretion in Greenland proceeded by assemblage and collage of terranes (Nutman and Collerson, 1991; Nutman et al., 2004). All these arguments militate in favour of plate tectonic like processes operating since the Early Archaean.
- Investigation performed on the mafic magmatic components of Isua and Akilia greenstone belts, and based on lead (Kamber et al., 2003) or short-life isotopes (Boyet et al., 2003; Caro et al., 2003) conclude to a very early differentiation of a proto-crust, probably during the magma ocean stage and its long-lived preservation during the Hadean.

4.3.1.3. *Uivak gneisses and Inukjuak (Porpoise Cove) greenstone belt (Canada)*
Along the northern Labrador coast, metamorphic rocks very similar to Amitsôq and Isua gneisses are exposed; there, they are called the Uivak

gneisses. Zircons analysed in these tonalitic (TTG) gneisses yield ages of
3.733 ± 0.009 Ga, however they also contain rounded cores dated at
3.863 ± 0.012 Ga (Schiøtte et al., 1989). However, in Labrador these rock
suffered a NeoArchaean granulite facies metamorphism (Collerson and
Bridgwater, 1979) which in not the case in Greenland. Very recently, a vol-
cano sedimentary formation very similar to Isua was discovered in Hudson
Bay (Canada) at Inukjuak (it was earlier called Porpoise Cove) and dated at
3.825 ± 0.016 Ga (David et al., 2002; Stevenson, 2003). The wide repartition
of these early Archaean crustal components militates in favour of the existence of a big continent (or of many small) older than 3.8 Ga in this part of
the northern hemisphere.

4.3.2. THE 4.0 GA ARCHAEAN CRUST

At Acasta, in Northern Territories (Canada) small amounts (20 km^2) of
Archaean rocks are exposed. They mainly consist in banded tonalite and
granodiorite (TTG) associated with subordinate amphibolite and ultramafic
rocks. Most of these outcrops are strongly deformed, such that in these high
strain zones all lithologies are juxtaposed and parallelized, resulting in banded rocks. However, small low strain zones exist; there, zircons measured in
TTG samples gave ages of 4.031 ± 0.003 Ga (Bowring and Williams, 1999).
Moreover, in a tonalitic sample, zircons contain cores that have been dated at
4.065 ± 0.008 Ga. Bowring and Williams (1999) interpret these cores as
reflecting the incorporation in the tonalitic magma of remnants of older
continental crust. Consequently, the Acasta TTG, not only represent the
oldest rocks (continental crust) so far discovered, but they also demonstrate
the existence of even older continental crust.

Ages measured on Amitsôq and Uivak gneisses, lead to the conclusion
that the mechanisms of TTG genesis by hydrous basalt melting were active at
3.87 Ga. In the light of Acasta data, this conclusion can be extended back in
time until 4.03 Ga and very probably until 4.06 Ga.

4.3.3. THE PRE-4. GA HADEAN CRUST

Until recently, Acasta rocks were the oldest known terrestrial materials, and the
question was to know if continental crust existed or not before 4.0 Ga. Consequently, isotopic compositions of the oldest rocks were used in order to indirectly discuss the possible existence and nature of the Hadean continental crust.

4.3.3.1. *Indirect evidences*
The basic idea on which are based these hypothesis is that continental crust
has been extracted from the mantle through partial melting processes. As the

composition of the crust is drastically different of that of the mantle, its extraction would have modified the mantle composition such that greater the continental crust volume greater the mantle composition change. The primordial Earth mantle is assumed to have the same Nd isotopic ratios as chondritic meteorites. In Figure 4.14, ε_{Nd} represents the difference between the $^{143}Nd/^{144}Nd$ of a rock and that of chondrites; consequently a $\varepsilon_{Nd} = 0$ will reflect a mantle source not affected by the extraction of the continental crust (CHUR = Chondritic Uniform Reservoir), whereas $\varepsilon_{Nd} > 0$ indicates a mantle impoverished by crust extraction; $\varepsilon_{Nd} < 0$ would rather reflect crustal source or contamination.

Even the oldest known rocks already display positive εNd, thus demonstrating that important volumes of continental crust were formed before 4.0 Ga. Simple calculations (McCulloch and Bennet, 1993; Bowring and Houst, 1995) showed that about 10% of the volume of the present day continental crust formed during Hadean times. Investigations performed on Hf (Vervoort et al., 1996; Albarède et al., 2000) and Pb (Kamber et al., 2003) isotopes led to the same conclusion. If all authors agree that continental crust differentiation started early in Earth history, they diverge about the rate and efficiency of this mechanism. For instance, Moorbath (1977) and Vervoort et al. (1996) estimate a relatively low degree of differentiation during the Hadean which could imply a constant crustal growth rate throughout the

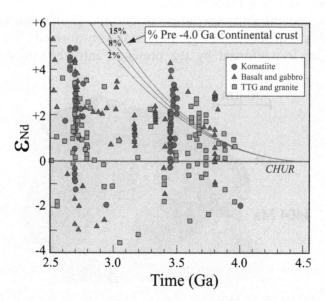

Figure 4.14. εNd vs. time diagram showing that even the oldest known rocks possess εNd > 0 characteristic of mantle composition impoverished by extraction of continental crust prior to 4.0 Ga. Red curves are theoretical mantle Nd isotopic composition assuming 2, 8 and 15% extraction of Hadean continental crust. Data are from McCulloch and Bennet (1993), Jahn (1997) and personal data. CHUR (Purple line) = Chondritic Uniform Reservoir.

whole Earth history (Moorbath, 1977). On the contrary, more recent re-
searches militate in favour of important rates of Hadean crust extraction
(Collerson et al., 1991; Bennet et al., 1993; McCulloch and Bennet, 1993;
Albarède et al., 2000). Similarly, studies of ^{142}Nd (produced by decay of
^{146}Sm, half-life = 0.103 Ga) points to an important crust differentiation
during the very early stages of Earth history, probably during the magma
ocean stage (Boyet et al., 2003; Caro et al., 2003). It has been proposed
(McCulloch and Bennet, 1994) that the early depletion of the mantle has been
buffered after 3.75 Ga by changes in the mantle convective regime resulting
in mixing between depleted mantle and underlying less depleted lower
mantle. Results presented by Albarède et al. (2000) are in agreement with this
model.

4.3.3.2. *Jack Hills zircons*

Recently, zircon crystals, extracted from Jack Hills meta-quartzites in
Australia gave an age of 4.404 ± 0.008 Ga (Wilde et al., 2001), which is the
oldest age so far obtained on terrestrial material (Figure 4.15). However, it
must be noted that many detrital zircons from Jack Hills and Mont Narryer
in Australia already gave a great variety of ages ranging between 4.3 and
4.0 Ga (Froude et al., 1983; Compston and Pidgeon, 1986; Cavosie et al.,
2004), which demonstrates that Hadean crust existed but also that it developed and grew all along Hadean times.

- Zircons predominantly crystallize in granitic (*s.l.*) melts, so it can be
  reasonably concluded that granitoids already formed at 4.4 Ga. This
  conclusion is reinforced by the presence, into the zircon crystals of

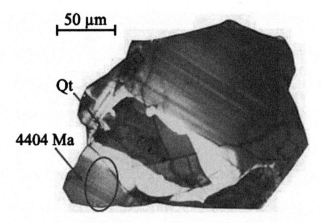

Figure 4.15. Cathodoluminescence image of zircon W74/2-36 from the Jack Hills metacon-
glomerate (Western Australia). An age of 4.404 Ga has been determined by ion microprobe
from the place shown by an ellipse. This crystal also contains quartz (Qt) inclusions indicating
that it crystallized in a granite-like magma (Photo: John Valley, University of Wisconsin,
Madison).

inclusions of quartz, feldspars, hornblende, biotite and monazite; these minerals crystallized together with the zircon and are typical of granitoids (Maas et al., 1992; Wilde et al., 2001; Cavosie et al., 2004). Consequently, as granitoids are the main components of continental crust, it can be concluded that continental crust existed at 4.4 Ga. Based on a hafnium isotope studies in these zircon crystals, Harrison et al. (2005) concluded that both continental crust genesis and plate tectonics began during the first 100–150 Ma of Earth history.

- From Rare Earth Element (REE) content in zircons, and based on partition coefficients between felsic melts and zircon, it is possible to estimate the REE content of the host magma. The melt compositions calculated from older zircons have REE patterns enriched in Light REE (LREE: La to Sm) and depleted in Heavy REE (HREE: Gd to Lu) (Wilde et al., 2001); which are characteristics of Archaean granites (TTG, Martin et al., 2005). Similarly, *Type-1* zircons from Jack Hills have the same REE patterns as Acasta gneiss ones that are TTG in composition (Hoskin, 2005). Consequently, very probably Jack Hills crystallized in TTG-type felsic magma.

- The population of Jack Hills zircons not only recorded ages of 4.4 Ga, but on the contrary, they recorded several continental crust genesis events. Cavosie et al. (2004), evidenced several picks of crustal growth at 4.4, 4.25, 4.2, 4.0 Ga (Figure 4.16). It can be concluded that the Hadean continental crust grew over a time range of about 400 Ma, even if the process could have been episodic (as it is since 4.0 Ga).

Jack Hills zircons also possess rims that indicate that during the 4.4–4.0 time interval, they underwent reworking (re-melting). Such mechanism signifies that this Hadean continental has been stable enough to be re-worked, by both

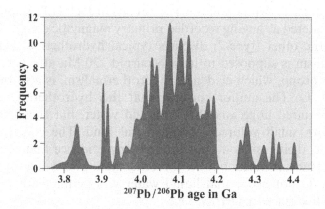

Figure 4.16. Histogram of age frequency of 88 zircons from Jack Hills, Mont Narryer and Barlee terrane (Cavosie et al., 2004).

magmatic and sedimentary processes. Similarly, the fact that these zircons exist implies that the early continental crust was enough stable (and consequently with a sufficiently huge volume) to resist at least partly to the Late Heavy Bombardment.

- Oxygen isotope measurements performed on Jack Hills zircons ($\delta^{18}O$ = 5.4 to 15) allow to calculate the isotopic composition of their host magma which ranges between $\delta^{18}O$ = 7 and $\delta^{18}O$ = 11 (Mojzsis et al., 2001; Peck et al., 2001; Wilde et al., 2001). These authors conclude that the zircons contain crustal material that had interacted with liquid water under surface or near surface conditions. These data are extremely important as they imply that liquid water (ocean?) was probably available on Earth surface as early as 4.4 Ga. In addition, the genesis of TTG like magmas results of hydrous melting of metamorphosed basalts, in the stability field of garnet (Martin, 1986; Martin et al., 2005); which, in turn, also implies the existence of an hydrosphere.

In conclusion, Jack Hills zircons, demonstrate that a TTG like continental crust existed as early as 4.4 Ga ago and that it more or less regularly continued to be generated and re-worked during the whole Hadean. This crust can be considered as stable, such that it can resist to meteoritic bombardment. In addition, Hadean crust composition (TTG) as well as zircons oxygen isotopic composition militate in favour of liquid water on Earth surface during Hadean, which could constitute a strong argument for discussing possible life existence and development.

4.3.3.3. *Discussion of zircon data*

Jack Hills zircons are subject to several extensive studies and recent papers (Utsunomiya et al., 2004; Hoskin, 2005) seem to moderate the enthusiasm provoked by the oxygen isotopes data. Indeed, based on REE, Hoskin (2005) showed that it exists two populations of zircons in Jack Hills population; one group is considered as having recorded primary magmatic composition (type-1), whereas the other (type-2) displays typical hydrothermal features. The hydrothermalism is supposed to have occurred 130 Ma after zircon crystallization. The group, which evidenced hydrothermalism, is also the one with the higher $\delta^{18}O$. The author consider that this hydrothermalism did not necessarily required huge amounts of liquid water, but that it could have developed with small volumes of ephemeral fluid. The conclusion of the author is that, their results do not provide any evidence that ocean did not exist during Hadean times, but on the other hand, most high $\delta^{18}O$ in Jack Hills zircons do not prove that liquid water was permanently available on Hadean Earth surface or near surface.

4.4. Late Heavy Bombardment (LHB)

PHILIPPE CLAEYS, ALESSANDRO MORBIDELLI

4.4.1. THE LATE HEAVY BOMBARDMENT (LHB)

The oldest known terrestrial rocks are represented by the gneisses outcropping in Amitsôq and Isua (Greenland) that are dated between 3.82 and 3.85 Ga and by small enclaves of older (4.03 Ga) gneisses in Acasta, Canada (Martin, 2001). An older formation age of 4.4 Ga has been measured in the cores of a few detrital zircon crystals recovered from Jack Hill (Australia) (Wilde et al., 2001; Valley et al., 2002). On Earth, the period comprised between the end of the accretion and the first occurrence of terrestrial rocks is commonly referred to as the Hadean Eon (Harland et al., 1989). The definition of this first division of the stratigraphic timescale is somewhat ambiguous because it is not sustained by a chronostratigraphic unit, which is a body of rock established to serve as reference for the lithologies formed during this span of time. The Hadean is not even represented on the recently published International Stratigraphic Chart that considers the Archaean as the first Eon, which "lower limit is not defined" (Gradstein et al., 2004). The Hadean covers the interval between the end of the accretion and 4.0 Ga. If there is almost no rock of Hadean age on Earth, this time interval is well represented by the Pre-Nectarian period and the Nectarian period on the Moon (Wilhems, 1987) or by the Early Noachian period on Mars (Figure 4.17). This stratigraphy places the LHB period in the beginning of the Archaean (in the Eoarchaean Era according to (Gradstein et al., 2004), coeval or slightly older than the oldest metamorphosed sediments found in Isua (Greenland).

The lack of lithological record on Earth between 4.5 and 3.9 Ga is explained by the LHB, which almost completely resurfaced the planet (Ryder et al., 2000) coupled with active geological processes such as plate tectonic and erosion, which continuously recycled the most ancient lithologies. The ancient surface of the Moon provides the best evidence for intense collisions in the period centred around 3.9 Ga. Its anorthositic (only made of feldspars) crust crystallized around 4.45 Ga, and the morphology of the highlands recorded a dense concentration of impact craters excavated prior to the emplacement of the first volcanic flows of the mare plains around 3.8 Ga (Wilhems, 1987; Snyder et al., 1996) (Figure 4.18). The LHB is now clearly established all over the inner solar system (Kring and Cohen, 2002). However, the magnitude and frequency of the collisions between 4.5 and 4 Ga remains a topic of controversy. Two explanations have been proposed (Figure 4.19). The frequency of impacts declined slowly and progressively since the end of

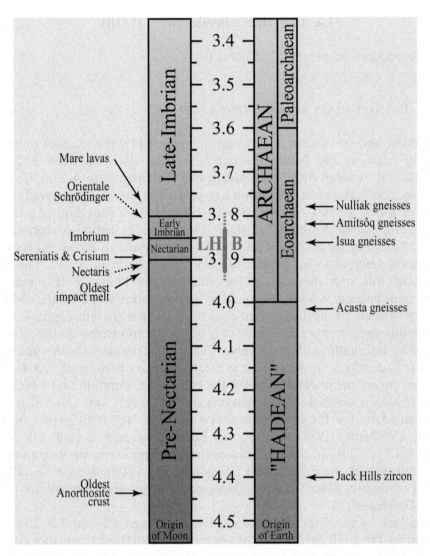

Figure 4.17. Comparative stratigraphy of the first 1.5 Ga on Earth and on the Moon. The full arrows are isotopic dates, the dotted-line arrows are dates based on relative chronology (modified after Ryder et al., 2000).

the accretion period explaining why the bombardment is still important around 3.9 Ga (Hartmann, 1975; Wilhems, 1987; Hartmann et al., 2000). According to these authors, the so-called LHB is not an exceptional event. Rather it is the tail end of a 600 million year period, during which the Earth was subject to major and devastating collisions, with seminal consequences such as the chronic melting of the crust and the vaporization of the oceans. Another view advocates a rapid decline in the frequency of impacts after the formation of the Moon to a value only slightly higher than today (∼2X). The

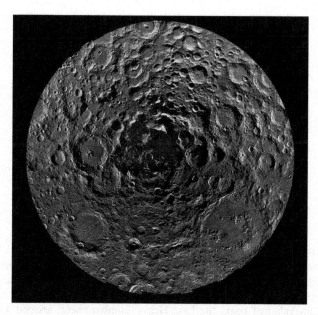

Figure 4.18. View of the Moon's South Polar Region (mosaic of 1500 images) showing a great number of impact craters of all sizes (Image NASA, Clementine mission) mostly formed during the LHB. The Schrödinger crater (320 km in diameter) is clearly visible on the lower right.

period between 4.0 and 3.80 represents thus an exceptional and cataclysmic event marked by an extraordinarily high rate of collisions (Tera et al., 1974; Ryder, 1990; Cohen et al., 2000; Ryder et al., 2000; Ryder, 2002).

Today, most authors favour the LHB cataclysmic scenario around 3.9 Ga. It is supported by a series of arguments; one of these arguments being that some 600 million years of continual impacts should have left an obvious trace on the Moon. So far such trace has not been found. The isotopic dating of both the samples returned by the various Apollo and Luna missions revealed no impact melt-rock older than 3.92 Ga (Ryder, 1990; Ryder et al., 2000; Stoeffler and Ryder, 2001). The lunar meteorites confirm this age limit; they provide a particularly strong argument because they likely originated from random locations on the Moon (Cohen et al., 2000). A complete resetting of all older ages all over the Moon is possible but highly unlikely taking into account the difficulties of a complete reset of all isotopic ages at the scale of a planet (Deutsch and Schrarer, 1994). The U–Pb and Rb–Sr isochrones of lunar highland samples indicate a single disruption by a metamorphic event at 3.9 Ga and between 3.85 and 4 Ga, respectively, after the formation of the crust around 4.4 Ga (Tera et al., 1974). There is no evidence for a re-setting of these isotopic systems by intense collisions between 4.4 and 3.9 Ga. The old upper crustal lithologies of the Moon do not show the expected enrichment in siderophile elements (in particular the

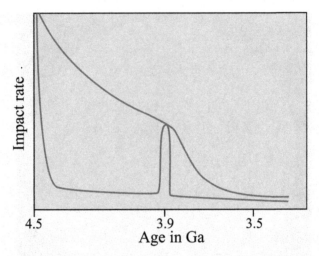

Figure 4.19. Schematic representation of the flux of projectile on Earth between 4.5 and 3.5 Ga (modified after Kring, 2003). The green curve shows a slow decrease of the bombardment, the red curve illustrates the leading hypothesis of a rapid decrease of the flux followed by an unexplained peak, the Late Heavy Bombardment, around 4.0–3.9 Ga. The duration of this high flux period varies (20 and 200 Ma); the slope of the decrease after 3.9 Ga is also poorly constrained. It cannot be excluded that other cataclysmic events took place between 4.5 and 4.0 Ga, but their traces must have been fully erased by the complete resetting of the Earth surface caused by the last event.

Platinum Group Elements, PGE) implied by an extended period of intense collisions (Ryder et al., 2000). Moreover, if the elevated mass accretion documented in the period around 3.9 Ga is considered to be the tail end of an extended period of collisions, the whole Moon should have accreted at about 4.1 Ga instead of 4.5 Ga (Ryder, 2002; Koeberl, 2004). On Earth, the oxygen isotopic signature ($\delta^{18}O$) of the oldest known zircons (4.4 Ga) indicates formation temperatures compatible with the existence of liquid water (Valley et al., 2002). This argument seems contradictory with an extended period of intense collisions. It can thus be concluded that today, there is strong evidence for a cataclysmic LHB event in the inner solar system around 3.9 Ga; in other words a short and intense peak in the cratering rate rather than a prolonged post-accretionary bombardment lasting ~600 Ma.

On planets, whose surface was not re-modelled by erosion, sedimentation and plate tectonics (Moon, Mars, Venus...) the ancient impact structures frequently exceed 1000 km in diameter (Stoeffler and Ryder, 2001). The battered old surfaces of these planets witness past collisions which frequency and scale were 2 to perhaps 3 orders of magnitude higher than the Phanerozoic values. It seems highly unlikely that a target the size of Earth would have been spared, even if active geological processes have obliterated the record of these past impacts. Their traces are perhaps to be found in the existence of elevated concentrations of shocked minerals, tektites or PGE in sedimentary sequences at the very base of the Archaean. So far, most searches have failed to yield convincing evidence (Koeberl et al., 2000; Ryder et al., 2000). Assuming that the bombardment was proportional with that on the Moon, several explanations account for the lack of impact signatures in Earth's oldest lithologies. First, it can be due to the relative small number and the limited type of samples available. Second, a high sedimentation rate could have diluted the meteoritic signal. Third, the LHB and the rocks from Isua may not overlap. A much younger age of 3.65 Ga has also been proposed by some authors for the Amitsôq gneisses (Rosing et al., 1996; Kamber and Moorbath, 1998). Such younger age based on whole rock Pb–Pb, Rb–Sr and Sm–Nd dating seems less accurate than the chronology obtained on single zircon crystals (Mojzsis and Harrison, 2000). Even considering the commonly accepted age of 3.80 Ga for the metasediments in Isua, it is possible that the LHB had already ended when deposition occurred. This is particularly likely if the bombardment rate declined rapidly after a short peak period. However, based on $^{182}W/^{183}W$ isotopic ratios, (Schoenberg et al., 2002) have detected a possible meteoritic signal in four out of six samples of metamorphosed sediments from Isua (Greenland) and Nulliak in Northern Labrador (Canada) dated between 3.8 and 3.7 Ga. These analyses should be replicated but they probably provide the first evidence of the existence of the LHB on Earth (Figure 4.20). However, recently a chromium isotope study carried out on similar lithologies at Isua failed to detect an extraterrestrial component (Frei and Rosing, 2005).

On the Moon, where the record of these collisions is best preserved, some 1700 craters larger than 20 km in diameter are known, among them 15 reach sizes between 300 and 1200 Km. All of them are dated between 4.0 and 3.9 Ga, ages that correspond to the end of the Nectarian and the beginning of the Imbrian periods according to the Lunar stratigraphy (Figure 4.17), (Wilhems, 1987). About 6400 craters with diameters >20 km should have been produced on Mars (Kring and Cohen, 2002). In comparison, the Earth because of its size and larger gravitational cross section represents an easier target to hit. Based on scaling the Moon flux, it would have been impacted by between 13 and 500 times more mass than the Moon according to the size distribution of the projectiles (Zahnle and Sleep, 1997; Hartmann et al., 2000). Using conservative values Kring (2003) estimates the formation of ~22,000 craters with a diameter over 20 km. The number of structures larger than 1000 km would vary between 40 and 200, and it is not impossible that some would reach 5000 km, that is the dimension of an entire continent (Grieve and Shoemaker, 1994; Kring and Cohen, 2002). The duration of the LHB cataclysmic event is difficult to estimate. Based on the cratering record of the Moon, it varies between 20 and 200 Ma, depending on the mass flux estimation used in the calculation. According to the lowest estimation on Earth a collision capable of

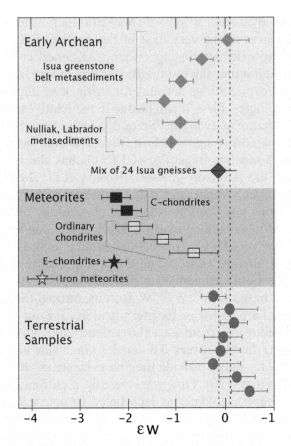

Figure 4.20. Values of εW for terrestrial rocks, meteorites and early Archaean samples relative to the composition of a tungsten standard, ACQUIRE-W (Figure modified from Schoenberg et al., 2002). Blue diamonds represent early Archaean metasediments, which plot outside the terrestrial values. Dashed lines show 2σ of mean uncertainty envelope of the standard (*n* = 46). Terrestrial samples are represented by red circles. Open star is an average of published iron meteorites; filled star an average of bulk rock enstatite chondrites; open squares are bulk rock ordinary chondrites; filled squares represent Allende carbonaceous chondrite. Dark blue diamond is an average of early Archaean crust samples.

generating a 20 km crater could take place every 10,000 years (Kring, 2003). This would correspond to the impact of a ~1 km projectile, an event that presently occurs once every 0.350–1 Ma. If the highest estimation is considered, the frequency increases to such an impact every 20 years!

Based on the lunar cratering record, it seems that around 3.8 Ga the impact frequency had decreased significantly and stabilized close to the present values. However, slope of the decrease is difficult to estimate. The ^{40}Ar–^{39}Ar dating of spherules formed by lunar impacts shows that this decrease in the frequency is perhaps more progressive (by factor 2 or 3 between 3.5 and 1 Ga), and that it has increased slightly again some 500 Ma ago (Culler et al., 2000). This interesting observation based on the dating of spherules collected by the Apollo 14 mission needs to be confirmed by applying the same approach to other sites sampled by Lunar missions.

4.4.2. ORIGIN OF THE LHB

The cause of such a bombardment has remained a puzzle for 30 years. Indeed, to have a cataclysmic spike in the bombardment rate, it is necessary that a massive reservoir of planetesimals remains intact during the planet formation epoch and the subsequent ~600 Ma and then, all of a sudden, is destabilized. This seems to be in conflict with the classical view according to which the solar system did not undergo substantial modifications since soon after its formation.

In a recent paper, (Gomes et al., 2005) designed a scenario that explains how the LHB might have been triggered. In their model, the giant planets formed on quasi-circular, coplanar orbits, between approximately 5.5 and 15 AU. This orbital configuration was much more 'compact' than the current one, where the planetary orbits are placed from 5.2 to 30 AU. A massive disk of planetesimals surrounded the planetary system, extending from a few AUs beyond Neptune outwards. The perturbations of the planets allowed the planetesimals from the inner parts of the disk to slowly develop planet-crossing orbits. The encounters of these planetesimals with the planets drove a slow, progressive migration of the planets. Computer simulations show that Saturn, Uranus and Neptune slowly moved outwards, and Jupiter slowly moved inwards. The rate of this migration was governed by the rate at which the planetesimals could leak out of the disk, which was slow because the disk was quite far from the closest planet.

As a consequence of Jupiter and Saturn's migrations in divergent directions, the ratio of the orbital periods of these two planets increased. This ratio is currently slightly smaller than 2.5, and thus it had to be smaller in the past. Gomes et al. (2005) proposed that the orbital period ratio was initially smaller than 2. Their simulations show that with reasonable initial conditions for the planets and the disk, the ratio of the orbital periods could reach the exact value of 2 (the 2:1 mean motion resonance in the jargon of celestial mechanics) quite late: on a time ranging from 350 Ma and 1.2 Ga, an interval that embraces the LHB time, ~600 Ma after planet formation.

The 1:2 resonance crossing triggered the LHB. Indeed, when the ratio of orbital periods passed through the value of 2, their orbits of Jupiter and Saturn became eccentric. This abrupt transition temporarily destabilized the system of the 4 giant planets, leading to a short phase of close encounters among Saturn, Uranus and Neptune. Because of these encounters, the orbits of the ice giants became eccentric and the furthest planet penetrated into the disk. This destabilized the disk completely. All planetesimals were scattered in sequence by the planets. A fraction of them temporarily penetrated into the inner solar system. About 10^{-7} of the original planetesimals hit the Moon and 10^{-6} hit the Earth. About 10^{-3} of the original planetesimals probably remained trapped in the Kuiper belt.

The interaction of the planets with the destabilized disk enhanced planet migration and damped the planetary eccentricities and inclinations. A large number of simulations of this process, made by Tsiganis et al. (2005) show that, if the planetesimal disk contained about 35 Earth masses at the time of the LHB, the planets statistically had to end on orbits very close to those currently observed. A more massive disk would have driven Jupiter and Saturn too far apart, and a less massive disk would not have moved them enough. Interestingly, given this total disk mass, the total mass of the planetesimals hitting the Moon turns out to be consistent with that inferred from the counting of basins formed at the LHB time.

The distant planetesimals were not the sole cause of the LHB though. The migration of Jupiter and Saturn from the 1:2 resonance to their current position destabilized also the asteroid belt. About 90% of the asteroids could develop orbits crossing those of the terrestrial planets, although this fraction is poorly estimated because it depends on the dynamical state of the asteroid belt before the LHB, which is unknown. Given the current mass of the asteroid belt (5×10^4 Earth masses), this implies that $\sim 5 \times 10^3$ Earth masses of asteroids became Earth crossers. Of these, about 2×10^{-4} hit the Moon, supplying again a mass of projectiles consistent with the LHB constraints. Thus, according to Gomes et al. (2005) model, the projectiles causing the LHB would have been a mixture of comets (from the distant disk) and asteroids, in roughly equal proportions. The exact relative contribution of asteroids and comets to the LHB cannot be stated with precision from the model, at least at the current state. Chemical analyzes on lunar samples collected in the vicinity of basins seem to indicate that asteroids of enstatite or more likely ordinary chondritic type participated in the process (Kring and Cohen, 2002; Tagle, 2005).

The strength of Gomes et al. (2005) model on the origin of the LHB is that it explains not only the main characteristics of the LHB (intensity, duration, abrupt start) but also several other puzzling properties of the solar system. It accounts for the orbital architecture of the giant planets (for what concerns both the mutual spacing, the eccentricities and the mutual inclinations of the orbits, Tsiganis et al., 2005), the population of Jovian Trojans (for what concerns their orbital distribution and total mass, Morbidelli et al., 2005), the absence of numerous asteroid families formed at the LHB time (Gomes et al., 2005).

There do not seem to be many alternatives to Gomes et al. (2005) model. The destabilization of a small body reservoir requires a change in planetary orbits. Planet migration gives such a change, and once it is done, the solar system has acquired the current state. Thus, to make a cataclysmic LHB planet migration had effectively to start at that time. Two conditions need to be fulfilled in order for a compact planetary system to stay quiet for more than half a billion years. The first is that the planetary orbits had to be quasi-circular. The second is that the region among the planets was essentially depleted of planetesimals. If the planetary orbits had to be quasi-circular, then an event had to excite the eccentricities. Resonance crossing provides eccentricity excitation, but only the 1:2 resonance between Jupiter and Saturn provides the required eccentricity values. Thus, Jupiter and Saturn had to cross this resonance. And these are the basic ingredients of Gomes et al. (2005) model.

References

Abe, Y.: 1993, *Lithos* **30**, 223–235.

Abe, Y., Drake, M., Ohtani, E., Okuchi, T. and Righter, K.: 2000, in K. Righter and R. Canup (eds.), Origin of the Earth and Moon (University of Arizona Press, Tucson, Arizona), pp. 413–433.

Albarède, F., Blichert-Toft, J., Vervoort, J. D., Gleason, J. D. and Rosing, M. T.: 2000, *Nature* **404**, 488–490.

Aléon, J., Engrand, C., Robert, F. and Chaussidon, M.: 2001, *Geochim. Cosmochim. Acta* **65**, 4399–4412.

Aleon, J., Robert, F., Chaussidon, M. and Marty, B.: 2003, *Geochim. Cosmochim. Acta* **67**(19), 3773–3783.

Amelin, Y., Krot, A. N., Hutcheon, I. D. and Ulyanov, A. A.: 2002, *Science* **297**, 1679–1683.

Anders, E.: 1989, *Nature* **342**, 255–257.

Appel, P. W. U., Moorbath, S. and Touret, J.: 2002, *Precambrian Res.* **176**, 173–180.

Arpigny, C., Jehin, E., Manfroid, J., Hutsemékér, D., Schulz, R., Stüwe, J. A., Zucconi, J. M. and Ilyin, I.: 2003, *Science* **301**, 1522–1524.

Baadsgaard, H., Nutman, A. P., Bridgwater, D., Rosing, M. T., McGregor, V. R. and Allaart, J. H.: 1984, *Earth Planet. Sci. Lett.* **68**, 221–228.

Ballentine, C. J., Marty, B., Lollar, B. S. and Cassidy, M.: 2005, *Nature* **433**(7021), 33–38.

Barnes, I. and O'Neil, J. R.: 1969, *Geol. Soc. Amer. Bull.* **80**, 1947–1960.

Bennet, V. C., Nutman, A. P. and McCulloch, M. T.: 1993, *Earth Planet. Sci. Lett.* **119**, 299–317.

Bizzarro, M., Baker, J. A. and Haack, H.: 2004, *Nature* **431**(7006), 275–278.

Black, L. P., Gale, N. H., Moorbath, S., Pankhurst, R. J. and McGregor, V. R.: 1971, Earth Planet. Sci. Lett. **12**, 245–259.

Bland, P. A., Smith, T. B., Jull, A. J. T., Berry, F. J., Bewan, A. W. R., Cloudt, S. and Pillinger, C. T.: 1996, *Mon. Not. R. Astron. Soc.* **283**, 551–565.

Boato, G.: 1954, *Geochim. Cosmochim. Acta* **6**, 209–220.

Bolhar, R., Kamber, B. S., Moorbath, S., Fedo, C. M. and Whitehouse, M. J.: 2004, *Earth Planet. Sci. Lett.* **222**(1), 43–60.

Bolhar, R., Kamber, B. S., Moorbath, S., Whitehouse, M. J. and Collerson, K. D.: 2005, *Geochim. Cosmochim. Acta* **69**(6), 1555–1573.

Bouvier, A., Blichert-Toft, J., Vervoort, J. D. and Albarède, F.: 2005, *Earth Planet. Sci. Lett.* **240**(2), 221–233.

Bowring, S. A. and Houst, T. B.: 1995, *Science* **269**, 1535–1540.

Bowring, S. A. and Williams, I. S.: 1999, *Contrib. Mineral. Petrol* **134**, 3–16.

Boyet, M., Blichert-Toft, J., Rosing, M., Storey, M., Telouk, P. and Albarede, F.: 2003, *Earth Planet. Sci. Lett.* **214**(3–4), 427–442.

Boyet, M. and Carlson, R. W.: 2005, *Science* **309**, 576–581.

Bridgwater, D., McGregor, V. R. and Myers, J. S.: 1974, *Precambrian Res.* **1**, 179–197.

Brownlee, D. E.: 1985, *Ann. Rev. Earth Planet. Sci. Lett.* **13**, 147–173.

Brownlee, D. E., Joswiak, D. J., Love, S. G., Nier, A. O., Schlutter, D. J. and Bradley, J. P.: 1993, *Lunar Planet. Sci.* **XXIV**, 205–206.

Canup, R. and Asphaug, E.: 2001, *Nature* **412**, 708–712.

Caro, G., Bourdon, B., Birck, J.-L. and Moorbath, S.: 2003, *Nature* **423**(6938), 428–432.

Cavosie, A. J., Valley, J. W. and Wilde, S. A.E. I. M. F.: 2005, *Earth Planet. Sci. Lett.* **235**, 663–681.

Cavosie, A. J., Wilde, S. A., Liu, D., Weiblen, P. W. and Valley, J. W.: 2004, *Precambrian Res.* **135**(4), 251–279.

Cherniak, D. J.: 2001, *Chem. Geol.* **177**(3–4), 381–397.

Cherniak, D. J., Lanford, W. A. and Ryerson, F. J.: 1991, *Geochim. Cosmochim. Acta* **55**, 1663–1673.

Chyba, C.: 1990, *Nature* **343**, 129–133.

Chyba, C. and Sagan, C.: 1992, *Nature* **355**, 125–132.

Chyba, C. F.: 1991, *Icarus* **92**, 217–233.

Clarke, W. B., Beg, M. A. and Craig, H.: 1969, *Earth Planet. Sci. Lett.* **6**, 213–220.

Clayton, R. N.: 1993, *Ann. Rev. Earth Planet. Sci. Lett.* **21**, 115–149.

Cohen, B. A., Swindle, T. D. and Kring, D. A.: 2000, *Science* **290**(5497), 1754–1756.

Collerson, K. D. and Bridgwater, D.: 1979, in F. Barker (ed.), Trondhjemites, Dacites and Related Rocks (Elsevier, Amsterdam), pp. 206–273.

Collerson, K. D., Campbell, I. H., Weaver, B. L. and Palacz, Z. A.: 1991, *Nature* **349**, 209–214.

Compston, W. and Pidgeon, R. T.: 1986, *Nature* **321**, 766–769.

Conrad, C. P. and Hager, B. H.: 1999, *J. Geophys. Res.* **104**, 17551–17571.

Culler, T. S., Becker, T. A., Muller, R. A. and Renne, P. R.: 2000, *Science* **287**(5459), 1785–1788.

Dauphas, N.: 2003, *Icarus* **165**(2), 326–339.

Dauphas, N. and Marty B.: 2002, *J. Geophys. Res. Planets* **107**, E12-1–E12-7.

Dauphas, N., Robert, F. and Marty, B.: 2000, *Icarus* **148**(2), 508–512.

David, J., Parent M., Stevenson R., Nadeau P. and Godin L.: 2002. La séquence supracrustale de Porpoise Cove, région d'Inukjuak; un exemple unique de croûte paléo-archéenne (ca. 3.8 Ga) dans la Province du Supérieur. 23 éme Séminaire d'information sur la recherche géologique, Ministère des ressources naturelles du Québec.(session 2).

de Ronde, C. E. J., Channer, D. M. d., Faure, K., Bray, C. J. and Spooner, T. C.: 1997, *Geochim. Cosmochim. Acta* **61**(19), 4025–4042.

Delsemme, A. H.: 1999, *Planet. Space Sci.* **47**, 125–131.

Deming, D.: 1999, *Palaeogeogr. Palaeoclimatol. Palaeoecol.* **146**, 33–51.

Deutsch, A. and Schrarer, U.: 1994, *Meteoritics* **29**, 301–322.

Dreibus, G. and Wanke, H.: 1989, in S. K. Atreya, J. B. Pollack and M. S. Matthews (eds.), Origin and Evolution of Planetary and Satellite Atmospheres (University of Arizona Press, Tucson), pp. 268–289.

Engrand, C., Deloule, E., Robert, F., Maurette, M. and Kurat, G.: 1999, *Meteor. Planet. Sci.* **34**, 773–786.

Engrand, C. and Maurette, M.: 1998, *Meteorit. Planet. Sci.* **33**, 565–580.

Farley, K. A., Montanari, A., Shoemaker, E. M. and Shoemaker, C.: 1998, Science 1250–1253.

Fedo, C. M.: 2000, *Precambrian Res.* **101**(1), 69–78.

Fedo, C. M., Myers, J. S. and Appel, P. W. U.: 2001, *Sediment. Geol.* **141–142**, 61–77.

Fedo, C. M. and Whitehouse, M. J.: 2002, *Science* **296**, 1448–1452.

Frank, L. A., Sigwarth, J. B. and Craven, J. D.: 1986, *Geophys. Res. Lett.* **13**, 303–306.

Frank, L. A., Sigwarth, J. B. and Craven, J. D.: 1986, *Geophys. Res. Lett.* **13**, 303–306.

Frei, R. and Rosing, M. T.: 2005, *Earth Planet. Sci. Lett.* **236**, 28–40.

Froude, D. O., Ireland, T. R., Kinny, P. D., Williams, I. S., Compston, W., Williams, I. R. and Myers, J. S.: 1983, *Nature* **304**, 616–618.

Gast, P. W.: 1960, *J. Geophys. Res.* **65**, 1287–1297.

Genda, H. and Abe, Y.: 2005, *Nature* **433**, 842–844.

Gomes, R., Levison, H. F., Tsiganis, K. and Morbidelli, A.: 2005, *Nature* **435**, 466–469.

Göpel, C., Manhès, G. and Allègre, C. J.: 1994, *Earth Planet. Sci. Lett.* **121**, 153–171.

Gounelle, M., Zolensky, M. and Liou, J. C.: 2002, *Geochim. Cosmochim. Acta* **67**, 507–527.

Gradie, J. C., Chapman, C. R. and Tedesco, E. F.: 1989, in T. G. M. S. M. R. P. Binzel (ed.), Asteroids II (Univ. Arizona Press, Tucson), pp. 316–335.

Gradstein F. M., Ogg J. G., Smith A. G., Agterberg F. P., Bleeker W., Cooper R. A., Davydov V., Gibbard P., Hinnov L., (†) M. R. H., Lourens L., Luterbacher H.-P., McArthur J., Melchin M. J., Robb L. J., Shergold J., Villeneuve M., Wardlaw B. R., Ali J., Brinkhuis H., Hilgen F. J., Hooker J., Howarth R. J., Knoll A. H., Laskar J., Monechi S., Powell J., Plumb K. A., Raffi I., Röhl U., Sanfilippo A., Schmitz B., Shackleton N. J., Shields G. A., Strauss H., Dam J. V., Veizer J., Kolfschoten T.v. and Wilson D.: 2004, *A Geological Time Scale 2004*. Cambridge University, Cambridge, 610 pp.

Gray, C. M. and Compston, W.: 1974, *Nature* **251**, 495–497.

Grieve, R. A. F. and Shoemaker, E. M.: 1994, in T. Gehrels (ed.), Hazards Due to Comets and Asteroids (Univ. Arizona Press, Tucson), pp. 417–462.

Halliday, A. N., Wänke, H., Birck, J. L. and Clayton, R. N.: 2001, *Space Sci. Rev.* **96**, 1–34.

Harland, W. B., Armstrong, R. L., Cox, A. V., Craig, L. E., Smith, A. G. and Smith, D. G.: 1989, *A Geologic Time Scale*, Cambridge University Press, Cambridge, 263 pp.

Harrison, C. G. A.: 1999, *Geophys. Res. Lett.* **26**, 1913–1916.

Harrison, T. M., Blichert-toft, J., Müller, W., Albarede, F., Holden, P. and Mojzsis, S. J.: 2005, *Science* **310**, 1947–1950.

Hartmann, W. K.: 1975, *Icarus* **24**, 181–187.

Hartmann, W. K., Ryder, G., Dones, L. and Grinspoon, D.: 2000, in K. Righter and R. Canup (eds.), Origin of the Earth and Moon (University f Arizona Press, Tucson, Arizona), pp. 493–512.

Hashizume, K., Chaussidon, M., Marty, B. and Robert, F.: 2000, *Science* **290**(5494), 1142–1145.

Hashizume, K., Kase, T. and Matsuda, J. I.: 1997, *Kazan* **42**, S293–S301.

Hashizume, K., Marty, B. and Wieler, R.: 2002, *Earth Planet. Sci. Lett.* **202**, 201–216.

Honda, M., McDougall, I., Patterson, D. B., Doulgeris, A. and Clague, D. A.: 1991, *Nature* **349**, 149–151.

Hood, L. L. and Horanyi, : 1993, *Icarus* **106**, 179–189.

Hoskin, P. W. O.: 2005, *Geochim. Cosmochim. Acta* **69**(3), 637–648.

Hunten, D. M., Pepin, R. O. and Walker, J. C. B.: 1987, *Icarus* **69**, 532–549.

Jacobsen, S. B. and Yin, Q.: 2003, *Lunar Planet. Sci.* **XXXIV**, 1913 .

Jahn, B. M.: 1997, in R. Hagemann and M. Treuil (eds.), Introduction à la Géochimie et ses Applications (Editions Thierry Parquet), pp. 357–393.

Javoy, M.: 1995, *J. Geophys. Res. Lett.* **22**, 2219–2222.

Javoy, M.: 1998, *Chem. Geol.* **147**, 11–25.

Javoy, M.: 2005, *C. R. Geosci.* **337**(1–2), 139–158.

Jewitt, D. C., Matthews, H. E., Owen, T. and Meier, R.: 1997, *Science* **278**(5335), 90–93.

Kamber, B. S., Collerson, K. D., Moorbath, S. and Whitehouse, M. J.: 2003, *Contrib. Mineral. Petrol.* **145**, 25–46.

Kamber, B. S. and Moorbath, S.: 1998, *Chem. Geol.* **150**, 19–41.

Keays, R. R., Ganapathy, R., Laul, J. C., Anders, E., Herzog, G. F. and Jeffery, P. M.: 1970, *Science* **167**, 490–493.

Kerridge, J. F.: 1985, *Geochim. Cosmochim. Acta* **49**, 1707–1714.

Kleine, T., Münker, C., Mezger, K. and Palme, H.: 2002, *Nature* **418**, 952–955.

Koeberl, C.: 2004, *Earth Moon Planets* **92**, 79–87.

Koeberl, C., Reimold, W. U., McDonald, I. and Rosing, M.: 2000, in I. Gilmour and C. Koeberl (eds.), Impacts and the Early Earth (Springer, Berlin), pp. 73–97.

Kortenkamp, S. J., Dermott, S. F., Fogle, D. and Grogan K.: 2001, in: B. P.-E.a.B. Schmitz (ed.), Accretion of Extraterrestrial Matter Throughout Earth's History (Kluwer Academic), pp. 13–30.

Kring, D. A.: 2003, *Astrobiology* **3**(1), 133–152.

Kring, D. A. and Cohen, B. A.: 2002, *J. Geophys. Res.* **107**(E2), 10.1029/2001JE001529.

Kurat, G., Koeberl, C., Presper, T., Branstätter, F. and Maurette, M.: 1994, *Geochim. Cosmochim. Acta* **58**, 3879–3904.

Kyte, F. T. and Wasson, J. T.: 1986, *Science* **232**, 1225–1229.

Lee, T., Papanastassiou, D. A. and Wasserburg, G. J.: 1976, *Geophys. Res. Lett.* **3**, 109–112.

Lopez-Garcia, P.: 2005, in M. Gargaud, B. Barbier, H. Martin and J. Reisse (eds.), Lectures in Astrobiology (Springer-Verlag, Berlin), pp. 657–676.

Love, S. G. and Brownlee, D. E.: 1993, *Science* **262**, 550–553.

Maas, R., Kinny, P. D., Williams, I. S., Froude, D. O. and Compston W.: 1992, Geochim. Cosmochim. Acta **56**, (1281–1300).

Mahaffy, P. R., Donahue, T. M., Atreya, S. K., Owen, T. C. and Niemann, H. B.: 1998, *Space Sci. Rev.* **84**(1–2), 251–263.

Mamyrin, B. A., Tolstikhin, I. N., Anufriev, G. S. and Kamensky, I. L.: 1969, *Dokkl. Akad. Nauk. SSSR.* **184**, 1197–1199 (In Russian).

Martin, H.: 1986, *Geology* **14**, 753–756.

Martin, H.: 2001, in M. Gargaud, D. Despois and J.-P. Parisot (eds.), L'environnement De la Terre Primitive (Presses Universitaires de Bordeaux, Bordeaux), pp. 263–286.

Martin, H., Smithies, R. H., Rapp, R., Moyen, J.-F. and Champion, D.: 2005, *Lithos* **79**(1–2), 1–24.

Marty, B.: 1989, *Earth Planet. Sci. Lett.* **94**, 45–56.

Marty, B.: 1995, *Nature* **377**, 326–329.

Marty, B. and Dauphas, N.: 2003, *Earth Planet. Sci. Lett.* **206**, 397–410.

Marty, B., Jambon, A. and Sano, Y.: 1989, *Chem. Geol.* **76**, 25–40.

Marty, B., Robert, P. and Zimmermann, L.: 2005, *Meteorit. Planet. Sci.* **40**, 881–894.

Maurette, M.: 1998, *Orig. Life Evol. Biosph.* **28**, 385–412.

Maurette, M.: 2001, in M. Gargaud, D. Despois and J.-P. Parisot (eds.), L'environnement de la terre primitive (Presses Universitaires de Bordeaux), pp. 99–130.

Maurette, M., Duprat, J., Engrand, C., Gounelle, M., Kurat, G., Matrajt, G. and Toppani, A.: 2000, *Planet. Space Sci.* **48**(11), 1117–1137.

Mazor, E., Heymann, D. and Andus, E.: 1970, *Geochim. Cosmochim. Acta* **34**, 781–824.

McCulloch, M. T. and Bennet, V. C.: 1993, *Lithos* **30**, 237–255.

McCulloch, M. T. and Bennet, V. C.: 1994, *Geochim. Cosmochim. Acta* **58**, 4717–4738.

McGregor, V. R.: 1968, *Rapport Gronlands Geol. Unders* **19**, 31.

McGregor, V. R.: 1973, *Philos. Trans. R. Soc. Lond. A* **A-273**, 243–258.

Messenger, S.: 2000, *Nature* **404**, 968–971.

Messenger, S., Keller, L. P., Stadermann, F. J., Walker, R. M. and Zinner, E.: 2003, *Science* **300**, 105–108.

Messenger, S., Keller, L. P. and Walker, R. M.: 2002, *Discovery of Abundant Silicates in Cluster IDPs, LPS XXXIII*, LPI, Houston, 1887 pp.

Mizuno, H., Nakasawa, K. and Hayashi, C.: 1980, *Earth Planet. Sci. Lett.* **50**, 202–210.

Mojzsis, S. J., Arrhenius, G., Keegan, K. D., Harrison, T. H., Nutman, A. J. and Friend, C. L. R.: 1996, *Nature* **384**, 55–59.

Mojzsis, S. J., Harrison, M. T. and Pidgeon, R. T.: 2001, *Nature* **409**, 178–181.

Mojzsis, S. J. and Harrison, T. M.: 2000, *GSA Today* **10**, 1–6.

Moorbath, S. (ed.), 1977, *Aspects of the geochronology of ancient rocks related to continental evolution. The continental crust and its mineral deposits, 20*. Geological Association of Canada Special Paper, pp. 89–115.

Moorbath, S., O'Nions, R. K., Pankhurst, R. J., Gale, N. H. and McGregor, V. R.: 1972, *Nature* **240**, 78–82.

Morbidelli, A.: 2002, *Ann. Rev. Earth Planet. Sci.* **30**, 89–112.

Morbidelli, A., Chambers, J., Lunine, J. I., Petit, J. M., Robert, F., Valsecchi, G. B. and Cyr, K.E.: 2000, *Meteorit. Planet. Sci.* **35**, 1309–1320.

Morbidelli, A., Levison, H. F., Tsiganis, K. and Gomes, R.: 2005, *Nature* **435**(7041), 462–465.

Murakami, M., Hirose, K., Yurimoto, H., Nakashima, S. and Takafuji, N.: 2002, *Science* **295**, 1885–1887.

Myers, J. S.: 2001, *Precambrian Res.* **105**(2–4), 129–141.

Nutman, A. J., Bennet, V. C., Friend, C. R. L. and McGregor, V. R.: 2000, *Geochim. Cosmochim. Acta* **64**(17), 3035–3060.

Nutman, A. J. and Bridgwater, D.: 1986, *Contrib. Mineral. Petrol.* **94**, 137–148.

Nutman, A. J. and Collerson, K. D.: 1991, *Geology* **19**, 791–794.

Nutman, A. J., McGregor, V. R., Friend, C. R. L., Bennett, V. C. and Kinny, P. D.: 1996, *Precambrian Res.* **78**, 1–39.

Nutman, A. P., Friend, C. R. L., Barker, S. L. L. and McGregor, V. R.: 2004, *Precambrian Res.* **135**(4), 281–314.

Nyquist, L. E., Wiesman, H., Bansal, B., Shih, C.-Y., Keith, J. E. and Harper, C. L.: 1995, *Geochim. Cosmochim. Acta* **59**, 2817–2837.

Owen, T., Mahaffy, P. R., Niemann, H. B., Atreya, S. and Wong, M.: 2001, *Astrophys. J.* **553**, L77–L79.

Ozima, M. and Podosek, F. A.: 2002. *Noble Gas Geochemistry*, Cambridge University Press, Cambridge, 286 pp.

Peck, W. H., Valley, J. W., Wilde, S. A. and Graham, C. M.: 2001, *Geochim. Cosmochim. Acta* **65**(22), 4215–4229.

Pepin, R. O.: 1991, *Icarus* **92**, 1–79.

Pinti, D. L.: 2002, *Trends Geochem.* **2**, 1–17.

Pinti, D. L.: 2005, in M. Gargaud, B. Barbier, H. Martin and J. Reisse (eds.), Lectures in Astrobiology. Advances in Astrobiology and Biogeophysics (Springer-Verlag, Berlin), pp. 83–107.

Podosek, F. A. and Cassen, P.: 1994, *Meteorit. Planet. Sci.* **29**, 6–25.

Porcelli, D. and Halliday, A. N.: 2001, *Earth Planet. Sci. Lett.* **192**(1), 45–56.

Porcelli, D. and Pepin, R. O.: 2000, in R. M. Canup and K. Righter (eds.), Origin of the Earth and Moon (The University of Arizona Press, Tucson), pp. 435–458.

Robert, F.: 2001, *Science* **293**(5532), 1056.

Robert, F.: 2003, *Space Sci. Rev.* **106**(1–4), 87.

Robert, F. and Epstein, S.: 1982, *Geochim. Cosmochim. Acta* **46**, 81–95.
Robert, F., Gautier, D. and Dubrulle, B.: 2000, *Space Sci. Rev.* **92**, 201–224.
Rose, N. M., Rosing, M. T. and Bridgwater, D.: 1996, *Am. J. Sci.* **296**, 1004–1044.
Rosing, M. T. and Frei, R.: 2004, *Earth Planet. Sci. Lett.* **217**(3–4), 237–244.
Rosing, M. T., Rose, N. M., Bridgwater, D. and Thomsen, H. S.: 1996, *Geology* **24**, 43–46.
Ryder, G.: 1990, *Eos Trans AGU* **71**, 313–323.
Ryder, G.: 2002, *J. Geophys. Res. Planets* **107**, 6–14.
Ryder, G., Koeberl, C. and Mojzsis, S. J.: 2000, in R. M. Canup and K. Righter (eds.), Origin of the Earth and Moon (University of Arizona Press, Tucson, Arizona), pp. 475–492.
Sano, Y. and Pillinger, C. T.: 1990, *Geochem. J.* **24**, 315–325.
Sarda, P., Staudacher, T. and Allègre, C. J.: 1988, *Earth Planet. Sci. Lett.* **91**, 73–88.
Sasaki, S. and Nakasawa, K.: 1988, *Earth Planet. Sci. Lett.* **89**, 323–334.
Sasaki, S. and Nakazawa, K.: 1984, *Icarus* **59**, 76–86.
Schiøtte, L., Compston, W. and Bridgwater, D.: 1989, *Can. J. Earth Sci.* **26**, 1533–1556.
Schoenberg, R., Kamber, B. S., Collerson, K. D. and Moorbath, S.: 2002, *Nature* **418**, 403–405.
Shukolyukov, A. and Lugmair, G. W.: 1993, *Science* **259**, 1138–1142.
Sleep, N. H., Meibom, A., Fridriksson, T., Coleman, R. G. and Bird, D. K.: 2004, *Proc. Nat. Acad. Sci.* **101**, 12818–12823.
Sleep, N. H., Zahnle, K. and Neuhoff, P. S.: 2001, *Proc. Natl. Acad. Sci. U. S. A.* **98**(7), 3666–3672.
Snyder, G., Hall, C. M., Lee, D. C., Taylor, L. A. and Halliday, A. N.: 1996, *Meteorit. Planet. Sci.* **31**, 328–334.
Stevenson, R. K.: 2003, Geochemistry and isotopic evolution (Nd, Hf) of the 3.825 Ga Porpoise Cove sequence, Northeastern Superior Province, Québec., Vancouver 2003 GAC – MAC – SEG Meeting, Vancouver, pp. GS6.
Stoeffler, D. and Ryder, G.: 2001. *Stratigraphy and Isotopic Ages of Lunar Geologic Units: Chronological Standard for the Inner Solar System., The Evolution of Mars. Space Science Reviews*, International Space Science Institute, Bern, Switzerland, 7–53 pp.
Tagle, R.: 2005, *LL-Ordinary chondrite impact on the Moon: Results from the 3.9 Ga impact melt at the landing site of Apollo 17*, Lunar and Planetary Science Conference, Houston Texas, pp. CD-ROM Abstract # 2008.
Taylor, S. R.: 2001. *Solar System Evolution: A New Perspective*, Univ. Press, Cambridge, 484 pp.
Tera, F., Papanastassiou, D. A. and Wasserburg, G. J.: 1974, *Earth Planet. Sci. Lett.* **22**, 1–21.
Tolstikhin, I. N. and Hofmann, A. W.: 2005, *Phys. Earth Planet. Int.* In press.
Tolstikhin, I. N. and Marty, B.: 1998, *Chem. Geol.* **147**, 27–52.
Trieloff, M., Kunz, J., Clague, D. A., Harrisson, C. J. and Allègre, C. J.: 2000, *Science* **288**, 1036–1038.
Trull, T.: 1994, in J. I. Matsuda (ed.), Noble Gas Geochemistry and Cosmochemistry (Terra Sci. Pub. Co., Tokyo), pp. 77–88.
Tsiganis, K., Gomes, R., Morbidelli, A. and Levison, H. F.: 2005, *Nature* **435**(7041), 459–461.
Utsunomiya, S., Palenik, C. S., Valley, J. W., Cavosie, A. J., Wilde, S. A. and Ewing, R. C.: 2004, *Geochim. Cosmochim. Acta* **68**(22), 4679–4686.
Valley, J. W., Peck, W. H., King, E. M. and Wilde, S. A.: 2002, *Geology* **30**(4), 351–354.
Vervoort, J. D., Patchett, P. J., Gehrels, G. E. and Nutman, A. J.: 1996, *Nature* **379**, 624–627.
Wasserburg, G. J. G. R. and Busso, M.: 1998, *Astroph. J.* **500**, L189–L193.
Westall, F. and Folk, R. L.: 2003, *Precambrian Res.* **126**, 313–330.
Wilde, S. A., Valley, J. W., Peck, W. H. and Graham, C. M.: 2001, *Nature* **409**, 175–178.

Wilhems, D. E.: 1987, *Geologic history of the Moon. Professional Paper, 1348*. US Geological Survey, Reston VA.

Wood, J. A., Dickey, J. S., Marvin, U. B. and Powell, B. N.: 1970, *Proc. Apollo* **11**(Lunar Sci. Conf. 1), 965–988.

Yada, T., Nakamura, T., Takaoka, N., Noguchi, T., Terada, K., Yano, H., Nakazawa, T. and Kojima, H.: 2004, *Earth Planets Space* **56**, 67–79.

Yin, Q., Jacobsen, S. B., Yamashita, K., Blicchert-Toft, J., Télouk, P. and Albarède, F.: 2002a, *Nature* **418**, 949–952.

Yin, Q., Jacobsen, S. B., Yamashita, K., B.-T., J., Télouk, P. and Albarède, F.: 2002b, *Nature* **418**, 949–952.

Yin, Q. Z. and Ozima, M.: 2003, *Geochim. Cosmochim. Acta* **67**(18), A564–A564.

Yokochi, R., Marty, B., Pik, R. and Burnard, P.: 2005. Geochem. Geophys. Geosyst. 6, Q01004.

Zahnle, K. and Dones L.: 1998. Source of terrestrial volatiles, Origin of the Earth and Moon. LPI Contribution, pp. 55–56.

Zahnle, K. J. and Sleep, N. H.: 1997, in P. J. Thomas, C. F. Chyba and C. P. McKay (eds.), Comets and the Origin and Evolution of Life (Springer, New York), pp. 175–208.

Zhang, Y. X. and Zindler, A.: 1993, *Earth Planet. Sci. Lett.* **117**(3–4), 331–345.

Zolensky, M. E., Weisberg, M. K., Buchanan, P. C. and Mittlefehldt, D. W.: 1996, *Meteorit. Planet. Sci.* **31**, 518–537.

Earth, Moon, and Planets (2006) 98: 153–203
DOI 10.1007/s11038-006-9089-3

5. Prebiotic Chemistry – Biochemistry – Emergence of Life (4.4–2 Ga)

ROBERT PASCAL and LAURENT BOITEAU
Département de Chimie, Université Montpellier II, Montpellier, France
(E-mails: rpascal@univ-montp2.fr; laurent.boiteau@univ-montp2.fr)

PATRICK FORTERRE
Institut de Génétique et Microbiologie, Université Paris-Sud, Orsay, France
(E-mail: fortere@igmors.u-psud.fr)

MURIEL GARGAUD
Observatoire Aquitain des Sciences de l'Univers, Université Bordeaux1, Bordeaux, France
(E-mail: gargaud@obs.u-bordeaux1.fr)

ANTONIO LAZCANO
Universidad Nacional Autónoma de México (UNAM), Mexico City, Mexico
(E-mail: alar@correo.unam.mx)

PURIFICACIÓN LÓPEZ-GARCÍA and DAVID MOREIRA
Unité d'Ecologie, Systématique et Evolution, Université Paris-Sud, Orsay, France
(E-mails: puri.lopez@ese.u-psud.fr; david.moreira@ese.u-psud.fr)

MARIE-CHRISTINE MAUREL
Institut Jacques Monod, Université Paris 6, Paris, France
(E-mail: marie-chistine.maurel@ijm.jussieu.fr)

JULI PERETÓ
Institut Cavanilles de Biodiversitat i Biologia Evolutiva, Universitat de València, València, Spain
(E-mail: Juli.Pereto@uv.es)

DANIEL PRIEUR
Laboratoire de Microbiologie des Environnements Extrêmes, Université de Bretagne Occidentale, Brest, France
(E-mail: Daniel.Prieur@univ-brest.fr)

JACQUES REISSE
Faculté des Sciences Appliquées (CP 165/64), Université Libre de Bruxelles, Brussels, Belgium
(E-mail: jreisse@mach.ulb.ac.be)

(Received 1 February 2006; Accepted 4 April 2006)

Abstract. This chapter is devoted to a discussion about the difficulties and even the impossibility to date the events that occurred during the transition from non-living matter to the first living cells. Nevertheless, the attempts to devise plausible scenarios accounting for the emergence of the main molecular devices and processes found in biology are presented including the role of nucleotides at early stages (RNA world). On the other hand, hypotheses on the development of early metabolisms, compartments and genetic encoding are also discussed in relation with their role in extant living organisms. The nature of the Last Common Ancestor is also presented as well as hypotheses on the evolution of viruses. The following sections constitute a collection of independent articles providing a general overview of these aspects.

Keywords: Origin of life, chemical evolution, origin of genetic information, metabolism, membrane

5.1. A Word of Caution about Chronology

JACQUES REISSE, LAURENT BOITEAU, PATRICK FORTERRE, MURIEL GARGAUD, ANTONIO LAZCANO, PURIFICACIÓN LÓPEZ-GARCÍA, MARIE-CHRISTINE MAUREL, DAVID MOREIRA, ROBERT PASCAL, JULI PERETÓ, DANIEL PRIEUR

Compared to astronomers and geologists, chemists and biochemists find themselves in a very difficult situation when asked to participate to a collective work on the dating of significant events in astrobiology. Little information is available that can allow them to date in detail the events that took place when the protosolar nebula started to collapse and eventually the young Earth was formed and life first appeared in our planet. Dating the origin of the constituents of living matter is in itself a huge problem. Some of these molecules were probably already present in the interstellar cloud long before it splitted into various nebulae. One of these nebulae was the proto-solar nebula, which therefore must have contained a vast ensemble of organic molecules. It is generally accepted that during the accretion of the solar system, these organic molecules were probably destroyed in the inner part of the system, but some of them may have remained intact or with little modifications in small volatile-rich bodies like comets or, perhaps, even in the parent bodies of carbonaceous chondrites. During post-accretional processes these preformed organic compounds, which may have included amino acids or nucleic bases, were delivered to the young Earth together with water molecules and other simple volatiles. Therefore, it could be argued that some of the components of living systems were probably formed before the solar system itself. Obviously, it can also be argued that the hydrogen atoms found in past and extant life forms were formed very soon after the Big Bang, and that the life story started many billions year before the origin of the solar system itself! The choice of an origin is *always* arbitrary and requires a careful definition of what is the starting point.

Although there are some dissenting views, it is has been generally assumed that life appeared on Earth once the physico-chemical conditions of the primitive environment were compatible with the presence of organic polymers like nucleic acids and polypeptides. Most researchers agree that one of these conditions was the presence of liquid water, and that therefore the prebiotic stage may have started around 4.4–4.2 billions year ago. On the other hand, although the identification of the oldest traces of life remains a contentious issue, it is generally agreed that a microbial biosphere had already developed on Earth 3.5 billions years ago. In between, chemical evolution took place. Any attempt to put exact dates on particular steps may be futile at the time being. Of course, and as we will see in the following pages, chemists and biochemists are able to make some reasonable assumptions about the sequence of some steps leading from the synthesis and accumulation of biochemical monomers to the first cells. Let us consider two examples: given the chemical lability of RNA, it is possible that oligopeptides may have existed before polyribonucleotides began to accumulate, and that RNA in turn evolved prior to DNA. Nevertheless, no one is able to state if the accumulation of a necessary (whatever "necessary" may have been) amount of polypeptides required 1 year or 1 million years. Similarly, nobody is ready to claim that the synthesis of a "sufficient" amount of polynucleotides took a "short time" or a "very long time", considering that even "short" and "very long" cannot be defined!

Individually, chemical reactions can be fast or slow depending on the rate constants values (which themselves are strongly depending on the temperature, pH, ionic strength, and so on) but also on the reactant concentrations. Given our poor understanding of primitive Earth conditions, it could be argued that it is impossible to estimate, even roughly, the time necessary, in an unknown place on the young Earth to go through the various steps required for the emergence of a living cell. However, simulation experiments can provide important insights on the rate of chemical syntheses and/or degradations under various conditions.

In order to avoid the problem related to the degradation of organic molecules and of the supramolecular systems into which they may have evolved, it is tempting to suggest that the multi-steps syntheses which led to the formation of the first living cell were fast processes. It is difficult to estimate the rate of self-organization of the precursors of life into replicating systems, because the chemical steps are unknown. Whatever the time scale required for the appearance of an informational polymer, once formed it must have persisted at least long enough to allow its replication. If polymers formed by a slow addition of monomers, this process must have been rapid compared to rates of hydrolysis, especially if a considerable amount of genetic information was contained in the polymer. Self-replicating systems capable of undergoing Darwinian evolution must have emerged in a period shorter than the destruction rates of their components; even if the backbone of primitive genetic polymers was highly stable, the nitrogen bases themselves would decompose over long periods of time. In fact, it can be argued that the accumulation of all components of the primitive soup will be limited by destructive processes, including the pyrolysis of the organic compounds in the submarine vents (large amounts of the entire Earth's oceans circulate through the ridge crests every 10 million years facing temperatures of 350 °C or more).

However, nobody knows the number of aborted "attempts" before the final result was reached and life appeared and persisted. What is sure is the fact that presently, prebiotic chemistry is unable to put dates on steps between a system that is definitively non-living and a system which could be recognized as living. It is doubtful that the situation will change in the near future: as we said previously, our poor knowledge on the actual conditions of the primitive environment, the absence of molecular relics from the prebiotic organic world, the dependence of organic reactions rates on external conditions and, last but not least, the fast degradation of organic molecules and supramolecular organic assemblies will perhaps preclude for ever to date prebiotic events. The only thing that can be done (with great care!) is to suggest a plausible sequence of events. This is what the authors of the following chapter have tried to do.

Following this approach, chemical evolution and the first stages of biochemical evolution may be thought as a succession of stages corresponding to chemical or biochemical structures of increasing complexity. However, this view is probably biased by our current description of the basic properties of extant life. Then early chemical and biochemical stages may have followed one another gradually and probably involved the coexistence of a large number of pathways, the majority of which disappeared and might be considered as dead branches of the evolution tree. It is also possible that what we call the "RNA world" (or the additional hypotheses suggested by those that follow other alternatives) simply corresponded to a subset of molecules that we consider as *qualitatively* significant among a large number of systems that came to a dead end, so that these stages may have occurred simultaneously. For these reasons, the chapter on chemical and early biological evolution has been divided into several parts corresponding to structures of increasing complexity, which does not mean that evolution proceeded following this sequence: for instance, it will never be possible to know if the confinement within vesicles preceded or not the appearance of primitive replicating genetic polymers, or if they coexisted from the very beginning.

Evolutionary biologists interested in very early evolution are largely in front of similar problems to those faced by chemists and biochemists. They could be helped in principle by some imposed constraints coming from geology, micropaleontology, and geochemistry, including bona-fide microfossils, isotopic data and molecular fossils. However, sometimes these signatures are ambiguous and controversial (see chapter 7.1), and in most cases do not allow any inference about the lifestyle, the metabolism or the phylogenetic affiliation of the corresponding living species. Biologists are obliged to use indirect arguments based on what they know about modern microorganisms, comparing genes and proteins involved in metabolic pathways to suggest a plausible evolutionary route that explains their observed contemporary patterns of distribution in organisms and, eventually, a series of possible sequential events towards the past that leads to speculate about the nature of the earliest metabolic strategies. However, in some cases this depends strongly on the model of evolutionary reconstruction that is chosen. One or a few parsimonious scenarios can be identified but, ultimately, evolution does not necessarily follow the most parsimonious way among all those that are possible, so that absolute certainty about the succession of early evolutionary steps is unattainable. Nobody knows for sure how the last unicellular common ancestor looked like or even when it became the dominant form of life on the young Earth. So, for different reasons, chemists interested in prebiotic chemistry and biologists interested in the first living organisms must accept that they have few things to say about chronology but they have to explain why it is so. The works that give the impression that the "how" and "when" questions concerning the origin of life are solved, except for few details, do not contribute to the development of astrobiology. It is much better to list the problems for which plausible explanations exist but also those which remain without solution. In these last cases, it is sometimes necessary to question the question and to try to find another way to formulate the questioning. In science, the impossibility to find a solution to a problem can be the proof that the problem is not well formulated.

5.2. A Scenario Starting from the First Chemical Building Blocks

LAURENT BOITEAU, ROBERT PASCAL

Although chronology is impossible in prebiotic chemistry, building plausible scenarios, linking the possibilities of simple abiotic processes to what the most ancient biochemical pathways are supposed to be, is the central goal of this field of science. In other words, there is no definitive answer to the question: how long did it take for life to be present? But there may be an (or several) answer(s) to: what is the sequence of stages that were covered for living beings to emerge in the abiotic environment of the early Earth, where simple organic compounds were synthesized in the early atmosphere, in other places (hydrothermal vents, volcanic plumes) or delivered from the outer space by meteoritic bombardment? The object of prebiotic chemistry is then to solve a problem that depends on the definition of what was the environment of the early Earth (unsteady over the 4.4–2 Ga b.p. era assigned to the origin of life) and on the definition of what could have been the biochemistry of early living organisms inferred from the sciences of evolution (see part 5.7). It is then necessary to analyze what occurred in the black box separating the two stages to get a continuous sequence of incremental steps as usually observed in the evolution of living organisms, the state of the system at the entrance and at the exit being not clearly known.

5.2.1. AVAILABILITY OF ORGANIC MATTER AND ENERGY

The origin of organic matter and energy on the early Earth is still debated mainly because of the lack of precise indications establishing the composition of the atmosphere (and especially the relative content in H_2 and CO_2, see chapter 6.2.1) and the surface and ocean temperatures (that may have been comprised within a range of less than 0 °C to above 80 °C, see chapters 4.4 and 6.2). Moreover, as pointed out by Miller (1998), the conditions under which organic molecules can be formed in an equilibrium state or in a stationary state (thanks to a flux of energy or of reactive species with a dehydration or oxidation state away from equilibrium) are usually those in which these molecules become instable. Then, any synthetic process in a given area (or under defined conditions) must also involve a quenching step to protect the activated mixture from further degradation by transferring it into a milder environment. This is true for any kind of energy sources including lightning, ultraviolet irradiation of the atmosphere, hydrothermal processes of synthesis and even for the delivery of exogenous organic matter by bombardment. Therefore, the claim of a hypothetical synthetic or deleterious character of a given energy source should only be considered after taking into account the efficiency of the quenching process. Anyway, it is unlikely that a single energy source may have been responsible for the formation of the organic matter needed by the origin of life process and the early stages of life evolution.

Two different kinds of processes have contributed to feed the early Earth with biogenic molecules, biomolecules and the building blocks needed for polymer synthesis: the exogenous delivery and the endogenous formation as a result of energy input in the atmosphere or in the oceans. The relative importance of the two processes was principally dependant on the degree of oxidation of the primitive atmosphere and thus on its hydrogen content (Chyba and Sagan, 1992; Chyba, 2005). The initial Miller experiment (Miller, 1953) was carried out in a mixture with a high content in H_2, CH_4, and NH_3 that is favourable for the synthesis of organic molecules. Then, it has been considered that the escape of hydrogen from the atmosphere to the outer space had been so fast that the Earth atmosphere rapidly reached a composition based on CO_2 and N_2 with a low content in H_2 (Kasting, 1993), much less favourable for synthesis. As a result, the amount of biogenic molecules synthesized on the early Earth may have been much lower and life depended on extraterrestrial delivery. Indications have been reported supporting the formation of organic molecules in the solar nebula and then their delivery to the early Earth since a high organic content can be found in meteorites and comets (Mullie and Reisse, 1987; Cronin and Chang, 1993). Then, it is obvious that the delivery of biogenic compounds from the outer space played a role at that stage but its relative contribution mainly depended on the local production, which can only be deduced from hypothetical models of the early atmosphere. A recent model of the evolution of atmosphere supports a hydrogen escape that could have been much slower than previously believed (Tian et al., 2005), so that the amount of organic matter produced on Earth could have been sufficient for the emergence of life. It has also been proposed that hydrothermal synthesis, by which organic molecule could have been synthesized by heating taking advantage of the presence of minerals (see for example Holm and Andersson, 1998; Russell and Martin, 2004), may have contributed to endogenous production. The catalogue of molecules produced includes high-energy biogenic basic species such as HCN, cyanate, formaldehyde and other aldehydes (Miller, 1998). These simple molecules can undergo different processes under conditions simulating the primitive Earth capable of yielding the building blocks of biochemistry: amino acids (Strecker reaction of aldehydes), nucleic bases, and sugars (formose reaction from formaldehyde). But it is more difficult to define a scenario of chemical evolution by which the system became more complex allowing the formation of macromolecular and supramolecular components of life and their combination into metabolic processes.

5.2.2. FAVOURABLE AREAS FOR PREBIOTIC CHEMICAL PROCESSES

Since the knowledge of the Earth formation suggests that an habitable ocean may have been present as early as 4.4 Ga (see chapter 4.2.2), it is possible to speculate that life emerged within the first few hundred million years of the history of the planet and may have survived the Late Heavy Bombardment (see chapters 4.4 and 6.2.1). But we have no indication about the occurrence of this event. Hypothetical places that would have been favourable can be inferred from the processes leading to the accumulation of organic matter in a specific environment. For instance, as soon as emerged lands were present, it can be devised that ocean tides (induced by the vicinity of the Moon) may have allowed the formation of pools concentrating monomers and may have triggered wetting/drying cycles capable of inducing their polymerization possibly with the assistance of minerals (Rohlfing, 1976, Lahav et al., 1978, Rode et al. 1999). They may also have induced solid-gas reactions that have been shown to induce peptide bond formation (Commeyras et al., 2002). Volcanic areas, rich in reduced and sulphur-derived compounds, or hydrothermal vents in the ocean may be considered as favourable locations for these processes as well.

5.2.3. CHEMICAL EVOLUTION THROUGH A STEPWISE PROCESS

Prebiotic chemistry is customarily divided into different stages corresponding to an increasing degree of complexity of the entities involved:

- The synthesis of building blocks (amino acids, nucleic bases, nucleo-sides, nucleotides...).
- The formation of polymers (nucleic acids, peptides).
- The emergence of supramolecular architectures including the formation of the membrane and hence of individual cells.

This partition corresponds to one of the first hypothesis on the origin of life suggesting that life began in a "primordial soup" containing all the chemical components needed for feeding the first living organism. Since Miller's experiment was performed, prebiotic chemistry has demonstrated the capacity of making a wide range of building blocks available (amino acids, carbohydrates, nucleic bases) under favourable abiotic environments (Miller, 1998). The next degree of organization, the formation of peptides and nucleic acids, may have been the result of interactions (involving possibly activating agents) in a pool of inactivated building blocks. Then, the origin of life is considered in this hypothesis as a sequence of events resulting from more or less improbable encounters of building blocks leading finally to a system capable of self-replication. As it does not involve any driving force, this hypothesis suggests that the origin of life was highly improbable. As a result, it would have been less unlikely starting from high concentrations of building blocks maintained during a long period of time in a stable environment required for monomers to accumulate and polymers to have a sufficient lifetime.

5.2.4. CHEMICAL EVOLUTION THROUGH A DYNAMIC PROCESS

The partition of the origin-of-life problem into several stages corresponds to an approach that could rather be considered as based on the present day biochemistry way of thinking and teaching. This partition may however be misleading since the capacity to evolve and to promote the emergence of new properties must have been the most important feature of the chemical system from which life arose. This behaviour is observed for chemical systems maintained in states far away from equilibrium by a constant or erratic flux of matter and energy (Eigen, 1971; Nicolis and Prigogine, 1977). The origin of life may therefore be considered as the emergence of individual structures with new properties (and incorporating an information content capable of being reproduced) from a prebiotic network of chemical reactions linking high-energy species to inactivated products. Then the supply of energy and activated reactants may be considered as the driving force for chemical evolution. Additionally, the process is likely to be dependant on physico-chemical constraints governing the reactivity of activated biogenic compounds. Increasing the strength of constraints and of the driving force would have made the emergence of life less improbable so that an unstable environment may have been less deleterious than in a process purely governed by chance. In other words, these constraints may have influenced the rate of evolution of the system and, consequently, the chronology of events even though the process remained non-deterministic and historically unique (as far as we know). The synthesis of building blocks, polymers and supramolecular structures may then have been associated in a single process so that no activation of monomers was needed, the energy needed for polymerization being carried by precursors. From high-energy biogenic compounds, any early chemometabolic pathways leading to more complex species required a sequence of kinetically and thermodynamically spontaneous reactions linking substrates to products (Weber, 2002). Catalytic abilities to overcome activation barriers were probably limited (Weber, 2002). Of course, thermodynamics would never by itself identify the pathway leading to the first replicating molecule or supramolecular edifice (Kuhn and Kuhn, 2003). But, as any other chemical reactions, chemometabolic pathways are governed by thermodynamic constraints, the quantitative analysis of which can be used to rule out or to validate these processes.

5.2.5. CATALYTIC ACTIVITY AND INFORMATION STORAGE

A minimal form of life would have needed the association of a carrier of information content and the chemical activity needed for its replication. The form of life that we are presently familiar with on Earth mainly developed around two classes of biopolymers: protein and peptides carrying catalytic activity and nucleic acids carrying information. But noteworthy exceptions (ribozymes, nucleotide-like enzyme cofactors) have been considered as strong indications that the situation may have been quite different at early stages of evolution (see part 5.5). More generally, we have no indication that the different classes of biomolecules played the same role at early stages as in modern biochemistry. Three main hypotheses can be devised concerning the development of this process. Two of them correspond to a stepwise process in which one class of polymers could have acquired a replicating ability and then, in a later stage, the translation process was discovered leading to the modern protein-nucleic acid world. The advantage of this process from a chronological point of view is that it can be described by a sequence of stages involving an increasing complexity, but the driving force leading to the emergence of new properties is not obvious. On the contrary, the last process, corresponding to coevolution, may have developed from a probably quite complicated network of organic reactions, maintained far away from equilibrium through the feeding with activated biogenic compounds.

5.2.5.1. *Emergence of life in a peptide world:*
Since there is no accepted abiotic pathways leading to nucleosides and nucleotides, and since amino acids are the more easily synthesized building blocks in prebiotic experiments and have been detected in extraterrestrial organic matter

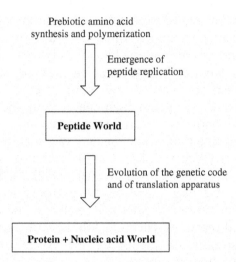

Prebiotic amino acid
synthesis and polymerization

Emergence of
peptide replication

Peptide World

Evolution of the genetic code
and of translation apparatus

Protein + Nucleic acid World

Figure 5.1. The sequence of stages corresponding to the origin of life from self-replicating peptides.

(meteorites), it is tempting to consider that life developed from a peptide only world. Peptide bonds can be formed thermodynamically from free amino acids by heating or under dehydrating conditions (generally in the presence of catalysts) or under the effect of activating agents. Additionally, these peptides may have been subject to hydrolysis so that elongation, by addition of activated monomers at the N-terminus, was taking place at the same time as peptide bonds were cleaved. These simultaneous processes may have led to some kind of selection through peptide protometabolisms (Commeyras et al., 2002; Huber et al., 2003, Plankensteiner et al., 2005). Peptides with sequences capable of catalytic activities may have been formed (Barbier and Brack, 1992) and others may have become prone to self-replication through a selection process in a population of continuously growing and disappearing random sequences. However, peptide self-replication is a highly improbable process since there are for instance 10^{13} (20^{10}) different decamers starting with the modern set of twenty amino acids. This hypothesis is unlikely because efficient catalysis usually requires peptides having several secondary structure domains (α-helices or β-sheets) associated to each other to ensure a properly defined fold (Corey and Corey, 1996). This requirement is achievable only for peptides having a sufficient length[1] (ca. 50 residues) that need an encoding system for the sequence to be

---

[1] In this chapter, (usually short) random poly-amino acid sequences, which generally do not fold into definite structures, are called (poly) *peptides*. This is a major difference with *proteins*, the catalytic or recognition abilities of which are the result of stable three-dimensional structures determined by their genetically encoded sequences in living organisms (Fersht, 1999).

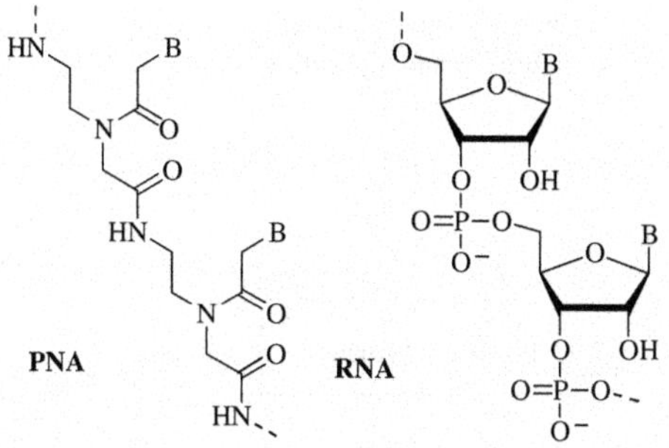

Figure 5.2. Peptide Nucleic Acid (PNA) as hypothetical early information carriers, their structure compared to that of RNA.

reproduced accurately from monomers. The next difficulty with this hypothesis is the need of a subsequent process that would have converted the amino acid sequences into genetic information. An alternative would be the existence of a first set of residues capable of carrying information such as Peptide Nucleic Acids involving a peptide-like backbone and nucleic bases on the side chains (Bohler et al., 1995; Nielsen, 1999; Nelson et al., 2000). In this way, peptides could have played both the role of information carriers and of catalysts in a pre-RNA world (Figure 5.2).

A replacement of this early genetic information system by the modern nucleic acid-based system is then needed. But the rationale for a radical change like this is not clear and it would require that no remnant of this former information storage system have been preserved by evolution.

5.2.5.2. *Emergence of life in an RNA world:*
According to the RNA world hypothesis (Gesteland et al., 1999) nucleic acid played both the role of information storage and the role of catalysts at an initial or intermediate stage of evolution (see part 5.5). Although there is no consensus on the absence of peptides in an RNA world, their presence is usually considered as non-essential at that stage (Figure 5.3). However, the direct emergence of a self-replicating RNA sequence is usually considered as unlikely because there is presently no prebiotically plausible pathway for the synthesis of mononucleotides so that the RNA world may have been pre-ceded by another system of replicating molecules called the pre-RNA world (Orgel, 2004). Moreover, there is no obvious driving force that would have led to the selection of the translation apparatus in an RNA world, though it has been suggested that RNA folding could have been improved by short coded peptides (Noller, 2004).

5.2.5.3. *Emergence of life from an RNA–peptide coevolution process:*
The former hypotheses suppose inherently that most biochemical evidences of the initial stage have been lost since a switch of information support or of

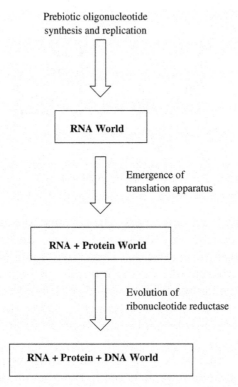

Figure 5.3. The sequence of stages corresponding to the direct emergence of an RNA world.

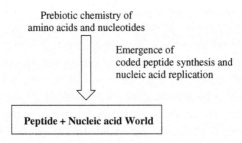

Figure 5.4. Coevolution process for the emergence of life leading directly to an RNA protein world.

catalytic molecules has erased records of the initial process. There is an alternative possibility that life evolved using directly both systems, which seems more complex on a first sight. But if life developed from an RNA–peptide coevolution process, the translation machinery could be considered as a metabolic remnant of the initial stage (Figure 5.4). Indeed, covalent bonds between amino acids and AMP or tRNAs are formed in the bio-

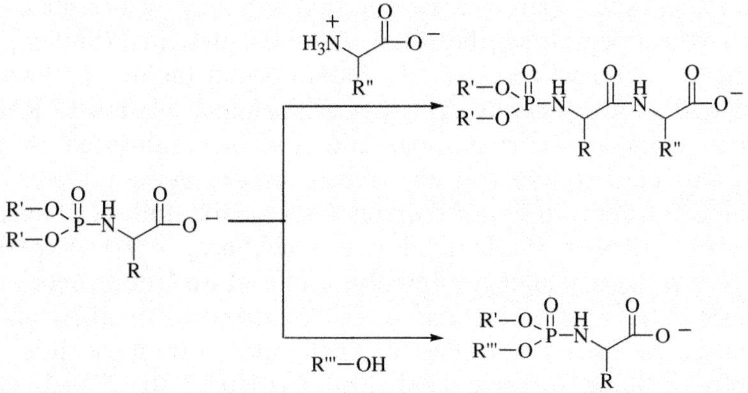

Figure 5.5. Ester exchange and peptide formation from *N*-(dialkylphosphoryl)amino acids.

chemical activation of amino acids and are essential in the reliability of the translation process. It is also supported by the catalytic activity of the ribosome (Nissen et al., 2000) on peptide bond formation.[2] Several chemical processes compatible with a coevolution scenario have been reported. For instance, it could have been involved as early as in the prebiotic synthesis of essential building blocks, which is supported by the discovery of a catalytic activity of amino acids (Weber, 2001) on the formose reaction, a likely process for the formation of sugars from formaldehyde. Actually, there is increasing evidence that amino acids and peptides are capable of very interesting stereoselective catalytic activities in the formation of carbon-carbon bonds through aldol-forming reaction, which may have been of importance for the emergence of homochirality (Pizzarello and Weber, 2004; Cordóva et al., 2005).

N-(dialkylphosphoryl)amino acids display interesting capabilities for the prebiotic syntheses of peptides and polynucleotides (Cheng et al., 2004). An intramolecular phosphoric–carboxylic mixed anhydride has been proposed to explain the specific behavior of *N*-(dialkylphosphoryl)amino acids that spontaneously give oligopeptides upon standing in various solvents (Figure 5.5). Peptide formation was accompanied by diester exchange on the phosphoryl group via a reaction that may have been useful for nucleotide ligation (the key step of RNA synthesis). Because of their abilities in both peptide and nucleic acid oligomerization, *N*-phosphoryl amino acids could have played an important role in prebiotic chemistry on condition that a plausible pathway of synthesis of these compounds has been made available.

---

[2] However, it has been recently determined that the rate increase brought about by the ribosome is consistent with a role of entropy trap so that it could be the result of binding both reacting tRNAs at the convenient position for reaction, without need for additional catalysis by the ribosomal RNA (Sievers et al., 2004).

The idea that life began in a system linking nucleic acid replication and genetically coded peptide synthesis has also been presented (Sutherland and Blackburn, 1997; Borsenberger et al., 2004), so that the two pathways may not be viewed as two separate processes. The aminoacylation of RNA may have arisen from an other purpose and was then subverted by protein synthesis; this may explain why the translation process developed whereas any advantage from translation requires a reasonably full set of specifically aminoacylated tRNAs (Sutherland and Blackburn, 1997). Experimental support in favour of a linkage of amino acid and nucleotide chemistries at early stages is the reaction of amino acid N-carboxyanhydrides with inorganic phosphate (Biron and Pascal, 2004) and with nucleotides (Biron et al., 2005), leading to mixed anhydrides (Figure 5.6). This is the first demonstrated abiotic pathway that may have led to the most activated forms of amino acids found in biology, aminoacyl phosphates and aminoacyl adenylates, which are involved in ribosomal (Arnez and Moras, 2003) and non-ribosomal peptide syntheses (Marahiel et al., 1997; Healy et al., 2000). Since there is increasing evidence that NCAs can be considered as unexpectedly common prebiotic molecules (Taillades et al., 1999; Maurel and Orgel, 2000; Leman et al., 2004), it is possible that, at early stages, amino acid activation was not dependent on the energy provided by phosphoanhydrides. Since aminoacyl adenylates and amino acid anhydrides with other nucleotides are mixed anhydrides, they could be considered as activated nucleotides as well as activated amino acids; i.e. the activation of nucleotides at that stage may have been dependent on amino acid chemistry (Pascal et al., 2005).

A coevolution process is also supported by the behavior of 3′-phosphonucleosides (Biron et al., 2005) that undergo two different intramolecular reactions with NCAs through the mixed anhydride leading either to a cyclic phosphodiester or to an amino acid ester that is reminiscent of aminoacylated tRNA (Figure 5.7).

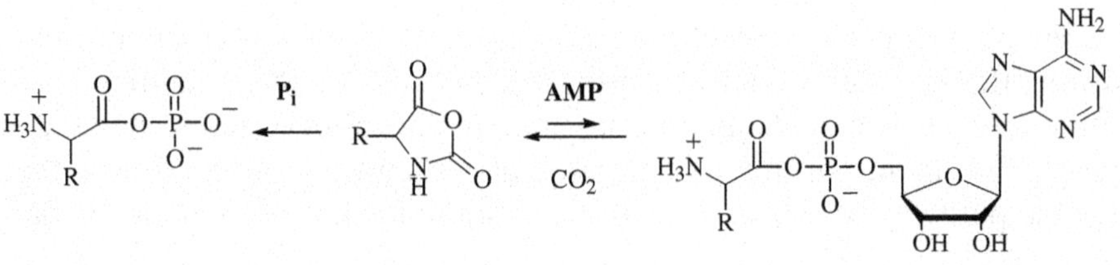

AA-P   **NCA**   **AA-AMP**

Figure 5.6. The formation of aminoacyl phosphates (AA-P) and aminoacyl adenylates (AA-AMP) from amino acid N-carboxyanhydrides and inorganic phosphate (P_i) or adenosine-5′-monophosphate (AMP), respectively.

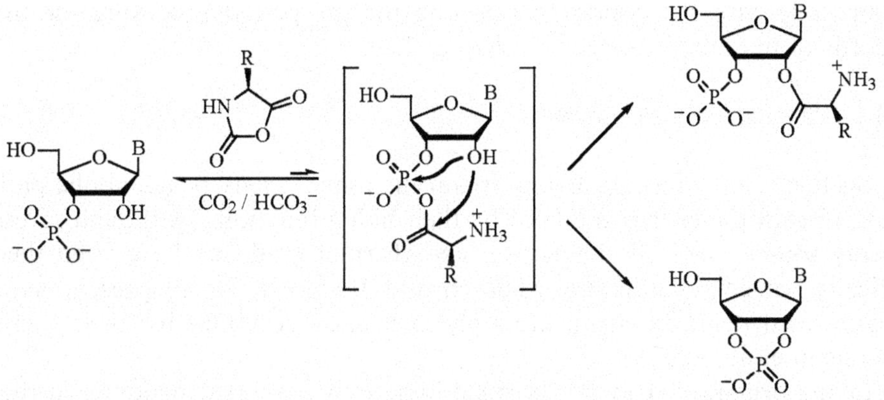

Figure 5.7. Intramolecular reactions of mixed anhydride derived from 3'-phosphorylated nucleosides.

5.3. Hypothesis about Early Metabolisms

DANIEL PRIEUR

All organisms living on Earth to day (with the exception of viruses, see part 5.8) are organized on a cellular basis. The cell is the fundamental unit of living matter, and is an entity isolated from its environment by a membrane (mostly made of lipids). Within this membrane are gathered molecules and sub-cellular structures required for cell life, and particularly macromolecules such as proteins, lipids, polysaccharides and nucleic acids. A cell may be single or associated to others, forming tissues, organs and finally complex organisms. But in any case, a living cell possesses five major functions: metabolism, growth (reproduction), differentiation, chemical communication and evolution. Consequently, the universal common ancestor of all living organisms (LUCA, see part 5.7), was almost probably also organized on a cellular basis.

The first cell function listed above is metabolism, which is the sum of all biochemical reactions occurring within the cell. These reactions aim to synthesize macromolecules (anabolism) or to obtain the energy required for all cellular functions (catabolism). A quick look at the biochemical pathways described for the metabolisms used by contemporary cells, and particularly prokaryotic cells, shows that these pathways (for catabolism and anabolism) are very complicated in the sense they require a variety of transporters, electron carriers, enzymes, co-enzymes, and a complex genome encoding all the required proteins. The first entities certainly used a rather simple process to gain their energy from their environment, and consequently to synthesize their components. No traces, no signatures, no dates are known for these early metabolisms, and the biologist can only imagine something simple, and compatible with the conditions existing for the period preceding the first records of microbial fossils (see chapter 7.1).

5.3.1. PRELIMINARY DEFINITIONS

Altogether, contemporary living organisms use a variety of metabolic pathways to gain the energy required by their living functions. Although several energy sources such as magnetism and thermal gradients have been theoretically considered (Schulze-Makuch and Irwin, 2002), all present living organisms depend on chemical or photochemical reactions for their energy (Madigan et al., 2003).

Living organisms can be classified in several groups, depending on their energetic metabolism. Those using light as energy source are called *photo-trophs*. Those gaining their energy from chemical reactions are named *chemotrophs*. In this case, those using organic molecules as energy sources are *chemo-organotrophs*, while those using inorganic molecules as energy sources are called *chemo-lithotrophs*.

In order to build their components, cells uptake in their environments small amounts of micro-nutrients (Cr, Co, Cu, Mn, Mo, Ni, etc) but larger amounts of macro-nutrients (C, N, H, O, P, S, K, Mg, Ca, Na, Fe). Among them, carbon is particularly important since it is present in all macromolecules. Cells using organic carbon molecules for biosynthesis are named *heterotrophs*, while those using inorganic carbon (carbon dioxide) are called *autotrophs*.

5.3.2. HOW THE FIRST ENTITIES PRESUMABLY GAINED THEIR ENERGY AND CARBON?

Despite the report of microfossils showing prokaryotic morphotypes in very old rocks (Westall et al., 2001), the metabolism displayed by these organisms cannot be deduced from these observations. If a photosynthetic (most probably anoxygenic) metabolism (based on morphological similarities with extant stromatolites) has been suggested for putative microorganisms occurring in stromatolite fossils aged of 3.5 Ga or younger (Schopf, 1993), this is not finally proved (Brasier et al., 2002), and the question of metabolism is still open for other old microfossils reported from Barberton rocks in Australia (see chapter 7.1).

The scenario given by Madigan et al. (2003) is rather convincing and explained below.

Whatever the exact dating for the first cellular-like entities with an energetic metabolism, it is obvious that this event occurred under anoxic conditions. From our knowledge of energetic metabolisms used by contemporary prokaryotes, several anaerobic metabolisms can be hypothesized: anoxic photosynthesis, fermentation and anaerobic respiration. All require a variety of enzymes, electron carriers, and for phototrophs, photosynthetic pigments, all involved in rather complex pathways that one cannot easily imagine for early energy generating systems. What kind of simple mechanism could be considered? If anoxic photosynthesis is excluded because of its complex pathways and because photosensitive pigments are already evolved molecules, a simple chemolithotrophy must be considered seriously. On the primitive Earth, likewise nowadays, a variety of reduced inorganic molecules (putative electron donors) did exist, and among them, molecular hydrogen represents an excellent candidate. Molecular hydrogen is a common energy source for prokaryotes living in geothermal (terrestrial and marine) areas (Prieur, 2005), and it has been demonstrated that this compound could drive hyperthermophilic ecosystems in Yellowstone National Park (Nealson, 2005; Spear et al., 2005). Molecular hydrogen may be a product of interactions between hydrogen sulphide and ferrous iron, or between protons and ferrous iron in the presence of UV radiation as an energy source (Spear et al., 2005). Whatever its origin, molecular hydrogen belongs with protons to a redox couple whose reduction potential ($E_0' = -0.42$ V) is very favourable for electron donation. With such an electron donor, there is a wide choice of putative electron acceptors in the absence of molecular oxygen. Among those (still inorganic) used by extant prokaryotes, elemental sulphur (S°) represents a good candidate. Elemental sulphur and hydrogen sulphide form a redox couple whose reduction potential ($E_0' = -0.28$) is favourable for free energy generation, but does not require a long series of electron carriers. As shown on Figure 5.8, a primitive hydrogenase would have been the single enzyme

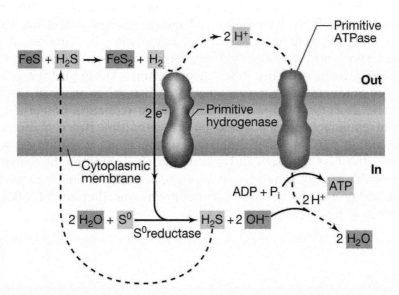

Figure 5.8. A hypothetical energy-generating metabolism for primitive cells (Madigan et al., 2003).

required for uptaking molecular hydrogen, producing protons and electrons, and consequently forming a proton gradient and a proton motive force. A primitive ATPase (this enzyme is present for all living organisms) would have used this proton motive force for ATP generation. This hypothesis is rather convincing, taking into account that the mechanism hypothesized is rather simple and requires only two primitive enzymes, and that the inorganic compounds involved were most probably abundant on the primitive Earth.

If such an energy generating mechanism is suggested, one must now consider the question of the carbon source (s).

An organism which uses molecular hydrogen as electron donor and elemental sulphur as an electron acceptor is called a chemolithotroph carrying out anaerobic respiration (see definitions above). Such organisms are, for most of those living nowadays, also autotrophs and use carbon dioxide as a single carbon source. Autotrophy is a property of various organisms that do not use organic carbon for their energy generation: anoxygenic and oxygenic photosynthesizers and chemolithotrophs. To transform carbon dioxide into organic carbon, they use rather complex pathways such as the Calvin cycle, but also the reverse citric acid or the hydroxypropionate cycles. These pathways require a variety of enzymes, which are again difficult to imagine working all together in primitive organisms. Since organic compounds (whatever their origins) existed on the primitive Earth, it is more probable that the primitive chemolithotrophic cells utilized already formed organic carbon as carbon sources, in the same way as extant mixotrophic organisms (such as certain sulphur-oxidizers) are doing today. If that was the case, although it is more difficult to consider it as a common and unique energy generating mechanism, a kind of simple fermentation[3] could also be considered as a possible hypothesis. Actually, various fermentations are nowadays supporting the growth of many different organisms. Although they generate rather low amounts of energy compared with respirations using electron donors and acceptors with a great difference of reduction potentials, they cannot be eliminated as a possible early metabolism. But it is difficult to imagine which particular fermentation among many possibilities, would have been the first. Consequently, molecular hydrogen might remain the most probable energy generating compound for early living organisms, as it is still today used by a variety of mesophilic and thermophilic Bacteria and Archea (Prieur, 2005).

---

[3] In fermentation an organic compounds serves as electron donor and carbon source, an electron acceptor is temporarily generated from an intermediate compounds resulting from the degradation of the initial carbon and an energy source.

5.4. Origin and Evolution of Compartments

PURIFICACIÓN LÓPEZ-GARCÍA, DAVID MOREIRA AND JULI PERETÓ

Although impossible to date, and hard to place in a relative succession of events, compartments must have appeared very early during the emergence of life to enclose metabolism (self-maintenance) and information storage in entities that reproduced and could undergo natural selection. For many authors, the earliest life forms must have had boundaries: life only appeared when the state of self-reproducing compartments was reached (Varela et al., 1974; Morowitz et al., 1988; Deamer, 1997; Luisi, 1998; Peretó, 2005). Contemporary cells are surrounded by membranes that assure their integrity, facilitate the necessary exchange with the external environment (diffusion of gases, active transport of ions and metabolites) and harbour energy-transducing systems that take advantage of the ion gradients maintained across the membrane using primary energy sources (i.e. visible light or chemical reactions). The exploitation of electrochemical gradients across membranes to supply energy (the chemiosmotic theory) is a universal property of living cells. Any hypothesis on the origin of life must explicitly state the way to convert energy into complex structures and organization, something that is only achieved in terrestrial life through compartments (Harold, 2001). Cell membranes are essentially made out of phospholipids, amphiphilic molecules composed of a hydrophilic glycerol-phosphate head bound to long hydrophobic fatty acid or isoprenoid tails. They organise in bilayers and host different proteins involved in transport and energy-transducing processes (Figure 5.9). Of course, at early stages, compartments must have been defined by much simpler barriers and less functions, possibly only two: definition of a 'self' boundary and of a not-quite-impermeable one, i.e. one allowing the import of ions and metabolites but retaining most of the internally produced material. A remarkable *ab initio* problem would be the osmotic crisis generated by the enclosure of polymers. This situation could be mitigated by the coevolution of membranes and primitive ion exchange systems that necessarily could not be sophisticated protein structures but simplest molecular devices (e.g. the *Escherichia coli* non-proteinaceous calcium channel made out of polyphosphate and polyhydroxybutyrate; Reusch, 2000).

5.4.1. AMPHIPHILIC VERSUS NON-AMPHIPHILIC COMPARTMENTS

What was the nature of those early compartments? The fact that amphiphilic lipid bilayers are universal today suggests that protocellular compartments may have been founded on simpler molecular grounds with similar properties such as long single-chain monocarboxylic acids, alcohols or monoglycerides,

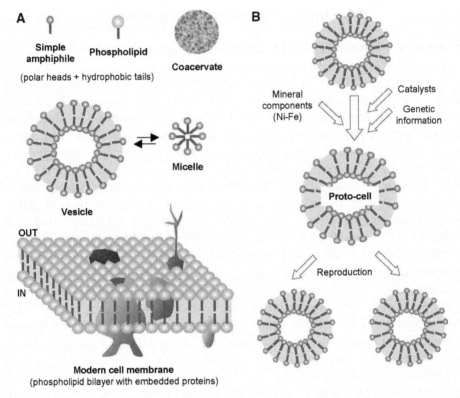

Figure 5.9. A. Various types of compartments and their components. B. A possible model of membrane evolution.

capable of self-assembly to form vesicles and available in prebiotic times (Deamer, 1986; Deamer et al., 2002; Monnard and Deamer, 2002). Oparin's coacervates (Oparin et al., 1976), spherical aggregates of macromolecular components, and micelles (Figure 5.9) are unlikely to have played a role in the origin of protocellular compartments. Amphiphilic vesicles are compatible with "prebiotic-soup" models, but they are also compatible with models of surface-metabolism (Wächtershäuser, 1988b) operating in early times. Surface metabolism likely played a role in the synthesis and accumulation of complex organic molecules (Lazcano, 2001; Monnard and Deamer, 2002) and perhaps in vesicle replication (Hanczyc and Szostak, 2004). Nevertheless, surface-based hypotheses are 'acellular' by definition. This inconvenience has been overcome by a proposal suggesting that the first compartments were tridimensional iron monosulphide bubbles that grew in hydrothermal environments, and that these mineral membranes persisted for a long time in actual biological evolution (Russell and Hall, 1997; Martin and Russell, 2003). Although such mineral compartments may have played also a role as initial chemical reactors, their persistence in relatively modern cells is highly improbable (see part 5.7). Additional evidence comes from contemporary membranes, which do not form *de novo* but grow and divide from pre-existing membranes. This membrane heredity view (Cavalier-Smith, 2001) together with the heuristic principle of continuity (Morowitz et al., 1991) would conform to the idea that some type of amphiphilic bilayer existed since early times and that a kind of continuum allowed its evolution to date. Within this framework, amphiphiles and vesicles would be key chemical intermediates during life emergence (Ourisson and Nakatani, 1994; Luisi et al., 1999; Segré et al., 2001).

5.4.2. SELF-ASSEMBLY AND EARLY EVOLUTION OF COMPARTMENTS

How and when did early compartments evolve? A possibility is that the earliest protocellular compartments resulted from the self-assembly of organic mixtures that were available on the planet. Long (C_{16-18}) fatty acids and alcohols assemble spontaneously, forming vesicles above a concentration threshold. Shorter fatty acids (C_{14}) would be even better candidates because they were easily synthesized and the bilayers formed are much more permeable, an advantage in times when membrane transport proteins had not yet evolved (Deamer, 1986; Monnard and Deamer, 2002). Interestingly enough, a selective passive incorporation of ribose into fatty acid vesicles has been shown (Sacerdote and Szostak, 2005), suggesting that preferential ribose uptake by primitive cells would play a role in the rise of a hypothetical RNA world. Long monocarboxylic acids were present in prebiotic conditions both by exogenous delivery and endogenous synthesis. The exogenous delivery is attested by the presence of mixed aliphatic and aromatic compounds in meteorites such as Murchison that, indeed, are able to form vesicles spontaneously (Monnard and Deamer, 2002). This kind of compounds can also be synthesized by Fischer-Tropsch-type reactions (Deamer et al., 2002). Deamer (1997) suggested that prebiotic conditions favoured saturated over unsaturated acids and considered that isoprenoid-type molecules were difficult to make abiotically. Interestingly, the presence of "impurities", for instance pyrrolic compounds (the first pigments) increases vesicle permeability (Deamer, 1997). Furthermore, it has been shown experimentally that the interaction and incorporation of minerals (montmorillonite) to fatty acids catalyze vesicle formation (Hanczyc et al., 2003). Such vesicles can grow by incorporating surrounding fatty acids and divide (Hanczyc et al., 2003). Moreover, the growth of vesicles can generate incipient ion gradients (Chen and Szostak, 2004). RNA and other macromolecular species can be encapsulated in vesicles leading to an experimental approach to the simplest life forms (Luisi et al., 1994; Chen et al., 2004). Therefore, it is plausible that prebiotic mixed short-chain fatty acids formed vesicles that encapsulated, sequentially, catalytic species and genetic systems to become self-reproducing compartments. Nevertheless, whether this occurred and how it occurred remains a matter of speculation. Attempts to re-create life *in vitro* by encapsulating macromolecules such as ribozymes (see part 5.5) and other catalytic species in replicative vesicles will certainly yield interesting results (Szostak et al., 2001). Hopefully, these experiments will at least contribute to test the feasibility of those hypotheses.

5.5. The Hypothesis of an RNA World

MARIE-CHRISTINE MAUREL

The time necessary to go from an habitable Earth to a protocell-like procaryote can be divided in three periods. During the first period molecular organics sources, from which building blocks of life could have appeared, accumulated on the early Earth. The next period led to macromolecular synthesis from small monomers and to the first metabolic steps including the formation of the first replicating polymers. Subsequently a scenario can be described for the development from random polymers to a replicative system, capable of evolving by mutation and natural selection. This last period, called the *RNA world period*, would have opened the door to evolutionary biology as we know it today, leading to organized and complex systems.

The question of how long did it take to go from prebiotic building blocks to the first living cell must be addressed according to the inherent constraints imposed by primeval conditions. Any extrapolations from results obtained in the laboratory to what may have occurred 4 billion years ago are tenuous. As a result we have to study the stability of all components in extreme conditions, that is the behaviour of monomers and macromolecules of life at high and cold temperatures (Schwartzman and Lineweaver, 2004; Vergne et al., 2006), with and without salt (Tehei et al., 2002), at low and high pH (Kühne and Joyce, 2003), at low and high pressure (Tobé et al., 2005; Di Giulio, 2005), in different redox conditions, radio-ionizating and cosmic conditions, solvent conditions etc., as well as in conditions simulating an ocean-boiling asteroid impact...

Diverse molecular ecosystems could potentially have arisen in these physico-chemical specifications and we have to take them into account especially if the hereditary criteria are retained as mandatory in designing life. Thus the half-lives for the decomposition of the components of life (amino acids, peptides, sugars, lipids) and of the first genetic materials, that are nucleobases must be measured and considered on the geological time scale. Also the balance between synthesis and degradation must lead to consistent concentrations.

Minerals and mineral surfaces, salt and crystals may help to stabilize macromolecules and monomers (Tehei et al., 2002; Cornée et al., 2004; Ricardo et al., 2004). Purines and pyrimidines have been found in sediment cores from both ocean and lake basins, some dating back as far as 25×10^6 year, but they may be the result of contamination or decomposition under anhydrous conditions. On the other hand, adenine, uracile, guanine, xanthine, hypoxanthine nitrogenous bases and several organics have been found in the Murchison meteorite (Stoks and Schwartz, 1982) and it is now possible to detect subpicomoles of purine bases trapped in a mineral or a colloidal supports (El Amri et al., 2003, 2004, 2005).

Finally, these materials, if applicable to any origin of life theory founded on Darwinian evolution, may have resisted to several extinctions where the survival of a single organism (in a micro- environment) would be sufficient to reestablish an entire ecosystem.

Speculations from results obtained in the laboratory, specially unconstrained sequences obtained by *in vitro* selection, to what may have occurred 4 billion years ago, are weak. Again, a valuable approach lies in the examination and the experimental test of the resistance of RNA molecules under the inherent constraints imposed by prebiotic geochemical and geophysical conditions.

Lastly as it seems likely that the RNA World may not have been the pristine nucleic acid-dominated ecosystem but simply a transient go-between during the evolution to the contemporary DNA–protein world, considerations above would apply to any alternative pre-RNA backbone before the emergence of standard ribonucleic acids.

5.5.1. THE RNA WORLD SCRIPT

A scenario of evolution postulates that an ancestral molecular world, *the RNA World*, existed originally before the contemporary DNA–RNA–Protein world meaning that the functional properties of nucleic acids and proteins as we see them today would have been produced by molecules of ribonucleic acids (Gilbert, 1986; Benner et al., 1989, 1993; Joyce, 1989; Orgel, 1989; Bartel and Unrau, 1999; Gesteland et al., 1999; Joyce and Orgel, 1999; McGinness et al., 2002; Joyce, 2002). RNAs occupy a pivotal role in the cell metabolism of all living organisms and several biochemical observations resulting from the study of contemporary metabolism should be stressed. For instance, throughout its life cycle, the cell produces the deoxyribonucleotides required for the synthesis of DNA from ribonucleotides, the monomers of RNA. Thymine, a base specific of DNA, is obtained by transformation (methylation) of uracil a base specific of RNA, and RNAs serve as obligatory primers during DNA synthesis. Finally, the demonstration that RNAs act as catalysts is an additional argument in favour of the presence during evolution of RNAs before DNA. Therefore it seems highly likely that RNA arose before DNA during biochemical evolution, and for this reason DNA is sometimes considered as modified RNA better suited for the conservation of genetic information. This genetic privilege would constitute a logical step in an evolutionary process during which other molecules could have preceded RNA and transmitted genetic information. The idea of an *RNA world* rests primarily on three fundamental hypotheses, developed by Joyce and Orgel (1999):

- during a certain period in evolution, genetic continuity was assured by RNA replication;
- replication was based on Watson–Crick type base pairing;
- early catalysis was performed by small non genetically coded peptides and by ribozymes.

Orgel and his coworkers showed that starting from activated monomers, it is possible in certain conditions to copy a large number of oligonucleotide sequences containing one or two different nucleotides in the absence of enzyme (Inoue and Orgel, 1983; Joyce and Orgel, 1986; Orgel, 1992; Hill et al., 1993). On the other hand, Ferris and his coworkers studied the assembly of RNA oligomers on the surface of montmorillonite (Ferris, 1987; Ferris and Ertem, 1992). Thus, experimental results demonstrated that minerals which serve as adsorbing surfaces and as catalysts (Paecht-Horowitz et al., 1970; Ferris et al., 1996), can lead to accumulation of long oligonucleotides, given that activated monomers are available. One can thus envisage that activated mononucleotides assembled into oligomers on the montmorillonite surface or on an equivalent mineral surface. The longest strands, serving as templates, direct the synthesis of complementary strands starting from monomers or short oligomers, leading double-stranded RNA molecules to accumulate. Finally, a double RNA helix – of which one strand is endowed with RNA polymerase activity –, would dissociate to copy the complementary strand and to produce a second polymerase that would copy the first to produce a second complementary strand, and so forth. The RNA world would thus have emerged from a mixture of activated nucleotides. However, a mixture of activated nucleotides would need to have been available! Finally, when either the first replicative molecule, the template or one of its elements (nucleotides) is to be synthesized from the original building blocks, in particular the sugars that are constituents of nucleotides, a certain number of difficulties are encountered (Sutherland and Whitfield, 1997). First, synthesis of sugars from formaldehyde produces a complex mixture in which ribose is in low amounts. Second, production of a nucleoside from a base and a sugar leads to numerous isomers, and no synthesis of pyrimidine nucleosides has so far been achieved in prebiotic conditions. Finally, phosphorylation of nucleosides also tends to produce complex mixtures (Ferris, 1987). Consequently, onset of nucleic acid replication is nearly inconceivable if one does not envisage a simpler mechanism for the prebiotic synthesis of nucleotides. Eschenmoser succeeded in producing 2,4-diphosphate ribose during a potentially prebiotic reaction between glycol aldehyde monophosphate and formaldehyde (Eschenmoser, 1999). It is thus possible that direct prebiotic nucleotide synthesis occurred by an alternative chemical pathway. Nevertheless, it is more likely that a certain organized form of chemistry preceded the RNA world, hence the notion of "genetic take-over". Since the ribose-phosphate skeleton is theoretically not indispensable for the transfer of genetic information, it is logical to propose that a simpler replication system would have appeared before the RNA molecule. During the evolutionary process, a first genetic material, mineral in nature would have been replaced by another totally distinct material of organic nature. The hypothesis of a precursor of nucleic acid (Cairns-Smith, 1966, 1982; Joyce et al., 1987) is a relatively ancient idea, but it is only within the last few years that research has been oriented towards the study of simpler molecules than present day RNAs, yet capable of auto-replication. In the *Peptide Nucleic Acids* (PNA) of Nielsen and coworkers, the ribofuranose-phosphate skeleton is replaced by a polyamidic skeleton on which purine and pyrimidine bases are grafted (see Figure 5.2). Indeed, PNAs form very stable double helices with an RNA or a complementary DNA (Egholm et al., 1993) and can serve as template for the synthesis of RNA, or vice versa (Schmidt et al., 1997). Moreover, PNA–DNA chimeras containing two types of monomers have been produced on DNA or PNA templates (Koppitz et al., 1998). Eventually, the information can be transferred from PNAs (achiral monomers) to RNA during directed synthesis; the double helical molecule with a single complementary RNA strand is stable. Transition from a "PNA world" to an "RNA world" is hence possible. The group of Eschenmoser recently replaced the ribose moiety of RNA by a four-carbon sugar, threose, whose prebiotic synthesis seems easier. The resulting oligonucleotides designated TNAs, $(3' \rightarrow 2')$-α-L-threose nucleic acid, can form a double helix with RNA (Schöning et al., 2000). TNA is capable of antiparallel Watson–Crick pairing with complementary DNA, RNA and TNA oligonucleotides. Finally, this leads us to a major conclusion, namely that a transition may have occurred between two different systems without loss of information.

From the point of view of evolution, the studies described previously demonstrate that other molecules capable of transmitting hereditary information may have preceded our present day nucleic acids. This is what Cairns-Smith coined the "take-over" (Cairns-Smith, 1982), the evolutionary encroachment or genetic take-over, or to some extent what François Jacob (1970) calls genetic tinkering, in other words, making new material from the old.

The role of cofactors at all steps of the metabolism and their distribution within contemporary groups of organisms suggests that a great variety of nucleotides was present in the ancestor common to all forms of life and before. Several authors have underscored the possible presence of coenzymes before the appearance of the translation machinery (White, 1976). Present-day coenzymes, indispensable cofactors for many proteins, would be living fossils of catalysts of primitive metabolism. Most coenzymes are nucleotides (NAD, NADP, FAD, coenzyme A, ATP...) or contain heterocyclic nitrogen bases and it is even possible to consider that catalytic groups that were part of nucleic enzymes were incorporated in specific amino acids rather than being "retained" as coenzymes. This could be the case of imidazole, the functional group of histidine, whose present synthesis in the cell is triggered by a nucleotide.

Work has been carried out based on the demonstration of esterase activity in a nucleoside analogue N^6-ribosyladenine (Fuller et al., 1972; Maurel and Ninio, 1987; Maurel, 1992). This activity which is due to the presence of an imidazolyl group that is free and available for catalysis, is comparable to that of histidine placed in the same conditions. We have studied the kinetic behaviour of this type of catalyst (Ricard et al., 1996) and have shown that the catalytic effect increases greatly when the catalytic element, pseudohistidine, is placed in a favourable environment within a macromolecule (Décout et al., 1995). Moreover, primitive nucleotides were not necessarily restricted to the standard nucleotides encountered today, and because of their replicative and catalytic properties, the N6 and N3 substituted derivatives of purines could have constituted essential links between the nucleic acid world and the protein world.

5.5.2. THE CASE OF ADENINE

Purine nucleotides, and in particular those containing adenine, participate in a large variety of cellular biochemical processes (Neunlist et al., 1987; Maurel and Décout, 1999; Nissen et al 2000). Also, the ease with which purine bases are formed in prebiotic conditions (Oró, 1960) suggests that these bases were probably essential components of an early genetic system. Furthermore, purines have also been found in the Murchison meteorite showing the range of resistance of this molecule. The first genetic system was probably capable of forming base pairs of the Watson–Crick type, Hoogsteen and other atypical associations, by hydrogen bonds as they still appear today in RNA. It probably contained a different skeleton from that of RNA, and no doubt it also modified bases, thereby adding chemical functions, but also hydrophobic groups, and functions such as amine, thiol, imidazole, etc. Wächtershäuser (1988a) also suggested novel pairings of the purine–purine type.

Today, from a vast combination of nucleic acids, one can isolate aptamers that possess catalytic properties (RNA ligation, cleavage or synthesis of a peptide bond, transfer of an aminoacyl group, etc.). The first nucleic acids could possess independent domains, separated by flexible segments, creating reversible conformational motifs, dependent on ions and bound ligands. Thus, a 10 amino acid-long peptide can recognize fine structural differences within a micro RNA helix (discrimination can be made between two closely-placed microhelices). Just as protein and antibodies, RNA molecules can present hollows, cavities, or slits that make these specific molecular recognitions possible. RNAs must "behave as proteins". Whatever the chronology and the order of appearance of the various classes of molecules, the importance lies in the shape, the scaffolding and the architecture that have allowed functional associations.

Starting from a heterogenous population of RNAs with 10^{15} variants (a population of 10^{15} different molecules), five populations of RNAs capable of specifically recognizing adenine after about ten generations have been selected (Meli et al., 2002). When cloned, sequenced and modelled, the best one among the individuals of these populations, has a shape reminiscent of a claw capable of grasping adenine. Is it the exact copy of a primitive ribo-organism that feeds on prebiotic adenine in prebiotic conditions? Functional and structural studies presently under way will highlight other activities, other conformations...

Following this line of investigation two adenine-dependent ribozymes capable of triggering reversible cleavage reactions have been selected. One of them is also active with imidazole alone. This result leads to very important perspectives (Meli et al., 2003).

A considerable amount of research has been focused on the selection of ribozymes *in vitro*. Recently, it was demonstrated that a ribozyme is capable of continuous evolution, adding successively up to 3 nucleotides to the initial molecule (McGinness, 2002). It is also possible to construct a ribozyme with only two different nucleotides, 2,6-diaminopurine and uracil (Reader and Joyce, 2002).

5.5.3. PROVISIONAL CONCLUSIONS

Very little is known to date about the behavior of macromolecules in "extreme" environments. How do structures behave? What are the major modifications observed? What are the conditions of structural and functional stability? How are the dynamics of the macromolecules and their interactions affected? What are the possibilities of conserving biological macromolecules in very ancient soils or in meteorites? Can we find traces of these macromolecules as molecular biosignatures, and if so in what form (Maurel and Zaccaï, 2001; Tehei et al., 2002)?

The selection of thermohalophilic aptamers, RNAs resistant to high temperatures (80 °C) in the presence of salt (halites 30 million years old) (Vergne et al., 2002, 2006), will maybe allow us to answer some of these questions, that are fundamental for the search and the date of past traces of life, and of life on other planets...

5.6. The RNA/DNA Transition and the Origin of the Genetic Code

5.6.1. THE ORIGIN OF THE GENETIC CODE

JULI PERETÓ

The hypothesis of an RNA world is widely accepted (see part 5.2 and part 5.5). Nevertheless, the different evolutionary paths emerging from such a scenario have attracted much less attention. Did the protein synthesis and the genetic code emerge in an RNA–peptide world and, afterwards, DNA was invented in an RNA/protein world? Or did DNA precede proteins? All these and other alternative possibilities have been critically discussed by Dworkin et al. (2003). Albeit a precise chronology of the different transitions is impossible, we can try to establish the most parsimonious order of events based in our chemical and biochemical knowledge. Therefore, and although we are conscious that alternative solutions can not be totally ruled out, the current consensus around a growing experimental evidence favours an emergence of the machinery for biosynthesis of coded peptides in a world of ribozymes helped by amino acids and short peptides as cofactors (i.e. an RNA–peptide coevolution process, see part 5.2; Szathmáry, 1999; Noller, 2004). The view of an early emergence – in any case, before the universal cenancestor, see part 5.7 – of the machinery for both transcription and translation is consistent with genomic and structural studies of the major macromolecular components of RNA polymerases and ribosomes (Fox and Naik, 2004).

5.6.1.1. *Origin of protein synthesis in an RNA world*
The biosynthesis of coded proteins in modern cells is performed by the ribosome, an extraordinarily complex assembly of proteins and RNA whose high-resolution structure has been recently elucidated – the large subunit described by Ban et al. (2000), the small subunit by Wimberly et al. (2000), and the complete ribosome by Yusupov et al. (2001). There are good structural (Nissen et al., 2000; Hoang and Noller, 2004) and chemical (Zhang and Cech, 1997) reasons to propose that peptidyltransferase activity resides in the 23S rRNA of the large ribosomal subunit, although this view is not fully consistent with some biochemical and genetic observations (see Fox and

Naik, 2004, and references therein). It is widely assumed that the ribozymic nature of the ribosomal peptidyltransferase is a molecular fossil from the RNA world.

Protein biosynthesis is one of the more complex, energy dependent, metabolic processes. To perform its function, the ribosome is assisted by many different molecular components: tRNAs – which both activate amino acids through an ester bond and carry the anticodon triplet complementary of a codon triplet in the mRNA –, aminoacyl-tRNA synthases – aaRS, which catalyze the synthesis of aminoacyl-tRNAs from each amino acid and its cognate tRNA throughout an aminoacyl-adenylate intermediate –, and many protein factors that participate in the different biosynthetic phases – initiation, elongation, translocation, and termination of the polypeptide chain.

In vitro selection experiments have shown the ability of ribozymes to catalyze the basic steps of translation: RNA aminoacylation – including the formation of the activated aminoacyl-adenylate – and peptide bond synthesis (see Joyce, 2002, and references therein). Thus, the emergence of this metabolic process seems chemically plausible in an RNA world (Lazcano et al., 1992; Maynard Smith and Szathmáry, 1995; Brosius, 2001; Joyce, 2002; Fox and Naik, 2004; Noller, 2004).

Some models try to present the evolutionary transitions during the establishment of all the macromolecular components of the translation machinery and the origin of the genetic code in the context of the RNA world hypothesis. Thus, the key role of tRNAs for the emergence of the primitive ribosome in an RNA world has been emphasized by several authors (Weiner and Maizels, 1987; Brosius, 2001). The coding coenzyme handle (CCH) hypothesis is a testable scheme for the origin of translation and the code (Szathmáry, 1999). In short, this proposal starts with the classical notion that present day coenzymes (carrying a ribonucleotide moiety) are molecular fossils of the RNA world (White 1976, 1982). In the primitive stage of an RNA world, ribozymic activities would be supplemented with amino acids and short peptides acting as cofactors, leading to further metabolically complex stages. The CCH hypothesis suggests that the binding of cofactors to the ribozymes was non-covalent, through base-pairing of short oligonucleotides recognizing some sequence of the ribozyme active site. The progressive replacement of ribozymes by polypeptides led to the transition to an RNA–protein world with an incipient genetic code initially established by the recognition of the oligonucleotide handles (precursors of anticodons) of the coenzymes and the handle binding sites (precursors of codons) of the ribozymes (for further details see Szathmáry and Maynard Smith, 1997; Szathmáry, 1999). In summary, translation might initially have evolved to rise the functional versatilities of the RNA world (Noller, 2004) but eventually also sparked off its fall (Joyce, 2002).

5.6.1.2. *Origin and evolution of the genetic code*

The origin of the genetic code – i.e. the assignment of base triplets to amino acids during protein biosynthesis – remains a mystery since its deciphering in the 1960s. One of the most impressive characteristics of the code is its universality: except for some recent evolutionary innovations, notably mitochondrial variants, all extant organisms use the same code (Santos and Tuite, 2004). This is one of the most compelling arguments favoring the existence of a universal cenancestor (see part 5.7). However, the classic hypothesis by Crick (1968) suggesting that the code is a *frozen accident* (i.e. an historical accident fixed in the universal cenancestor) has been challenged by models with different balance between chance and necessity during the origin and evolution of the code and, especially, by the very occurrence of code variants (Santos and Tuite, 2004).

5.6.1.3. *An expanding code*

In their classical paper on the hypercycle and the origin of genetic information Eigen and Schuster (1978) suggested that the primitive code did use units of less than three bases. It follows that the number of initial codons (and coded amino acids) was less than the current 64 (for the 20 amino acids universally found in proteins). During its early evolution the code would have increased both the number of codons and of coded amino acids, and the present code would reflect the pattern of this historical expansion (for recent hypotheses see: Patel, 2005 and Wu et al., 2005).

5.6.1.4. *The stereochemical hypothesis*

Pelc (1965) and Dunnill (1966) postulated the appearance of a primitive code based in the specific steric interaction between base triplets – codons or anticodons – and amino acids. Nowadays one experimental test of this model searches the synthetic RNA sequences that bind strongly an amino acid. The *in vitro* selection methods have originated several aptamers – i.e. specific RNA ligands – for some protein amino acids, like Phe, Ile, His, Leu, Gln, Arg, Trp, and Tyr, that contain the coding sequences (codon and/or anticodon) for those amino acids (Yarus et al., 2005), but this method does not work in other cases, like Val (Yarus, 1998). In general, it is assumed that stereochemical interactions played some role – more or less strong, more or less visible in present-day cells – during the origin and early evolution of the code. A critical analysis on the stereochemical hypothesis can be found in Ellington et al. (2000).

5.6.1.5. *Adaptive evolution*

The degeneracy of the code (i.e. the existence of synonymous codons) and the similarity of the codons (one base difference) for physicochemically similar amino acids suggest that some optimization process has sculpted most of the code to minimize the damage due to point mutations (Sonneborn, 1965) or mistranslations (Woese, 1965). Recent approaches to test this idea consist in the statistical comparison of the natural code with thousands of computationally generated random codes. The results suggest that nature's choice might be the best possible code (Freeland et al., 2003). Historical, stereochemical, and adaptive patterns have been combined into a coherent picture by Knight et al. (1999).

5.6.1.6. *The coevolution model*

Degeneracy aside, there are other patterns observed in the table of the genetic code (for a good summary, see Figure 1 in Szathmáry, 1999). Thus, amino acids adjacent in biosynthetic pathways cluster together in the code (Dillon, 1973; Jukes, 1973; Wong, 1975). Prebiotic chemistry on early Earth did not supply all 20 current protein amino acids (see part 5.2) and most likely some of them had a biosynthetic origin. The expanding amino acid repertoire might have therefore coevolved with the code, namely new metabolic products could usurp codons previously used by their metabolic precursors (for a recent review, see Wong, 2005).

5.6.1.7. *Clues from aminoacyl-tRNA synthases and tRNAs*

Aminoacyl-tRNA synthases (aaRS) are responsible of the actual decoding: in extant cells, each one of these 20 enzymes specifically recognizes each protein amino acid and their corresponding cognate tRNAs. A notable observation is the existence of a set of rules through which current aaRS recognize the tRNA molecules, mostly at the end of the acceptor stem – i.e. the end part of the tRNA that it is esterified by the amino acid, far from the anticodon stem, the part that recognizes the codon in the mRNA. Those rules – also known as the *operative code* – were established using minihelices, i.e. small fragments of RNA that mimic the aforementioned acceptor stems (Schimmel et al., 1993). On the other hand, structural studies of the active sites have revealed the existence of two classes of aaRS, each deriving from a different, non-homologous, ancient protein domain (Schimmel et al., 1993). Members of the two classes differ in the binding region on the acceptor stem of tRNAs, so that each tRNA can be potentially recognized by two binding sites, each one from one class, in a symmetric way. The specific pairings of the two aaRS classes to the different tRNAs have served to unveil a new pattern in the code. Ribas de Pouplana and Schimmel (2001) convincingly argued that the origin and coevolution of the aaRS-tRNA specific recognitions left their imprints in the present-day code. Thus, in a primitive stage each tRNA was recognized by two ancestral aaRS active sites, each from a different class, and each amino acid of the primitive – smaller – set had several codons assigned. The incorporation of new amino acids required the duplication of the elements (aaRS and tRNAs), leading to both the redistributide of codons and aaRS classes for the new couples amino acid/tRNA to originate the extant code. Subsequently, aaRS acquired new domains to better recognize both amino acid lateral chains and cognate tRNAs.

In summary, most of our current ideas about the origin of the genetic code are refinements of the early proposals made just after the code was established more than 30 years ago. The different hypotheses are complementary rather than mutually exclusive and the processes postulated could have been simultaneous. Although we may never know exactly how and when the universal code was established, experimental testing of the biochemical plausibility of the different proposals may eventually contribute to complete a coherent evolutionary narrative with component parts from all the hypotheses.

5.6.2. DATING GENETIC TAKEOVERS: HOW OLD ARE CELLULAR DNA GENOMES?

ANTONIO LAZCANO

As demonstrated by the numerous double-stranded polymeric structures, with backbones quite different from those of nucleic acids but held together by Watson–Crick base pairing, that have been synthesized in the past few years, a wide variety of informational molecules that have the potential for genetic information transfer is possible. These nucleic acid analogues provide useful laboratory models of the molecules that may have bridged the gap between the prebiotic soup and the earliest living systems. However, nowadays the only genetic polymers found in biological beings are RNA and DNA, two closely related nucleic acids. Which of them is older? Although the possibility of a simultaneous origin of DNA and RNA has been suggested (Oró and Stephen-Sherwood, 1974), many agree with the pioneering proposals made by A. N. Belozerskii and J. Brachet at the 1957 Moscow meeting of the origin of life, who suggested in an independent way that the high amounts of cellular RNA could be interpreted as evidence of a more conspicuous role during early biological evolution (cf. Dowrkin et al., 2003). This possibility is strongly supported by deoxyribonucleotide biosynthesis, a highly conserved pathway that forms the monomeric constituents of DNA via the enzymatic reduction of RNA's monomers. As reviewed elsewhere (Lazcano et al., 1988), this metabolic pathway can be interpreted as evidence that the transition from RNA to DNA genomes did not involve major metabolic changes but was facilitated by the evolutionary acquisition of one single enzymatic step (Lazcano et al., 1988).

During the past 15 years the possibility of early replicating and catalytic cell systems based on RNA and devoid of both proteins and DNA, as suggested in the late 1960's by Carl R. Woese, Francis H. Crick and Leslie E. Orgel, i.e. the so-called RNA world, has received considerable support with the demonstration of wide range of biochemical reactions catalyzed by ribozymes (Joyce, 2002). Although the nature of the predecessor(s) of the RNA world are completely unknown and can only be surmized, it is reasonable to assume that the extreme chemical instability of catalytic and replicative polyribonucleotides may have been the main limitation of such RNA world. In fact, the chemical lability of RNA implies that primordial ribozymes must have been very efficient in carrying out self-replication reactions in order to maintain an adequate inventory of molecules needed for survival. The instability of RNA may have been one of the primary reasons underlying the transition to the extant DNA/RNA/protein world where, due to the increased stability of the genetic molecules, survival would have been less dependent on polymer stability.

How did such transition take place? According to Joyce (2002), it is possible that in the RNA world ribozymes arose that could catalyze the polymerization of DNA; in this manner information stored in RNA could be transferred to the more stable DNA. On the other hand, since four of the central reactions involved in protein biosynthesis are catalyzed by ribozymes, their complementary nature suggest that they may have first appeared in an RNA world (Kumar and Yarus, 2001), i.e. double-stranded DNA genomes appeared once ribosome-catalyzed, nucleic acid-coded protein synthesis had evolved in an RNA-dominated primitive biosphere. Thus, the most likely explanation for DNA takeover could have been that because of increased stability much longer oligomers could have accumulated and this provided for an enhanced storage capacity of information that could be passed on to the next generation of living entities. Before long, RNA which once played the singular role of replication and catalysis was replaced by the more efficient and robust DNA/protein world wherein RNA was demoted to a role of messenger/transcriber of DNA stored information needed for protein biosynthesis.

When did the transition from RNA to DNA cellular genomes take place? As is often the case with metabolic innovations, such antiquity of such step cannot be documented from geological data. However, since all extant cells are endowed with DNA genomes, the most parsimonious conclusion is that this genetic polymer was already present in the last common ancestor (LCA) that existed prior to the divergence of the three primary domains, i.e., the Bacteria, Archaea, and Eucarya. As discussed elsewhere (Delaye et al., 2004), although there have been a number of suggestions that the LCA (or its equivalents) was endowed with genomes formed by small-sized RNA molecules or hybrid RNA/DNA genetic system, there are manifold indications that double-stranded DNA genomes of monophyletic had become firmly established prior to the divergence of the three primary domains. The major arguments supporting this possibility are:

(a) in contrast with other energetically favourable biochemical reactions (such as hydrolysis of the phosphodiester backbone, or the transfer of amino groups), the direct removal of the oxygen from the 2'-C ribonucleotide pentose ring to form the corresponding deoxy-equivalents is a thermodynamically much less-favoured reaction. This is a major constraint that strongly reduces the likelihood of multiple, independent origins of biological ribonucleotide reduction;

(b) the demonstration of the monophyletic origin of ribonucleotide reductases (RNR), which are the enzymes involved in the biosynthesis of deoxyribonucleotides from ribonucleotide precursors, is greatly complicated by their highly divergent primary sequences and the different mechanisms by which they generate the substrate 3'-radical species required for the removal of the 2'-OH group. However, sequence analysis and biochemical characterization of RNRs from the three primary biological domains has confirmed their structural similarities, which speaks of their ultimate monophyletic origin (see Freeland et al., 1999 and references therein);

(c) the sequence similarities shared by many ancient, large proteins found in all three domains suggest that considerable fidelity existed in the operative genetic system of their common ancestor (i.e., the LCA), but such fidelity is unlikely to be found in RNA-based genetic systems (Lazcano et al., 1992).

The nucleic acid replication machinery requires, at the very least, a set of enzymes involving a replicase, a primase, and a helicase. Quite surprisingly, when the corresponding enzymes are compared between the three primary domains, they appear to be of polyphyletic origins although there are indications of a closer phylogenetic relationship between the Eucarya and the Archaea. Given the central role that is assigned to nucleic acid replication in mainstream definitions of life, the lack of conservation and polyphyly of several of its key enzymatic components is somewhat surprising. However, there is structural evidence that some of components do have a single origin. It is reasonable to assume, for instance, that the oldest function of polymerases is the formation of the phosphodiester bond in a growing nucleic acid molecule. This reaction is catalyzed by the so-called palm domain of DNA polymerase I, and there is evidence of the ample phylogenetic distribution of this subdomain not only in the homologues of bacterial DNA polymerase I, but also among mitochondrial and viral DNA-dependent RNA polymerases (Steitz, 1999). These findings have been interpreted to imply that the palm domain is one of the oldest recognizable components of an ancestral cellular polymerase, that may have acted both as a replicase and a transcriptase during the RNA/protein world stage (Delaye et al., 2004). The presence of homologous palm domain in DNA- and RNA polymerases suggests that once the advent of double-stranded DNA took place, relatively few mutations would have been required for the evolution of this RNA replicase into a DNA polymerase, well before the divergence of the three domains. If this hypothesis is correct, then the palm domain of extant DNA polymerases is a silent but chemically active witness of the nature of biological systems older than the double stranded DNA cellular genomes. The biosphere never losses the traces of its evolutionary past.

5.7. The Last Common Ancestor

DAVID MOREIRA AND PURIFICACIÓN LÓPEZ-GARCÍA

In *The origin of species*, Darwin concluded that a logical outcome of the premises of descent with modification and natural selection is '*that probably all the organic beings which have ever lived on this earth have descended from some one primordial form, into which life was first breathed*' (Darwin, 1859). The strong development of biochemistry and molecular biology during the last century clearly demonstrated that all terrestrial life is based on common biochemical themes (Kornberg, 2000; Pace, 2001), lending solid ground to that idea. Since the basic structural constituents of the cell and its most fundamental metabolic reactions are shared by the three domains of life, Bacteria, Eucarya and Archaea (Woese and Fox 1977b), the hypothesis of a common ancestor from which modern organisms possessing already those traits made its way naturally.

Such hypothetical ancestor has received different names: the last common ancestor, the cenancestor (from the Greek *kainos* meaning recent and *koinos* meaning common) (Fitch and Upper, 1987) and the last universal common ancestor or LUCA (Forterre and Philippe, 1999). Although impossible to date, the cenancestor likely existed several billion years ago, predating the diversification of the three life domains. Therefore, the reconstruction of its model portrait is a difficult, speculative, task. Most authors favour the idea that the cenancestor was a single organism, an individual cell that existed at a given time and that possessed most of the features (and the genes encoding them) common to all contemporary organisms (Figure 5.10A–D). Others, on the contrary, envisage a population of cells which, as a whole, possessed all those genes, although no single individual did (Kandler, 1994; Woese, 1998; Woese, 2000). The level of gene exchange and spreading in this population should have been very high but, at some point, a particular successful combination of genes occurred in a subpopulation that became 'isolated' and gave rise to a whole line of descent. Kandler, for instance, proposed in his 'pre-cell theory' that

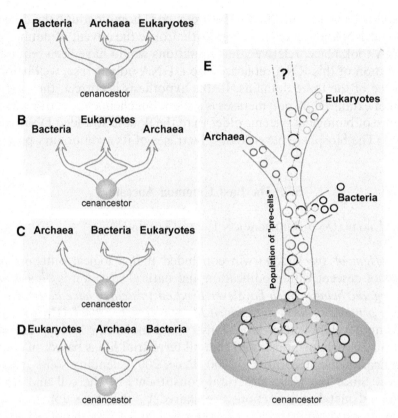

Figure 5.10. Different models of evolution of the three domains of life from a single-cell cenancestor (A–D) or a population of pre-cells (E).

bacteria, archaea, and eukaryotes emerged sequentially in this way (Figure 5.10E) (Kandler, 1994; Wächtershäuser, 2003). Despite the controversy between the 'single-cell' and 'population' hypotheses, the cenancestor is generally conceived as an already quite complex entity. This implies that life had already undergone a more or less long evolutionary pathway from the prebiotic times to the cenancestor stage. Therefore, both the origin of life and the nature of the cenancestor are different evolutionary questions. However, although rather complex, the cenancestor's level of complexity depends on the model. For Woese, it was a relatively primitive entity, a 'progenote', that *'had not completely evolved the link between genotype and phenotype'* (Woese and Fox, 1977a). For many other authors the progenote state occurred prior to the cenancestor, which was nearly a modern cell (Doolittle, 2000a, b).

Is it possible to build a model portrait of the cenancestor? At least, several of its fundamental characteristics can be inferred thanks to biochemistry and molecular biology comparative studies and, more recently, to the most powerful comparative genomics and molecular phylogeny. This approach has provided strong evidence for the very ancient origin of the ribosome-based protein synthesis (translation), a well-developed transcription machinery for the synthesis of structural and messenger RNAs, and the energy-obtaining process based on the generation of a proton gradient across membranes. All these are universal features in contemporary cells and, consequently, most likely present in the cenancestor. Other cenancestor's properties are, however, much more controversial, such as the existence of a DNA-based genome or even the possession of lipid-based membranes.

5.7.1. PROTEIN SYNTHESIS IN THE CENANCESTOR

When all available complete genome sequences of bacterial, archaeal and eukaryotic species are compared, only ~60 genes are found to be common to all of them. This is a small number knowing that prokaryotic species contain between 500 and 10,000 genes, and eukaryotic species between 2,000 and 30,000 genes. Interestingly, the set of 'universal' genes is almost entirely integrated by genes encoding ribosomal RNA and ribosomal proteins and other proteins involved in translation (aminoacyl-tRNA synthetases and translation factors) (Koonin, 2003). These genes are most likely ancestral, strongly suggesting that the cenancestor possessed a ribosomal-based translation machinery for protein synthesis comparable to that found in modern organisms. Protein synthesis by ribosomes is, therefore, the most universally conserved process and it appears to have remained practically unchanged since the cenancestor's times.

A few genes of that universal core encode RNA polymerase subunits, responsible for the synthesis of messenger and other RNAs from their DNA templates (genes). This suggests that the cenancestor possessed at least part of the transcription machinery that is found in contemporary organisms. Nevertheless, several RNA polymerase subunits and transcription factors are not universally distributed, so that the evolutionary conservation of the transcription machinery is not as high as that of translation.

5.7.2. THE CENANCESTOR'S GENOME: DNA OR RNA?

DNA is the molecule where genetic information is stored in all contemporary cells. However, only three out of the ~60 universal genes are related to DNA replication and/or repair: one DNA polymerase subunit, one exonuclease and one topoisomerase (Koonin, 2003). Today, archaea and eukaryotes share many genes involved in DNA replication that are absent from bacteria which, in turn, possess apparently unrelated genes encoding proteins performing equivalent functions. Various hypotheses try to explain this paradox. One of them postulates that the cenancestor did not possess a DNA, but an RNA, genome. DNA and its replication would have evolved twice independently, once in the bacterial line of descent and other in a lineage leading to archaea and eukaryotes (Leipe et al., 1999; Forterre, 2002). However, although few, universally conserved proteins and protein domains involved in DNA metabolism exist (Giraldo, 2003), suggesting that the cenancestor possessed indeed a DNA genome. Furthermore, due to a higher mutation rate, RNA is much more error-prone than DNA and, consequently, individual RNA molecules cannot exceed a certain size (Eigen limit) without falling into replicative catastrophe (Eigen, 1971; 2002). This size is small (~30–50 kb), so that a single RNA molecule can contain only a few dozens of genes. Since the cenancestor may have had over 600 genes (Koonin, 2003), its genome would have required many RNA molecules, entailing serious problems of stability and partition among daughter cells. Contemporary DNA and RNA viruses are good examples for these stability problems: DNA viruses can have very large genomes, up to ~1 Mbp (Raoult et al., 2004), but those of RNA viruses do not exceed ~30 kb (Domingo and Holland, 1997).

The remaining models postulate a DNA-based genome for the cenancestor. One of them hypothesises that transcription and translation were already well developed while DNA replication was still very primitive. DNA replication would have been refined as the two lines of descent leading to the bacteria and to the archaea/eukaryotes diverged (Olsen and Woese, 1997). An opposite possibility could be that the cenancestor had a very complex DNA-based metabolism, containing ancestral versions of the proteins found today in both the bacterial and the archaeal/eukaryotic replication machineries. One set of these proteins would be involved in replication, whereas the other would be specialized in DNA repair. During the separation of the two lineages, each line of descent would have retained only one of the two sets of proteins. In another hypothesis, the DNA replication machinery would be already developed in the cenancestor, but it evolved very fast in one or the two lineages descending from the cenancestor, so that the similarity between homologous genes in the two lineages are no longer recognisable (Olsen and Woese, 1997; Moreira, 2000). Yet in another model, the cenancestor would have had a DNA replication machinery, whose genes were inherited by archaea and eukaryotes, but that was replaced by viral genes in bacteria (Forterre, 1999). However, the directionality of these gene transfers is highly discussed, since present evidence demonstrates that viruses have acquired many, including replication-related, genes from their bacterial hosts (Moreira, 2000).

5.7.3. ENERGY AND CARBON METABOLISM IN THE CENANCESTOR

The question of how metabolism looked like in the cenancestor is rarely addressed, partly because the reconstruction of early metabolic pathways is extremely difficult. Genes involved in energy and carbon metabolism most often display a patchy distribution in organisms of the three domains of life and usually belong to large multigenic families whose members have been recruited for different functions in various metabolic pathways. In addition, horizontal gene transfer frequently affects metabolic genes, since they may confer an immediate adaptive advantage to the new host. Hence, the reconstruction of ancestral metabolic pathways is hampered by a complex history of gene duplications and losses, differential enzyme recruitment, and horizontal gene transfer (Castresana and Moreira, 1999). Nevertheless, the universal presence of a highly conserved membrane-bound ATPase in contemporary organisms indicates that the cenancestor was able to produce energy in the form of ATP by generating a proton gradient across the cell membrane. The energy source required to generate this proton gradient was likely chemical (oxido-reduction – redox-reactions), since phototrophy (light-based) evolved later and only in the bacterial line. The type of electron donors and acceptors involved in those redox reactions, whether they were organic, inorganic or both, is not known. It is possible that the cenancestor used a variety of oxidized inorganic molecules as electron acceptors (Castresana and Moreira, 1999), and that it carried out a simple heterotrophic metabolism, at least some kind of fermentation.

A purely heterotrophic cenancestor, needing to uptake organic molecules from the environment, would be logical if life originated in a prebiotic soup (Oparin, 1938; Broda, 1970). An autotrophic cenancestor, able to synthesise organic molecules from CO or CO_2, would be a logical outcome of models proposing that the first living forms were chemolithoautotrophic, deriving energy from redox reactions involving inorganic molecules such as H_2S, H_2 and FeS (Wächtershäuser, 1988b; Russell and Hall, 1997, see also part 5.3). However, this is not incompatible with soup-based models, since autotrophy might have evolved between life emergence and the cenancestor. If the cenancestor was autotrophic, which metabolic pathway to fix carbon did it use? Today, four different pathways of autotrophic carbon fixation are known: The Calvin–Benson cycle (reductive pentose-phosphate pathway), the Arnon cycle (reductive citric acid pathway), the Wood–Ljundahl cycle (reductive acetyl-CoA pathway), and the hydroxypropionate pathway, none of which is universal (Peretó et al., 1999). The complex Calvin–Benson cycle appeared relatively late during bacterial evolution, but any of the other, simpler, pathways might have been ancestral following different authors (Wächtershäuser, 1990; Peretó et al., 1999; Russell and Martin, 2004).

5.7.4. THE MEMBRANE OF THE CENANCESTOR

All contemporary cells are surrounded by a plasma membrane that is made out of phospholipids, generally organized in bilayers (see part 5.4). However there exist profound differences between the membrane lipids of archaea and those of bacteria and eukaryotes. In archaea, phospholipids are generally made out of isoprenoid lateral chains that are bounded by ether linkages to glycerol-1-phosphate, whereas in bacteria and eukaryotes they are made out of fatty acids bounded by ester linkages to glycerol-3-phosphate. The enzymes that synthesise the glycerol-phosphate stereoisomers are not homologous in archaea and bacteria/eukaryotes, belonging to different enzymatic families. To explain this profound difference, some authors even proposed that the cenancestor was not yet membrane-bounded and that membrane lipids together with membrane-bounded cells evolved independently to generate bacteria and archaea (Koga et al., 1998). Others proposed that the cenancestor was cellular but, instead of lipid membranes, it possessed iron monosulphide boundaries. Cells would be mineral compartments in an ever-growing hydrothermal chimney traversed by alkaline fluids, and membrane lipids would have been invented independently in bacteria and archaea (Martin and Russell, 2003). Another, less radical, option is that the cenancestor had a lipid membrane that was heterochiral, i.e. composed of a mixture of lipids built upon glycerol-1-phosphate and glycerol-3-phosphate (Wächtershäuser, 2003). The biosynthetic pathways to produce the two types of homochiral membranes would have evolved as archaea and bacteria diverged. This is supported by recent phylogenetic analyses of the enzymes involved suggesting that the cenancestor synthesized phospholipids via a non-stereospecific pathway (Peretó et al., 2004). In addition, the occurrence of universally conserved membrane-bound proteins, such as the proton-pump ATPases (Gogarten et al., 1989) and the signal recognition particle, SRP (Gribaldo and Cammarano 1998) fully supports the hypothesis of a cenancestor endowed with lipid membranes.

5.7.5. OTHER UNRESOLVED QUESTIONS

Many additional questions about the nature of a hypothetical cenancestor remain open, e.g. whether the cenancestor was hyperthermophilic or not, and whether it was 'simple' or 'complex'. The hypothesis of a hyperthermophilic cenancestor arose from the discovery of hyperthermophilic bacteria and archaea growing optimally at $> 80\ °C$ and from the proposals of a hot, autotrophic origin of life in a hotter early Earth (Achenbach-Richter et al., 1987; Pace, 1991). The first criticisms to it derived from the fact that RNA and other biomolecules have relatively short life-times at high temperatures so that a hot origin of life would be unlikely (Lazcano and Miller, 1996). However, the origin of life and the cenancestor might have occurred in different environmental conditions (Arrhenius et al., 1999). Hence, attempts to reconcile a hyperthermophilic cenancestor with a cold origin of life propose that only hyperthermophiles could survive the late heavy meteorite bombardment

~3.9 Ga ago (Gogarten-Boekels et al., 1995; about LHB, see chapter 4.4). The most decisive argument sustaining a hyperthermophilic cenancestor was that hyperthermophilic bacteria and archaea branched at the most basal positions in phylogenetic trees (Stetter, 1996). Two major objections followed. First, computer reconstruction of ancestral rRNAs suggests that the content of guanine and cytosine in the cenancestor's rRNA was incompatible with life at > 80 °C (Galtier et al., 1999), although the analysis of the same data by other methods favours, on the contrary, a hyperthermophilic cenancestor (Di Giulio, 2000; 2003). Second, refined phylogenetic analyses of rRNAs suggest that the basal emergence of hyperthermophilic bacteria was an artefact of phylogenetic tree reconstruction, and that they adapted secondarily to hyperthermophily (Brochier and Philippe, 2002). By contrast, the archaeal ancestor was most likely hyperthermophilic (Forterre et al., 2002). Therefore, if the bacterial ancestor was also a hyperthermophile, then the cenancestor was most likely hyperthermophilic too, but if the bacterial ancestor was not hyperthermophilic, the question remains open. At any rate, all models appear compatible with the occurrence of a thermophilic (60–80 °C) cenancestor (López-García, 1999).

Another controversial issue concerns the level of structural complexity of the cenancestor. For most authors, it was 'structurally simple', resembling today's prokaryotes, with the genetic material directly immersed in the cytoplasm. Such an ancestor would agree with the widely accepted bacterial rooting of the tree of life, between the bacteria and the archaea/eukaryotes (Figure 5.10A) (Woese et al., 1990), but would be also compatible with two alternative tree topologies (Figure 5.10B and C). However, other authors propose that the root lies between the eukaryotes and a branch leading to the two prokaryotic domains (Figure 5.10D) (Brinkmann and Philippe, 1999; Philippe and Forterre, 1999). This rooting would still be compatible with a prokaryote-like cenancestor, but opens the possibility that the cenancestor had some features of modern eukaryotes, such as a membrane-bound nucleus and many small RNAs claimed to be relics of a hypothetical RNA world (Poole et al. 1999). Despite the position of the root is indeed an open question, models proposing a eukaryotic-like cenancestor do not explain how such a complex entity was built from the prebiotic world. In this sense, a simpler, prokaryotic-like cenancestor appears much more parsimonious in evolutionary terms.

5.8. The Origin of Viruses

PATRICK FORTERRE

Viruses are often ignored in evolutionary scenarios. For many biologists, viruses are not even considered as genuine living beings, since they are absolutely dependent on cellular organisms (archaea, bacteria or eukaryotes) for their development. Nevertheless, viruses are very diverse and can be quite complex, as is revealed by recent observations including (i) the discovery of a giant virus (Mimivirus) whose genome doubles the size of certain cell genomes, (ii) the fact that, after the analysis of many viral genomes, many viral proteins have no known homologues in cells, (iii) the existence of a phylogenetic link between viruses and different evolutionary distant cellular hosts (e.g. man and bacteria) or (iv) the discovery of an unexpected morphological and functional diversity, particularly in the less-explored archaeal viruses.

5.8.1. VIRAL PROPERTIES

Viruses may be considered as living entities if we consider their chemical composition and their developmental cycle. Viruses are made out of the same molecules as cellular organisms: proteins, nucleic acids, and often, lipids and sugars. However, viruses exhibit a number of singularities. By contrast to cellular genomes, always made out of double-stranded DNA, viral genomes can be constituted by DNA or RNA, either single- or double-stranded. The essential difference between viruses and their cellular hosts is that viruses are unable to synthesize their own proteins and to produce their ATP. Viruses lack ribosomes as well as any form of energy and carbon metabolism. Therefore, viruses are obligatory parasites of cells (Villareal, 2005).

During their cycle, viruses exhibit two different forms. One form is stable, normally extracellular, the virion, which is unable to grow and reproduce, and the other (or others) is intracellular and can lead to the viral reproduction. Virions are constituted by a protective shell, the capsid, which encapsulates the viral genome that is generally associated with proteins forming a nucleoprotein filament. Virions can be observed only by electron microscopy and display very different morphologies (spherical, filamentous, head-and-tail, etc). Being very small, they were initially discovered by their capacity to traverse filters that retain bacteria. Nonetheless, their size can vary from a few dozens to various hundreds of nanometres, thus the largest viruses are comparable in size to the smallest known cells (Villareal, 2005). Viral genomes can also vary from a few kilobase pairs for the smallest RNA or DNA viruses to 1,200 kilobase pairs for the Mimivirus genome (Raoult et al., 2005). The simplest protein capsids are assembled following a precise geometry, but many virions have one or two additional envelopes, often from cellular origin. Virions have frequently been considered as passive forms simply allowing the transport and protection of the viral genome. However, an archaeal virus has been discovered recently, whose virion is able to change its morphology extracellularly (Haring et al., 2005).

The virion can either penetrate entirely within the cell cytoplasm or just inject its genetic material into it. These processes require more or less sophisticated mechanisms involving proteins that recognise the host cell surface, allow the fusion of the viral envelope with that of the cell, the capture of the virus by cellular endocytose or the transport of DNA through the cell plasma membrane. Once inside, the virus deviates the cellular metabolism to replicate and transcribe its own genome, make its capsid proteins and, in the end, produce new viruses. This cycle may trigger the death of infected cells with the sudden liberation of many virions (virulence), or may induce the continuous production of a limited amount of virions while allowing cell survival. Finally, the viral genome may stay within the cell without producing new viruses (lysogenic state). During the lysogenic state, viral genomes can become integrated in the host chromosome. In the case of retroviruses (e.g. AIDS virus), this integration is achieved by retrotranscription of the viral RNA to DNA. Lysogenic viruses can be reactivated when cells undergo different stress, which allows the virus to escape and initiate a new cycle.

All cellular organisms, archaea, bacteria or eukaryotes, can be infected by an extraordinary diversity of viruses. Viruses constitute indeed a major component of the biosphere and play, and have likely played, a determinant role in the evolution of their hosts (Villareal, 2005). Therefore, it is probable that viruses have also influenced early steps in biological evolution but, when and how did viruses originated? The origin of viruses is still a mystery. Many researchers think that viruses are polyphyletic, i.e. they have multiple origins. However, viruses share a number of properties that may suggest, on the contrary, a common mechanism for their emergence.

5.8.2. HYPOTHESES ON THE ORIGIN OF VIRUSES

Traditionally, three hypotheses to explain the origin of viruses have been put forward. First, viruses would be primitive entities that appeared before cells. Second, viruses would derive from ancient cells that parasitized other cells by a reduction process implying the loss of their ribosomes and energy and carbon metabolism. And third, viruses would be chromosome fragments from cells that became autonomous and began to parasitize (Balter, 2000, Forterre, 2006). The three hypotheses have been severely criticized. The first one was refuted because, being obligatory parasites, viruses could have never been emerged before cells. The second was also criticized because there are no known intermediates between viruses and cellular parasites, since even the most reduced cellular parasites (mycoplasma within the bacteria or microsporidia within the eukaryotes) have retained their basic cellular features. The third hypothesis was favoured by most biologists, although it does not explain how RNA or DNA fragments could have escaped from cells and acquire a capsid, and how the mechanisms that allow viruses to penetrate cells were developed. In addition, in its original form, this hypothesis postulated that viruses infected bacteria (bacteriophages) and those infecting eukaryotes had originated from bacterial and eukaryotic genomes, respectively. However, many viral proteins have no known homologues in their host genomes, which exclude, for some authors, a common origin. Furthermore, some viral proteins may have homologues in a different cellular domain to that being actually infected by the virus (Bamford, 2003).

Recent structural analyses of conserved viral proteins show that RNA polymerases from RNA bacteriophages are homologous to those from RNA viruses infecting eukaryotes, and that certain capsid proteins from some DNA bacteriophages are homologous to those from DNA viruses infecting eukaryotes and archaea. Although the possibility that viruses can "jump" and infect organisms from different domains cannot be excluded, various researchers think that viruses are very ancient, existing well before the divergence of the three cellular domains, that is, before the last universal common (or cellular) ancestor (Bamford, 2003, Forterre, 2005).

All the criticism raised against the above-mentioned hypotheses is difficult to overcome if viruses emerged in the cellular world that we know today (e.g. it would be difficult to imagine a regressive evolution from modern cells towards a viral form). However, the situation may be different if viruses originated prior to a last cellular ancestor during an RNA world. Thus, some authors that support the idea of a long acellular evolution during this RNA world period have hypothesized that viruses appeared before cells. Viruses would have been "hosted" primarily by a primitive semi-liquid soup or by mineral "cells" (Koonin and Martin, 2005). This idea is criticized by other authors favouring the idea that, on the contrary, free cells surrounded by a plasma membrane emerged very early. If this was the case, two different scenarios corresponding to the second and third hypotheses mentioned above can be envisaged: viruses emerged either by reduction of RNA cells that parasitized other RNA cells, or by the separation and independence of RNA fragments that would become autonomous and infectious. These scenarios, improbable in the modern cellular world, could be more realistic in the context of primitive RNA-based cells whose genomes were likely constituted by several linear RNA molecules (Forterre, 2006).

In any case, the first viruses may have been RNA viruses, although the question of when they emerged remains unanswered. The discovery of structural homologies between viral RNA and DNA polymerases suggests that DNA viruses evolved secondarily from certain RNA viruses. It could be even possible that DNA appeared for the first time in viruses, since a chemical modification in the viral genome could have represented an immediate selective advantage in order to escape from cellular defences designed to destroy the viral genome. This hypothesis would explain why many viruses encode their own ribonucleotide reductase (the enzyme that reduces the ribose into deoxyribose) and/or thymidylate synthase (responsible for the synthesis of thymine, the nucleotide that replaces uracile in DNA) (Forterre, 2002).

Although the mechanisms that led to the emergence of viruses remain hypothetical, viruses must have played a considerable role in biological evolution. Many viral genomes or genome fragments are present (cryptic) in contemporary cell genomes. Viruses can pick up genes from one organism and transfer them to another, thus serving as vehicles of gene transfer between cellular lines. Horizontal gene transfer is now recognized as an important motor in cellular evolution. The essential role of viruses in biological history is beginning to be fairly recognized.

References

Achenbach-Richter, L., Gupta, R., Stetter, K. O. and Woese, C. R.: 1987, *Syst. Appl. Microbiol.* **9**, 34–39.

Arnez, J. G. and Moras, D.: 2003, *Trends Biochem. Sci.* **22**, 211–216.

Arrhenius, G., Bada, J. L., Joyce, G. F., Lazcano, A., Miller, S. and Orgel, L. E.: 1999, *Science* **283**, 792.

Balter, M.: 2000, *Science* **289**, 1866–1867.

Bamford, D. H.: 2003, *Res. Microbiol.* **154**, 231–236.

Ban, N., Nissen, P., Hansen, J., Moore, P. B. and Steitz, T. A.: 2000, *Science* **289**, 905–920.

Barbier, B. and Brack, A.: 1992, *J. Am. Chem. Soc.* **114**, 3511–3515.

Bartel, D. P. and Unrau, P. J.: 1999, *Trends Biochem. Sci.* **24**, 9–13.

Benner, S. A.: 1993, *Science* **261**, 1402–1403.

Benner, S. A., Ellington, A. D. and Tauer, A.: 1989, *Proc. Natl. Acad. Sci. U.S.A.* **86**, 7054–7058.

Biron, J.-P. and Pascal, R.: 2004, *J. Am. Chem. Soc.* **126**, 9198–9199.

Biron, J.-P., Parkes, A. L., Pascal, R. and Sutherland, J. D.: 2005, *Angew. Chem. Int. Ed.* **44**, 6731–6734.

Bohler, C., Nielsen, P. E. and Orgel, L. E.: 1995, *Nature* **376**, 578–581.

Borsenberger, V., Crowe, M. A., Lehbauer, J., Raftery, J., Helliwell, M., Bhutia, K., Cox, T. and Sutherland, J. D.: 2004, *Chem. Biodivers.* **1**, 203–246.

Brasier, M. D., Green, O. R., Jephcoat, A. P., Kleppe, A. K., van Kranendonk, M., Lindsay, J. F., Steele, A. and Grassineau, N.: 2002, *Nature* **416**, 76–81.

Brinkmann, H. and Philippe, H.: 1999, *Mol. Biol. Evol.* **16**, 817–825.

Brochier, C. and Philippe, H.: 2002, *Nature* **417**, 244.

Broda, E.: 1970, *Prog. Biophys. Mol. Biol.* **21**, 143–208.

Brosius, J.: 2001, *Trends Biochem. Sci.* **26**, 653–656.

Cairns-Smith, A. G.: 1966, *J. Theor. Biol.* **10**, 53–88.

Cairns-Smith, A. G.: 1982, *Genetic Takeover and the Mineral Origins of Life*, Cambridge University Press, Cambridge, 477 pp.

Castresana, J. and Moreira, D.: 1999, *J. Mol. Evol.* **49**, 453–460.

Cavalier-Smith, T.: 2001, *J. Mol. Evol.* **53**, 555–595.

Chen, I. and Szostak, J. W.: 2004, *Proc. Natl. Acad. Sci. U.S.A.* **101**, 7965–7970.

Chen, I., Roberts, R. W. and Szostak, J. W.: 2004, *Science* **305**, 1474–1476.

Cheng, C. M., Liu, X. H., Li, Y. M., Ma, Y., Tan, B., Wan, R. and Zhao, Y. F.: 2004, *Origins Life Evol. Biosphere* **34**, 455 464.

Chyba, C.: 2005, *Science* **308**, 962–963.

Chyba, C. and Sagan, C.: 1992, *Nature* **355**, 125–132.

Commeyras, A., Collet, H., Boiteau, L., Taillades, J., Vandenabeele-Trambouze, O., Cottet, H., Biron, J.-P., Plasson, R., Mion, L., Lagrille, O., Martin, H., Selsis, F. and Dobrijevic, M.: 2002, *Polym. Int.* **51**, 661–665.

Cordóva, A., Engqvist, M., Ibrahem, I., Casas, J. and Sundén, A.: 2005, *Chem. Commun.* 2047–2049.

Corey, M. J. and Corey, E.: 1996, *Proc. Natl. Acad. Sci. U.S.A.* **93**, 11428–11434.

Cornée A., Vergne J., Guyot F., Goffe B., Rouchy J.-M. and Maurel M.-C.: 2004, in G. Camoin and P. Gautret (eds.), *Microbialites and Microbial Communities in Sedimentary Systems*, Publication Spéciale ASF, Paris, 31 p.

Crick, F. H. C.: 1968, *J. Mol. Biol.* **38**, 367–379.

Cronin J. R. and Chang S. 1993, in J. M. Greenberg, C. X. Mendoza-Gomez and V. Pirronello (eds.), *The Chemistry of Life's Origins*, Kluwer Dordrecht pp. 209–258.

Darwin, C.: 1859, *The Origin of Species by Means of Natural Selection*, J. Murray, London.

Deamer, D. W.: 1986, *Origins Life Evol. Biosphere* **17**, 3–25.

Deamer, D. W.: 1997, *Microbiol. Mol. Biol. Rev.* **61**, 239–261.

Deamer, D., Dworkin, J. P., Sandford, S. A., Bernstein, M. P. and Allamandola, L. J.: 2002, *Astrobiology* **2**, 371–381.

Décout, J.-L., Vergne, J. and Maurel, M.-C.: 1995, *Macromol. Chem. Phys.* **196**, 2615–2624.

Delaye L., Becerra A. and Lazcano A.: 2004, in L. Ribas de Pouplana (ed.), *The Genetic Code and the Origin of Life*. Landes Bioscience, Georgetown, pp. 34 –47.

Di Giulio, M.: 2000, *J. Theor. Biol.* **203**, 203–213.

Di Giulio, M.: 2003, *J. Theor. Biol.* **224**, 277–283.

Di Giulio, M.: 2005, *Gene* **346**, 7–12.

Dillon, L. S.: 1973, *Bot. Rev.* **39**, 301–345.

Domingo, E. and Holland, J. J.: 1997, *Annu. Rev. Microbiol.* **51**, 151–178.

Doolittle, R. F.: 2000a, *Res. Microbiol.* **151**, 85–89.

Doolittle, W. F.: 2000b, *Curr. Opin. Struct. Biol.* **10**, 355–358.

Dunnill, P.: 1966, *Nature* **210**, 1267–1268.

Dworkin, J. P., Lazcano, A. and Miller, S. L.: 2003, *J. Theor. Biol.* **222**, 127–134.

Egholm, M., Buchardt, O., Christensen, L., Behrens, C., Freier, S. M., Driver, D. A., Berg, R. H., Kim, S. K., Norden, B. and Nielsen, P. E.: 1993, *Nature* **365**, 566–568.

Eigen, M.: 1971, *Naturwissenschaften* **58**, 465–523.

Eigen, M.: 2002, *Proc. Natl. Acad. Sci. U.S.A.* **99**, 13374–13376.

Eigen, M. and Schuster, P.: 1978, *Naturwissenschaften* **65**, 341–369.

El Amri, C., Baron, M.-H. and Maurel, M.-C.: 2003, *Spectrochim. Acta Part A* **59**, 2645–2654.

El Amri, C., Baron, M.-H. and Maurel, M.-C.: 2004, *J. Raman Spectrosc.* **35**, 170–177.

El Amri, C., Maurel, M.-C., Sagon, G. and Baron, M.-H.: 2005, *Spectrochim. Acta Part A* **61**, 2049–2056.

Ellington, A. D., Khrapov, M. and Shaw, C. A.: 2000, *RNA* **6**, 485–498.

Eschenmoser, A.: 1999, *Science* **284**, 2118–2124.

Ferris, J. P.: 1987, *Cold Spring Harbor Symp. Quant. Biol.* **LII**, 29–39.

Ferris, J. P. and Ertem, G.: 1992, *Science* **257**, 1387–1389.

Ferris, J. P., Hill, A. R., Liu, R. and Orgel, L. E.: 1996, *Nature* **381**, 59–61.

Fersht, A. R.: 1999, *Structure ans Mechanism in Protein Science: A Guide to Enzyme Catalysis and Protein Folding*, Freeman, New York.

Fitch, W. M. and Upper, K.: 1987, *Cold Spring Harbor Symp. Quant. Biol.* **52**, 759–767.

Forterre, P.: 1999, *Mol. Microbiol.* **33**, 457–465.

Forterre, P.: 2002, *Curr. Opin. Microbiol.* **5**, 525–532.
Forterre, P.: 2005, *Biochimie* **87**, 793–803.
Forterre P.: 2006, *Virus Res.*, 5–16.
Forterre, P. and Philippe, H.: 1999, *Biol. Bull.* **196**, 373–375.
Forterre, P., Brochier, C. and Philippe, H.: 2002, *Theor. Popul. Biol.* **61**, 409–422.
Fox G. E. and Naik A. K. 2004, in Ll. Ribas de Pouplana (ed.), *The Genetic Code and the Origin of Life*, Landes Bioscience, Georgetown, pp. 92–106.
Freeland, S. J., Knight, R. D. and Landweber, L. F.: 1999, *Science* **286**, 690–692.
Freeland, S. J., Wu, T. and Keulman, N.: 2003, *Origins Life Evol. Biosphere* **33**, 457–477.
Fuller, W. D., Sanchez, R. A. and Orgel, L. E.: 1972, *J. Mol. Biol.* **67**, 25–33.
Galtier, N., Tourasse, N. and Gouy, M.: 1999, *Science* **283**, 220–221.
Gesteland, R. F., Cech, T. R. and Atkins, J. F., (eds.): 1999, *The RNA World* (2nd ed.), Cold Spring Harbor Laboratory Press, Cold Spring Harbor NY.
Gilbert, W.: 1986, *Nature* **319**, 618.
Giraldo, R.: 2003, *FEMS Microbiol. Rev.* **26**, 533–554.
Gogarten, J. P., Kibak, H., Dittrich, P., Taiz, L., Bowman, E. J., Bowman, B. J., Manolson, M. F., Poole, R. J., Date, T., Oshima, T., Konishi, J., Denda, K. and Yoshida, M.: 1989, *Proc. Nat. Acad. Sci. U.S.A.* **86**, 6661–6665.
Gogarten-Boekels, M., Hilario, E. and Gogarten, J. P.: 1995, *Origins Life Evol. Biosphere* **25**, 251–264.
Gribaldo, S. and Cammarano, P.: 1998, *J. Mol. Evol.* **47**, 508–516.
Hanczyc, M. M. and Szostak, J. W.: 2004, *Curr. Opin. Chem. Biol.* **8**, 660–664.
Hanczyc, M. M., Fujikawa, S. M. and Szostak, J. W.: 2003, *Science* **302**, 618–622.
Haring, M., Vestergaard, G., Rachel, R., Chen, L., Garrett, R. A. and Prangishvili, D.: 2005, *Nature* **436**, 1101–1102.
Harold, F. M.: 2001, *The Way of the Cell. Molecules, Organisms and the Order of Life*, Oxford University Press, Oxford.
Healy, V. L., Mullins, L. S., Li, X., Hall, S. E., Raushel, F. M. and Walsh, C. T.: 2000, *Chem. Biol.* **7**, 505–514.
Hill, A. R., Orgel, L. E. and Wu, T.: 1993, *Orig. Life Evol. Biosphere* **23**, 285–290.
Hoang, L., Fredrick, K. and Noller, H. F.: 2004, *Proc. Natl. Acad. Sci. U.S.A.* **101**, 12439–12443.
Holm, N. G. and Andersson, E. M.: 1998, in A. Brack (ed.), *The Molecular Origins of Life Assembling pieces of the Puzzle*, Cambridge University Press, Cambridge, pp. 86–99.
Huber, C., Eisenreich, W., Hecht, S. and Wächtershäuser, G.: 2003, *Science* **301**, 938–940.
Inoue, T. and Orgel, L. E.: 1983, *Science* **219**, 859–862.
Joyce, G. F.: 1989, *Nature* **338**, 217–224.
Joyce, G. F.: 2002, *Nature* **418**, 214–221.
Joyce, G. F. and Orgel, L. E.: 1986, *J. Mol. Biol.* **188**, 433–441.
Joyce, G. F. and Orgel, L. E.: 1999, in R. F. Gesteland, T. R. Cech and J. F. Atkins (eds.), *The RNA World*, Cold Spring Harbor Laboratory Press, Cold Spring Harbor NY, pp. 49–77.
Joyce, G. F., Schwartz, A. W., Miller, S. L. and Orgel, L. E.: 1987, *Proc. Natl. Acad. Sci. U.S.A.* **84**, 4398–4402.
Jukes, T. H.: 1973, *Biochem. Biophys. Res. Commun.* **53**, 709–714.
Kandler O.: 1994, in: S. Bengston (ed.), *Early Life on Earth*, Columbia University Press, pp. 152–160.
Kasting, J. F.: 1993, *Science* **259**, 920–926.
Knight, R. D., Freeland, S. J. and Landweber, L. F.: 1999, *Trends Biochem. Sci.* **24**, 241–247.
Koga, Y., Kyuragi, T., Nishihara, M. and Sone, N.: 1998, *J. Mol. Evol.* **46**, 54–63.

Koonin, E. V.: 2003, *Nat. Rev. Microbiol.* **1**, 127–136.
Koonin, E. V. and Martin, W.: 2005, *Trends Genet.* **21**, 647–654.
Koppitz, M., Nielsen, P. E. and Orgel, L. E.: 1998, *J. Am. Chem. Soc.* **120**, 4563–4569.
Kornberg, A.: 2000, *J. Bacteriol.* **182**, 3613–3618.
Kuhn, H. and Kuhn, C.: 2003, *Angew. Chem. Int. Ed.* **42**, 262–268.
Kühne, H. and Joyce, G. F.: 2003, *J. Mol. Evol.* **57**, 292–298.
Kumar, R. K. and Yarus, M.: 2001, *Biochemistry* **40**, 6998–7004.
Lahav, N., White, D. and Chang, S.: 1978, *Science* **201**, 67–69.
Lazcano A. 2001, in D. E. G. Briggs and P. R. Crowther (eds.), *Paleobiology II*, Blackwell Science Oxford, ch. 1.
Lazcano, A. and Miller, S. L.: 1996, *Cell* **85**, 793–798.
Lazcano, A., Guerrero, R., Margulis, L. and Oró, J.: 1988, *J. Mol. Evol.* **27**, 283–290.
Lazcano, A., Fox, G. E. and Oró, J.: 1992, in R. P. Mortlock (ed.), *The Evolution of Metabolic Function*, CRC Press, Boca Raton FL, pp. 237–295.
Leipe, D. D., Aravind, L. and Koonin, E. V.: 1999, *Nucleic Acids Res.* **27**, 3389–3401.
Leman, L., Orgel, L. and Ghadiri, M. R.: 2004, *Science* **305**, 283–286.
López-García, P.: 1999, *J. Mol. Evol* **49**, 439–452.
Luisi, P. L.: 1998, *Origins Life Evol. Biosphere* **28**, 613–622.
Luisi, P. L., Walde, P. and Oberholzer, T.: 1994, *Ber. Bunsenges. Phys. Chem.* **98**, 1160–1165.
Luisi, P. L., Walde, P. and Oberholzer, T.: 1999, *Curr. Opin. Colloid Interface Sci.* **4**, 33–39.
Madigan, M. T., Martinko, J. M. and Parker, J.: 2003, *Brock Biology of Microorganisms*, Prentice Hall Upper Saddle River, NJ, USA.
Marahiel, M. A., Stachelhaus, T. and Mootz, H. D.: 1997, *Chem. Rev.* **97**, 2651–2673.
Martin, W. and Russell, M. J.: 2003, *Philos. Trans. R. Soc. Lond., B Biol. Sci.* **358**, 59–83.
Maurel, M.-C.: 1992, *J. Evol. Biol.* **2**, 173–188.
Maurel, M.-C. and Décout, J.-L.: 1999, *Tetrahedron* **55**, 3141–3182.
Maurel, M.-C. and Ninio, J.: 1987, *Biochimie* **69**, 551–553.
Maurel, M. C. and Orgel, L.: 2000, *Origins Life Evol. Biosphere* **30**, 423–430.
Maurel, M.-C. and Zaccaï, G.: 2001, *BioEssays* **23**, 977–978.
Maynard Smith, J. and Szathmáry, E.: 1995, *The Major Transitions in Evolution*, Freeman, New York.
McGinness, K. E., Wright, M. C. and Joyce, G. F.: 2002, *Chem. Biol.* **9**, 585–596.
Meli, M., Vergne, J., Décout, J.-L. and Maurel, M.-C.: 2002, *J. Biol. Chem.* **277**, 2104–2111.
Meli, M., Vergne, J. and Maurel, M.-C.: 2003, *J. Biol. Chem.* **278**, 9835–9842.
Miller, S. L.: 1953, *Science* **117**, 528–529.
Miller, S. L.: 1998, in A. Brack (ed.), *The Molecular Origins of Life. Assembling pieces of the Puzzle*, Cambridge University Press, Cambridge, pp. 59–85.
Monnard, P. A. and Deamer, D. W.: 2002, *Anat. Rec.* **268**, 196–207.
Moreira, D.: 2000, *Mol. Microbiol.* **35**, 1–5.
Morowitz, H. J., Heinz, B. and Deamer, D. W.: 1988, *Origins Life Evol. Biosphere* **18**, 281–287.
Morowitz, H. J., Deamer, D. W. and Smith, T.: 1991, *J. Mol. Evol.* **33**, 207–208.
Mullie, F. and Reisse, J.: 1987, *Top. Curr. Chem.* **139**, 83–117.
Nealson, K. H.: 2005, *Proc. Natl. Acad. Sci. U.S.A.* **102**, 3889–3890.
Nelson, K. E., Levy, M. and Miller, S. L.: 2000, *Proc. Natl. Acad. Sci. U.S.A.* **97**, 3868–3871.
Neunlist, S., Bisseret, P. and Rohmer, M.: 1987, *Eur. J. Biochem.* **87**, 245–252.
Nicolis, G. and Prigogine, I.: 1977, *Self-Organization in Nonequilibrium Systems*, Wiley, New York.
Nielsen, P. E.: 1999, *Acc. Chem. Res.* **32**, 624–630.
Nissen, P., Hansen, J., Ban, N., Moore, P. B. and Steitz, T. A.: 2000, *Science* **289**, 920–930.

Noller, H. F.: 2004, *RNA* **10**, 1833–1837.
Olsen, G. J. and Woese, C. R.: 1997, *Cell* **89**, 991–994.
Oparin, A. I.: 1938, *The Origin of Life*, Mac Millan, New York.
Oparin, A. I., Orlovskii, A. F., Bukhlaeva, V. and Gladilin, K. L.: 1976, *Dokl. Akad. Nauk SSSR* **226**, 972–974.
Orgel, L. E.: 1989, in M. Grunberg-Manago et al. (eds.), *Evolutionary Tinkering in Gene Expression*, Plenum Press, London, pp. 215–224.
Orgel, L. E.: 1992, *Nature* **358**, 203–209.
Orgel, L. E.: 2004, *Crit. Rev. Biochem. Mol. Biol.* **39**, 99–123.
Oró, J.: 1960, *Biochem. Biophys. Res. Commun.* **2**, 407–412.
Oró, J. and Stephen-Sherwood, E.: 1974, in J. Oró, S. L. Miller, C. Ponnamperuma and R. S. Young (eds.), *Cosmochemical Evolution and the Origins of Life*, Reidel, Dordrecht, pp. 159–172.
Ourisson, G. and Nakatani, Y.: 1994, *Chem. Biol.* **1**, 11–23.
Pace, N. R.: 1991, *Cell* **65**, 531–533.
Pace, N. R.: 2001, *Proc. Natl. Acad. Sci. U.S.A.* **98**, 805–808.
Paecht-Horowitz, M., Berger, J. and Katchalsky, A.: 1970, *Nature* **7**, 847–850.
Pascal, R., Boiteau, L. and Commeyras, A.: 2005, *Top. Curr. Chem.* **259**, 69–122.
Patel, A.: 2005, *J. Theor. Biol.* **233**, 527–532.
Pelc, S. R.: 1965, *Nature* **207**, 597–599.
Peretó, J.: 2005, *Int. Microbiol.* **8**, 23–31.
Peretó, J. G., Velasco, A. M., Becerra, A. and Lazcano, A.: 1999, *Int. Microbiol.* **2**, 3–10.
Peretó, J., Lopez-Garcia, P. and Moreira, D.: 2004, *Trends Biochem. Sci.* **29**, 469–477.
Philippe, H. and Forterre, P.: 1999, *J. Mol. Evol.* **49**, 509–523.
Pizzarello, S. and Weber, A. L.: 2004, *Science* **303**, 1151–1151.
Plankensteiner, K., Reiner, H. and Rode, B. M.: 2005, *Curr. Org. Chem.* **9**, 1107–1114.
Poole, A., Jeffares, D. and Penny, D.: 1999, *Bioessays* **21**, 880–889.
Prieur, D.: 2005, in T. Tokano (ed.), *Water on Mars and Life*, Springer, Berlin, pp. 299–324.
Raoult, D., Audic, S., Robert, C., Abergel, C., Renesto, P., Ogata, H., La Scola, B., Suzan, M. and Claverie, J. M.: 2004, *Science* **306**, 1344–1350.
Reader, J. S. and Joyce, G. F.: 2002, *Nature* **420**, 841–844.
Reusch, R. N.: 2000, *Biochemistry (Moscow)* **65**, 280–295.
Ribas de Pouplana, Ll. and Schimmel, P.: 2001, *Trends Biochem. Sci.* **26**, 591–596.
Ricard, J., Vergne, J., Décout, J.-L. and Maurel, M.-C.: 1996, *J. Mol. Evol.* **43**, 315–325.
Ricardo, A., Carrigan, M. A., Olcott, A. N. and Benner, S. A.: 2004, *Science* **303**, 196.
Rode, B. M., Son, H. L., Suwannachot, Y. and Bujdak, J.: 1999, *Origins Life Evol. Biosphere* **29**, 273–286.
Rohlfing, D. L.: 1976, *Science* **193**, 68–70.
Russell, M. and Hall, A. J.: 1997, *J. Geol. Soc. Lond.* **154**, 377–402.
Russell, M. J. and Martin, W.: 2004, *Trends Biochem. Sci.* **29**, 358–363.
Sacerdote, M. G. and Szostak, J. W.: 2005, *Proc. Natl. Acad. Sci. U.S.A.* **102**, 6004–6008.
Santos M. A. S. and Tuite M. F. 2004, in Ll. Ribas de Pouplana (ed.), *The Genetic Code and the Origin of Life*, Landes Bioscience, Georgetown, pp. 183–200.
Schimmel, P., Giege, R., Moras, D. and Yokoyama, S.: 1993, *Proc. Natl. Acad Sci. U. S. A.* **90**, 8763–8768.
Schmidt, J. G., Nielsen, P. E. and Orgel, L. E.: 1997, *Nucleic Acids Res.* **25**, 4797–4802.
Schöning, K.-U., Scholz, P., Guntha, S., Wu, X., Krishnamurthy, R. and Eschenmoser, A.: 2000, *Science* **290**, 1347–1351.
Schopf, J. W.: 1993, *Science* **260**, 640–646.
Schulze-Makuch, D. and Irwin, L. N: 2002, *Astrobiology* **2**, 105–121.

Schwartzman, D. W. and Lineweaver, C. H.: 2004, *Biochem. Soc. Trans.* **32**, 168–171.

Segré, D., Ben-Eli, D., Deamer, D. W. and Lancet, D.: 2001, *Origins Life Evol. Biosphere* **31**, 119–145.

Sievers, A., Beringer, M., Rodnina, M. V. and Wolfenden, R.: 2004, *Proc. Natl. Acad. Sci. U.S.A.* **101**, 7897–7901.

Sonneborn T. M. 1965, in V. Bryson and H. J. Vogel (eds.), *Evolving Genes and Proteins*, Academic Press, New York, pp. 379–397.

Spear, J. R., Walker, J. J., McCollom, T. M. and Pace, N. R.: 2005, *Proc. Natl. Acad. Sci. U.S.A.* **102**, 2555–2560.

Steitz, T. A.: 1999, *J. Biol. Chem.* **274**, 17395–17398.

Stetter, K. O.: 1996, *FEMS Microbiol. Rev.* **18**, 149–158.

Stoks, P. G. and Schwartz, A. W.: 1982, *Geochim. Cosmochim. Acta* **46**, 309–315.

Sutherland, J. D. and Blackburn, J. M.: 1997, *Chem. Biol.* **4**, 481–488.

Sutherland, J. D. and Whitfield, J. N.: 1997, *Tetrahedron* **53**, 11493–11527.

Szathmáry, E.: 1999, *Trends Genet.* **15**, 223–229.

Szathmáry, E. and Maynard Smith, J.: 1997, *J. Theor. Biol.* **187**, 555–571.

Szostak, J. W., Bartel, D. P. and Luisi, P. L.: 2001, *Nature* **409**, 387–390.

Taillades, J., Collet, H., Garrel, L., Beuzelin, I., Boiteau, L., Choukroun, H. and Commeyras, A.: 1999, *J. Mol. Evol.* **48**, 638–645.

Tehei, M., Franzetti, B., Maurel, M.-C., Vergne, J., Hountondji, C. and Zaccaï, G.: 2002, *Extremophiles* **6**, 427–430.

Tian, F., Toon, O. B., Pavlov, A. A. and De Sterck, H.: 2005, *Science* **308**, 1014–1017.

Tobé, S., Heams, T., Vergne, J., Hervé, G. and Maurel, M.-C.: 2005, *Nucleic Acids Res.* **33**, 2557–2564.

Varela, F. G., Maturana, H. R. and Uribe, R.: 1974, *Biosystems* **5**, 187–196.

Vergne J. and Maurel M.-C. 2002, *Origins Life Evol. Biosphere* (abstract) **32**(5–6), 538.

Vergne J., Cognet J. A. H., Szathmáry E., and Maurel M.-C. 2006, *Gene*, **371**, 182–193.

Villarreal L. P. 2005, in L. P. Villarreal (ed.), *Viruses and the Evolution of Life*, ASM Press, Washington.

Wächtershäuser, G.: 1988a, *Proc. Natl. Acad. Sci. U.S.A.* **85**, 1134–1135.

Wächtershäuser, G.: 1988b, *Microbiol. Rev.* **52**, 452–484.

Wächtershäuser, G.: 1990, *Proc. Natl. Acad. Sci. U.S.A.* **87**, 200–204.

Wächtershäuser, G.: 2003, *Mol. Microbiol.* **47**, 13–22.

Weber, A. L.: 2001, *Origins Life Evol. Biosphere* **31**, 71–86.

Weber, A. L.: 2002, *Origins Life Evol. Biosphere* **32**, 333–357.

Weiner, A. M. and Maizels, N.: 1987, *Proc. Natl. Acad. Sci. U.S.A.* **84**, 7383–7387.

Westall, F., De Wit, M. J., Dann, J., Van der Gaast, S., De Ronde, C. and Gerneke, D.: 2001, *Precambrian Res.* **106**, 93–116.

White, H. B.: 1976, *J. Mol. Evol.* **7**, 101–104.

White H. B.: 1982, in J. Everse, B. Anderson and K. You (eds.), *The Pyridine Nucleotide Coenzymes*, Academic Press, New York, pp. 1–17.

Wimberly, B. T., Brodersen, D. E., Clemons, W. M. Jr, Morgan-Warren, R. J., Carter, A. P., Vonrhein, C., Hartsch, T. and Ramakrishnan, V.: 2000, *Nature* **407**, 327–339.

Woese, C. R.: 1965, *Proc. Natl. Acad. Sci. U.S.A.* **54**, 1546–1552.

Woese, C.: 1998, *Proc. Natl. Acad. Sci. U.S.A.* **95**, 6854–6859.

Woese, C. R.: 2000, *Proc. Natl. Acad. Sci. U.S.A.* **97**, 8392–8396.

Woese, C. R. and Fox, G. E.: 1977a, *J. Mol. Evol.* **10**, 1–6.

Woese, C. R. and Fox, G. E.: 1977b, *Proc. Natl. Acad. Sci. U.S.A.* **74**, 5088–5090.

Woese, C. R., Kandler, O. and Wheelis, M. L.: 1990, *Proc. Natl. Acad. Sci. U.S.A.* **87**, 4576–4579.

Wong, J. T.: 1975, *Proc. Natl. Acad. Sci. U.S.A.* **72**, 1909–1912.
Wong, J. T.: 2005, *BioEssays* **27**, 416–425.
Wu, H. L., Bagby, S. and van Elsen, D.: 2005, *J. Mol. Evol.* **61**, 54–64.
Yarus, M.: 1998, *J. Mol. Evol.* **47**, 109–117.
Yarus, M., Caporaso, J. G. and Knight, R.: 2005, *Annu. Rev. Biochem.* **74**, 179–198.
Yusupov, M., Yusupova, G., Baucom, A., Lieberman, K., Earnest, T. N., Cate, J. H. and Noller, H. F.: 2001, *Science* **292**, 883–896.
Zhang, B. and Cech, T. R.: 1997, *Nature* **390**, 96–100.

Earth, Moon, and Planets (2006) 98: 205–245
DOI 10.1007/s11038-006-9090-x

6. Environmental Context

HERVÉ MARTIN
Laboratoire Magmas et Volcans, Université Blaise Pascal, Clermont-Ferrand, France
(E-mail: martin@opgc.univ-bpclermont.fr)

PHILIPPE CLAEYS
DGLG-WE, Vrije Universiteit Brussel, Brussels, Belgium
(E-mail: phelaeys@vub.ac.be)

MURIEL GARGAUD
Observatoire Aquitain des Sciences de l'Univers, Université Bordeaux1, Bordeaux, France
(E-mail: gargaud@obs.u-bordeaux.fr)

DANIELE L. PINTI
GEOTOP-UQAM-McGill, Université du Québec à Montréal, Quebec, Canada
(E-mail: pinti.daniele@uqam.ca)

FRANCK SELSIS
Centre de Recherche Astronomique de Lyon and Ecole Normale Supérieure de Lyon, Lyon, France
(E-mail: franck.selsis@ens-lyon.fr)

(Received 1 February 2006; Accepted 4 April 2006)

Abstract. On Earth, the Archaean aeon lasted from 4.0 to 2.5 Ga; it corresponds to a relatively stable period. Compared with today, internal Earth heat production was several times greater resulting in high geothermal flux that induced the genesis of rocks such as komatiites and TTG suites, which are no more generated on Earth since 2.5 Ga. Similarly, the details of plate tectonic modalities (plate size, plate motion rate, plate thickness, tectonic style, irregular crustal growth, etc...) were different of modern plate tectonics. Both atmosphere and ocean compositions have been progressively modified and the greater heat production favoured the development of hydrothermalism and therefore created niches potentially favourable for the development of some forms of life. Catastrophic events such as giant meteorite falls or world-sized glaciations drastically and suddenly changed the environment of Earth surface, thus being able to strongly affect development of life. Even if specialists still debate about the age of the oldest indubitable fossil trace of life, Archaean can be considered as having been extremely favourable for life development and diversification.

Keywords: Archaean, continental growth, atmosphere and ocean evolution, meteoritic impacts, glaciations, young sun

After the Late Heavy Bombardment (see chapter 4.5), the Earth entered in a relative stability period that lasted until now. This does not mean that at 4.4 Ga, the Earth was similar to modern Earth, on the contrary, it was archaic and the internal as well as external processes were significantly different of those active on our modern Earth. For instance, if plate tectonics existed, the detail of its modality (plate size, plate motion rate, plate thickness, irregular crustal growth, etc...) changed until today. Both atmosphere and ocean compositions have been progressively modified. All these changes, even if slow and progressive certainly had significant influence on the development and evolution of life on Earth surface. In addition to these progressive modifications, more punctual and/or catastrophic events took place, such as giant meteorite falls or world-sized glaciations. These events drastically and suddenly changed the environment of Earth surface, and even if they lasted a short period of time (at geological time scale), they also strongly affected development of life. The target of this chapter is to identify and describe these changes and to try to replace them in their temporal frame.

6.1. Evolution of Geological Mechanisms: The 2.5 Ga Transition

HERVÉ MARTIN

6.1.1. EPISODIC CRUSTAL GROWTH

Since more than 15 years, it has been clearly established that, if continental crust growth has been continuous, it was also a highly irregular process (Condie, 1989; McCulloch and Bennet, 1993; Condie, 1998). Figure 6.1 is based on two complementary approaches: estimate of crustal growth rates (McCulloch and Bennet, 1993) and compilation of crustal zircon ages (Condie, 1989; Condie, 1998). It shows that throughout the whole Earth history, crustal growth proceeded by super-events (i.e. 3.8, 2.7, 1.8, 1.1 and

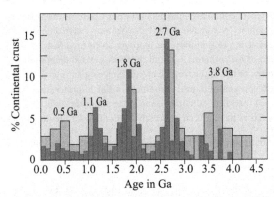

Figure 6.1. Histograms showing the estimated rates of crustal growth averaged over 200 Ma intervals (after McCulloch and Bennet, 1993, orange pattern) and the distribution frequency (100 Ma intervals) of U-Pb zircon ages (after Condie, 1989 and Condie, 1998, red pattern). This diagram clearly shows that the growth of continental crust has been a continuous but irregular process.

0.5 Ga). Typically each event lasted 250 to 350 Ma and is separated from the others by about 700–900 Ma.

Several authors consider that mantle plume activity could be responsible for the periodicity of Earth crust production (Stein and Hofmann, 1994; Albarède, 1998; Albarède, 2005). Albarède (1998) proposed that plume activity resulted in the emplacement of large oceanic plateaus, similar to the Mesozoic Ontong-Java plateau, where thick piles of plume basalts erupted on the oceanic floor. Indeed, when entering in subduction, these plateaus are able to undergo partial melting and to generate continental crust. For instance, today in Ecuador, the Carnegie ridge, which is generated by the Galapagos hot spot activity, is subducted under the South American plate since 5 Ma. There, the volcanic activity is more important than in other parts of Andes, the arc is larger and adakites are generated (Gutscher et al., 2000; Samaniego et al., 2002; Bourdon et al., 2003; Samaniego et al., 2005). This modern analogue testifies that increase of magmatic activity can result of basaltic plateau subduction.

Based on the Stein and Hofmann work (1994), Condie (1998) proposed an alternative model, which considers that in a subduction zone, the descending residual oceanic crust, transformed into eclogite, accumulates at the 660-km seismic discontinuity. When the amount of accumulated oceanic crust exceeds a threshold, it breaks the discontinuity and suddenly sinks into the mantle as a cold avalanche. These cold eclogites reach the mantle-core boundary (D″ layer) and take the place of the autochthonous warm mantle peridotite resulting in the uprising of a hot mantle plume. This catastrophic event can be considered as being a mantle and thermal overturn. As proposed by Albarède (1998) these uprising hot plumes can generate oceanic plateaus. Condie (1998) proposed that ascent of hot mantle mostly contributes to heat the upper mantle, which increases the rate of oceanic crust genesis, resulting in smaller and faster plates that enter more rapidly in subduction. Faster subduction of younger oceanic crust also results in increasing the rate of production continental crust (Martin and Moyen, 2002; Martin et al., 2005).

The downward motion of cold avalanches drags the overlying mantle, thus causing the development of "super subduction zones" above, which attract plates from great distances (Peltier et al., 1997; Condie, 1998). The convergence of the plates results in the growth of super continent. In fact, the development of super continents in these periods is due to two concomitant mechanisms: the collage and/or collision of pre-existing continental plates and the genesis of new juvenile continental crust due to subduction-related magmatic activity. Some of these supercontinents received names: Rodinia (1.1 Ga); Gondwana (0.5 Ga). Vaalbara has also been proposed for late Archaean (see: Condie, 1998; Zegers et al., 1998; Bourrouilh, 2001, for more details).

Moyen (1997) also evidenced the cyclic continent growth and evolution, which he divided into 4 stages. The first stage consists in arc complexes, where continental crust is generated by melting of subducted slab (TTG and adakites) or of overlying mantle wedge (classic calc-alkaline magmas). During a second episode, arcs accrete due to collision or collage; this can be correlated to the cold avalanche as described by Condie (1998). The third stage consists in the reworking of the newly accreted crust. Indeed, genesis of juvenile continental crust also results in transferring and concentrating strongly incompatible elements such as U, Th and K from the mantle into the crust. These elements have radiogenic isotopes (^{235}U, ^{238}U, ^{232}Th, ^{40}K), whose decay contributes to heat the crust and thus, can help in its melting and recycling. The last episode is a quiescence period (may be partly due to the temporal impoverishment of the upper mantle in radioactive elements); which can evolve towards a new cycle or towards cratonization and development of thick and cold lithospheric keel.

Existence of supercontinents has major influence not only in magmatic or deep geology, but also on climate and life. For instance, on a supercontinent fauna and flora will tend to homogenize and to follow a parallel evolution. After breakdown of such a continent, living being will evolve independently on their continental fragment.

6.1.2. Global evolution and changes (~2.5 Ga ago)

Taking into account the episodic and cyclic crustal growth could lead to the conclusion that the parameters of this cyclicity remained constant since the early Archaean. However, a rapid overlook of crustal composition through time points to major disparity between Early Archaean and Phanerozoic times. One of the more obvious differences consists in the temporal change in the nature and abundance of rocks. Some rocks are widespread in Archaean terrains whereas they are rare or inexistent after 2.5 Ga. This is the case for komatiites, TTG (Tonalite, Trondhjemite and Granodiorite) associations and Banded Iron Formations (BIF). Others, such as andesites, per-alkaline magmatic rocks and eclogites are abundant after 2.5 Ga and rare or unknown in Archaean terrains. In addition, others as high-Mg granodiorites (sanukitoids) are mainly known at the Archaean-Proterozoic boundary. Consequently, in superposition with cyclic crustal growth it exists an other secular evolution.

6.1.2.1. *Change in magma petrogenesis*
Typically, the main secular change in our planet history since its accretion is its progressive cooling. The more obvious evidence of this cooling is provided by komatiites (Figure 6.2). Komatiites are ultramafic lavas, which only exist in Archaean terrains. They were produced by high degrees of mantle melting ($\geqslant 50\%$) and they emplaced at temperatures ranging from 1525 to 1650 °C (Nisbet et al., 1993; Svetov et al., 2001) (today basalts are produced by 25 to 35% mantle melting at temperatures of about 1250–1350 °C). They

1 cm

Figure 6.2. Right: 3.445 Ga old komatiitic pillow lava from the Hooggenoeg formation in Barberton greenstone belt (South Africa); Left: Typical spinifex texture in a 3.2 Ga old komatiite flow from Umburanas greenstone belt (Bahia state, Brazil). (Photos H. Martin).

corroborate that during the first half of Earth history, upper mantle temperature was greater than today. Since its formation, Earth cools such that after 2.5 Ga, it was not anymore able to reach high temperatures and consequently it became unable to produce high degrees of mantle melting, thus accounting for the disappearance of komatiites after Archaean times.

When Earth accreted, it accumulated energy, such as residual accretion heat, heat release by exothermic core-mantle differentiation, radioactive element (^{235}U, ^{238}U, ^{232}Th, ^{40}K) disintegration heat, etc. Since 4.55 Ga, this potential energetic stock is gradually consumed and consequently Earth progressively cools (Figure 6.3, Brown, 1985).

Archaean TTGs lead to the same conclusion. Indeed TTGs are generated by hydrous basalt melting, very probably in a subduction–like environment, where, due to high geothermal gradients, the subducted oceanic crust melts instead of dehydrating as it is today (Martin, 1986; Martin and Moyen, 2002). After Archaean times, dehydration of slab resulted in calc-alkaline andesites whereas, in mantle plume environments, per-alkaline magmas were generated by low degrees of mantle melting. Different estimates consider that Early Archaean mantle temperature was 100 to 200 °C greater than today.

6.1.2.2. *Sanukitoids*

At the Archaean Proterozoic (~2.5 Ga) boundary, huge volumes of high-Mg magmas emplaced that were called sanukitoids by Shirey and Hanson (1984). They possess petrologic and chemical characteristics intermediate between Archaean TTGs and modern calc-alkaline andesites (Figure 6.4). Recent works demonstrated that sanukitoids formed by partial melting of a mantle

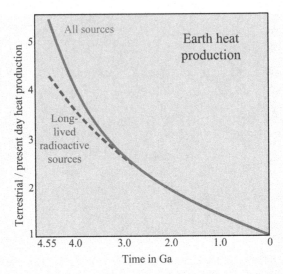

Figure 6.3. Temporal change of Earth heat production (after Brown, 1985).

Figure 6.4. 2.7 Ga old sanukitoids from Eastern Finland. Left: dyke of sanukitoid intrusive into the basic volcanics of the Kuhmo–Suomussalmi greenstone belt. Right: detail picture of sanukitoid, attesting of the abundance of feldspar megacrysts (white) in a matrix mainly made up of dark minerals (biotite and hornblende). (Photos H. Martin).

peridotite which composition has been modified by addition (metasomatism) of TTG magma (Smithies, 2000; Moyen et al., 2003; Rapp et al., 2003; Martin et al., 2005). In a subduction zone, the slab melts (TTG magma) have a low density and ascent through the overlying mantle wedge where they react with mantle peridotite. Rapp et al. (1999) established the concept of "effective melt/rock ratio": when this ratio is high, not all slab-melt is consumed during peridotite metasomatism and so, part of slab melt (TTG) reaches the continental crust. When the melt/rock ratio is low, all the slab

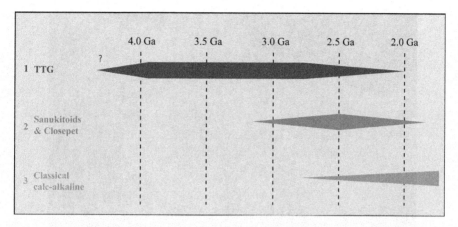

Figure 6.5. Schematic diagram illustrating the evolution of juvenile crustal magmatism in course of Earth history. The thickness of the colured domain is indicative of the volumetric importance of the magmatism. It clearly points to large domain of overlapping at the Archaean-Proterozoic transition (2.5 Ga).

melt is consumed in metasomatic reaction with the peridotite. Martin et al. (2005) proposed that sanukitoids represent transitional magmatism reflecting a major change in Earth thermal regime at about 2.5 Ga (Figure 6.5):

(1)   During Archaean times, Earth heat production was high leading to high degree of subducted slab melting. Consequently, during their ascent through the mantle wedge the melt/rock ratio was high such that part of slab melts reached the crust where they crystallized as TTG.

(2)   At the end of Archaean, Earth was significantly colder and consequently the degree of melting of subducted slab was low, resulting in a low melt/rock ratio. Therefore, the whole slab melts were consumed in reaction with mantle wedge, whose subsequent melting generated sanukitoids.

(3)   After 2.5 Ga, Earth was too cold to allow slab melting, so the subducted basalts cannot melt but only dehydrate and induce the melting of the overlying mantle peridotites, giving rise to the classical cal-alkaline magmas (andesites and granodiorites).

In other words, the sanukitoids are good markers of a major petrogenetic and thermal change in our planet. The fact that they mostly emplaced between 2.7 and 2.5 Ga allows to roughly dates this event.

6.1.2.3. *Change in tectonic style*
Changes did not occurred only in magma petrogenetic processes but also in global tectonic style. During Archaean times, modern-like horizontal

Figure 6.6. Satellite view of the 3.5–3.2 Ga old Archaean Pilbara block in Australia, showing the sagduction structures: the greenstone belts (dark green colour) are located in between TTG domes (white-yellow colours). The picture width is of about 450 km. (Photo Landsat).

tectonics operated and the large-scale horizontal structures (over-thrusts) resulted of continental collision events. However, vertical tectonics also appears in Archaean terrains whereas it is almost inexistent after 2.5 Ga. This tectonic style has been called sagduction by Gorman et al. (1978); it is a gravity driven tectonic, analogous to inverse diapirism (Figure 6.6). Indeed, when high density ($d = 3.3$) ultramafic rocks such as komatiites or even some iron-bearing sediments as BIF are emplaced over a low density ($d = 2.7$) continental crust made up of TTG, they generate a strong inverse density gradient leading to the down motion of high density greenstones into the TTG basement but also to the concomitant ascent of the surrounding low density TTG.

On Earth, the only high-density rocks able to emplace at the surface are komatiites and BIF. Today, most basalts have density that does not exceed 2.9 or 3.0. The resulting inverse density gradient is too low to initiate sagduction. As komatiites and BIF are restricted to Archaean times, vertical tectonics (sagduction) is also restricted to the primitive Earth crustal evolution.

6.1.2.4. *Juvenile vs. recycling*
Melting of the mantle or of a subducted oceanic crust generates felsic magmas that are extracted from the mantle; they contribute to increase the volume of the continental crust. This process is called a *juvenile* process. On the contrary, re-melting of older continental crust, generates new crustal magmas, but does not increase the continental crust volume. This mechanism is called *recycling*.

However, before to be able to recycle continental crust, it is necessary to create it. This is why, during Archaean times, juvenile processes were prominent and recycling was a subordinated process. Contrarily, after 2.5 Ga, the volume of continental crust was so great that recycling became prominent.

6.1.3. CONCLUSION

It appears that a very important period in Earth history has been the Archaean - Proterozoic boundary at about 2.7–2.5 Ga. It marks the change from archaic mechanisms before 2.5 Ga towards modern-like ones after. A single and simple cause can account for all the observed changes: the progressive cooling of our planet. However, the recorded change is relatively sudden and brutal whereas Earth cooling is a progressive mechanism. Consequently, we must invoke some threshold effect. This effect is well known in the case of felsic magmas (TTGs vs. calc-alkaline andesite). Indeed, a hydrous subducted oceanic slab melts at temperatures of about 700 °C; if it is dehydrated, it would only be able to melt for temperature greater than 1200 °C. Consequently, the threshold will be the geothermal gradient that will allow dehydration to occur for temperatures lower than 700 °C. Before 2.5 Ga, high geothermal gradients did not allow dehydration to occur prior to 700 °C, whereas today, lower geothermal gradients result in intense slab dehydration at temperatures lower than 700 °C.

The quick character of the change could have also been accentuated and reinforced by the genesis of a supercontinent at about 2.7 Ga. Figure 6.1 shows that this event was probably the more important ever recorder on our planet. Such a peak in crustal growth is accompanied by a huge transfer of heat producing radioactive elements from the mantle to the crust. One of the consequences is the warming of the crust leading to its intense recycling. However, another consequence is the important impoverishment of the upper mantle in these elements, which made it colder and significantly contributed in decreasing geothermal gradients, which could have accelerated the rapid character of the change.

Of course, these changes did not take place everywhere on Earth exactly at the same time. Indeed, for instance, today in subduction zones the geothermal gradient depends not only of the mantle temperature but also of the age of the subducted crust, the age of subduction, etc... (Martin, 1999). The value of these two last parameters can be highly variable, such that, for example, the age of oceanic crust when it enters in subduction ranges between 0 and 180 Ma. Consequently, locally calc-alkaline andesites were generated before 2.5 Ga whereas some TTG are known in Proterozoic terrains. However, it remains that most of the changes occurred after the 2.7 Ga supercontinent genesis, at about 2.5 ± 0.2 Ga.

6.2. Atmosphere and Ocean Physico-chemical Evolution

The presence of an atmosphere-hydrosphere early in the history of the Earth
has been essential to life, mainly supporting synthesis of organic molecules,
providing suitable ecologic niches (Nisbet and Sleep, 2001; Nisbet and
Fowler, 2003), and regulating the climate (Sleep and Zahnle, 2001a; Zahnle
and Sleep, 2002). Here, we report most recent hypothesis and the few direct
and indirect evidences for the evolution of the atmosphere and hydrosphere
during the Hadean and the Archaean eons. With the increasing geological
record available after the 2.5 Ga transition, a description of the chemical
evolution of the ocean and the atmosphere during Proterozoic and Phanerozoic times is a daunting work, well beyond the scope of this paper. A brief
and updated account of the entire geological history of seawater and of the
atmosphere can be found in (Holland, 2003).

6.2.1. ATMOSPHERE EVOLUTION

FRANCK SELSIS

6.2.1.1. *The prebiotic atmosphere*
By prebiotic, we refer to the era when life emerged on Earth. In order not to
eliminate the possibility of interplanetary or interstellar panspermia, we
consider here that life started to evolve on Earth during the prebiotic era,
whether the origin of life origin was endogenous or extraterrestrial. A "reference" composition of the prebiotic atmosphere is given by Kasting (1993).
This model describes the atmosphere around 3.8 Ga, a period closely
bounded by the Late Heavy Bombardment (3.9–3.8 Ga) and the oldest
plausible isotopic signatures of life (3.8 Ga) (Rosing, 1999). It is based on the
following assumption:
- The composition of volcanic gases is roughly the same as it is today.
  This is supported by the composition of the oldest (~3.9 Ga) magmatic
  rocks (ultramafic lavas?) (Delano, 2001; Li and Lee, 2004).
- Molecular nitrogen has reached its present atmospheric level (PAL).
  This is consistent with the outgassing timing inferred from isotopes, with
  a ~30% uncertainty on the level of N_2 due to recycling between the
  atmosphere, the crust and the mantle (see part 4.3.1.4.).
- Carbonate-silicate regulation of the mean surface temperature (T_S) is at
  work (Walker et al., 1981). This requires the weathering of emerged
  continents and imposes a CO_2 level of about 0.2 bars. This point was
  debated by Sleep and Zahnle (2001b) who suggested that such high level
  of atmospheric CO_2 would be unstable due to the carbonitization of

basalts, which would maintain the early Earth in a globally frozen state. Efficient carbonitization of seafloor seems however inefficient when occurring below a km-thick frozen ocean. As carbonate-silicate recycling was shown to be the mechanism allowing the Earth to recover from Snowball events (Hoffman et al., 1998), it is reasonable to assume that it was also a mechanism able to overcome carbonitization and to sustain T_S above 0 °C.

The abundance of molecular hydrogen is obtained by balancing volcanic outgassing and Jeans escape of atomic H to space. Assuming a diffusion-limited loss to space, as in Kasting's original work, yields a H_2 mixing ratio of 10^{-3}. However, escape of atomic hydrogen in the modern oxic atmosphere is only limited by its diffusion above the tropopause because the exosphere is sufficiently hot (~1000 K). The high temperature of Earth's exosphere is due to its high O_2 and low CO_2 contents (the exosphere of Venus, for instance, remains below 278 K). The prebiotic atmosphere was anoxic and the low exospheric temperature, rather than diffusion, limited the loss of hydrogen. Tian et al. (2005) estimated the hydrogen loss and showed that a volcanic rate equal to the present one would result in an H_2 mixing ratio 100 times higher (0.1 bar). For higher volcanic rates (which are likely), molecular hydrogen can become the dominant species in Earth prebiotic atmosphere.

In the prebiotic atmosphere, O_2 is produced only by photochemistry and escape of hydrogen following H_2O photolysis, and maintained to a very low level ($<10^{-10}$) by reaction with reducing volcanic gases.

The abundance of methane depends on the efficiency of its abiotic production, which is assumed to originate from reducing hydrothermal fluids produced after serpentinization of ultramafic rocks. Today's abiotic production (mainly at off-axis mid-ocean ridges) released in an anoxic atmosphere would lead to mixing ratio below 10^{-6}. As a more extensive hydrothermal activity was likely at 3.8 Ga and because meteoritic impacts also contributed to the methane production (Kress and McKay, 2004), significantly higher levels of CH_4 (10^{-5}–10^{-4}) could be sustained (Kasting, 2005).

The composition in other gases is much more speculative: CO is often pointed as a major prebiotic gas because of its formation during impact and its abundance in the protosolar nebula. However, no quantitative study on the sources and sinks of CO are available. Nitrogen-bearing prebiotic compounds like HCN and NO can be produced by lightning and impacts at a rate critically depending on the level of the other gases, among which H_2, CO_2 and CH_4 (Navarro-Gonzalez et al., 2001; Commeyras et al., 2004).

This model likely represents average conditions on Earth at about 3.8 Ga if life was absent or only marginally involved in the global geochemical cycles. However, recent revisions of the impact history question this dating of the prebiotic era at about 3.8 Ga. Earth became habitable after the last impact able to vaporize the whole ocean hit the Earth.[1] By triggering a magma ocean phase, such impact would not only annihilate the biosphere but also reset to simple molecules the complex organics that could have been formed before. By considering a monotonic decrease of the bombardment between the moon-forming impact and the dated lunar impacts, several authors estimated the rate of such catastrophes and the horizon for ancient life on Earth (Maher and Stevenson, 1988; Oberbeck and Fogelman, 1989; Sleep et al., 2001). They concluded that long lasting habitability started on Earth around 4.0 ± 0.1 Ga. As described in part 4.5, such monotonic decrease is not consistent with solar system formation models and the bombardment rate could have been comparable to the present one between the end of Earth accretion (~4.4 Ga) and the LHB (3.9 Ga; part 4.5). During the LHB itself, the Earth was bombarded by an integrated mass of asteroids lower than 1.8×10^{23}g (Gomes et al., 2005), which does not imply the occurrence of 500 km impactors (3×10^{12} Megatons) required to evaporate the entire oceans or 200 km bodies (10^{11} Megatons) able to warm it up to 100 °C. Therefore, and in the absence of constraints on the bombardment before 3.9 Ga, life could have emerged significantly earlier than previous hypotheses, as early as about 4.4 Ga. One of the main implications is that the prebiotic atmosphere may have been much more reducing that usually assumed. The composition of the atmosphere later than 4 Ga was the result of more than 500 Ma of irreversible oxidation through the escape of hydrogen to space, under the strong XUV irradiation of the early Sun (Selsis, 2004; Ribas et al., 2005), although hydrogen could still be abundant (Tian et al., 2005). H_2 and other reducing species including CH_4 could have dominated the very early atmosphere ~ 4.4 Ga ago (Sleep et al., 2004). It is important to insist that the uncertainty on the dating of the prebiotic atmosphere as well as on the dating of the oldest traces of life (see part 7.1) allows us to consider an origin of terrestrial life under a wide range of atmospheric composition: from H_2-, CH_4-rich to H_2, CH_4-poor.

6.2.1.2. The rise of O_2

Between 2.4 and 2.0 Ga, a change from global anoxic to oxic atmospheric conditions, the so-called "Great Oxidation Event" or G.O.E. (Holland, 2002), is revealed by several geological evidences (BIFs, red beds, palaeosols, detrital uraninite deposits) (Holland, 1994; Figure 6.7). It can now be dated with high precision thanks to sulphur isotopic enrichments recorded in rocks: Sulphur isotopes follow a "normal" mass-dependent fractionation for high

---

[1] This assumption is no longer true if life can survive such impact inside ejecta escaping the Earth and reseeding it after its recovery (Wells et al., 2003; Gladman et al., 2005).

levels of O_2 and "mass-independent" fractionation when this level is low. Mass-independent fractionation arises from the photolysis of SO_2 (Farquhar et al., 2000) in the absence of UV screening by O_3 and occurs at levels of O_2 below 10^{-5} PAL (Present Atmospheric Level) according to photochemical and radiative transfer modelling (Pavlov and Kasting, 2002). Mass-independent fractionation is found in all sediments between 3.8 and 2.32 Ga (Mojzsis et al., 2003). The partial pressure of O_2 increased from undetectable levels to more than 10^{-5} PAL at 2.32 Ga (Bekker et al., 2004). Palaeosols more recent than 2.2 Ga underwent oxidation at a level implying a minimum partial pressure of O_2 of 1-10 mbar, or 0.5–5% PAL (Rye and Holland, 1998). Although the quantitative inference of the partial pressure of O_2 from the composition of these palaeosols remains uncertain, palaeosols from 2.2–2.0 Ga seem to indicate a higher O_2 level (>5% PAL) than palaeosols from 1.8 to 0.8 Ga. One should keep in mind that these geological indicators may reveal local more than global conditions and can thus been affected by, for instance, locally enhanced biological release of O_2. Moreover, the average atmospheric level of O_2 is likely to have fluctuated due to complex biological, photochemical and climatic relationship between O_2, CO_2 and CH_4. Maybe more significant is the worldwide excursion of the $\delta^{13}C$ recorded between 2.25 and 2.05 Ga, which corresponds to the release of 10–20 times the present amount of atmospheric O_2 (Karhu and Holland, 1996). The existence of BIFs (Banded-Iron Formations; requiring reducing conditions) as late as 1.8 Ga, indicate that the deepest part of the ocean became oxic only 500 Ma

Figure 6.7. Banded Iron Formations (BIFs) from Sandur in India (Left) and from Kuhmo in Finland (Right); both are ~2.7 Ga old. The detail view on the right shows the alternations of quartz (white) and magnetite layers (black dark blue). (Photos H. Martin).

after the atmosphere (Holland, 1999). In fact, Anbar and Knoll (2002) showed that anoxic but sulphidic conditions in the deep ocean might still have prevailed after 1.8 Ga and as late as 0.6 Ga. Sulphur and carbon isotopes seem to indicate a second rise of O_2, from ~0.5% PAL to more than 5–20% PAL around 0.7 ± 0.1 Ga ago (Canfield and Teske, 1996; Catling and Claire, 2005). The present level of O_2 was at least reached before the explosion of complex life took place (543 Ma) and probably around 0.6 Ga. The partial pressure of O_2 remained above 50% PAL during the entire Phanerozoic with a notable excursion to 150–200% PAL (350–400 mbar) around 300 Ma (Permo-Carboniferous), associated with the rise of vascular land plants and the associated increase of carbon burial (Berner et al., 2003).

The question whether the appearance of oxygen producers coincided with the beginning of the great oxidation event or occurred much earlier is still open-ended. Molecular fossils of biogenic origins (hopanes and steranes) are a plausible indication for the existence of cyanobacteria and eukaryotes (the producer and consumer of O_2) at 2.7 Ga (Brocks et al., 2003b). Stromatolites built by phototrophs became abundant at about 2.8 Ga and exhibit similarities with modern stromatolites associated with cyanobacterial mats. Some of these stromatolites could result of non-oxygenic bacterial activity but Buick (1992) showed that at least some of them were not found in conditions allowing non-oxygenic photosynthesis. Isotopes of iron (Rouxel et al., 2005) and nitrogen (Shen et al., 2006) also exhibit changes at 2.7–2.8 Ga, possibly indicating that oxygenation of the Earth was taking place at that time. Local oxidizing conditions found in 3.7 Ga sediments (Rosing and Frei, 2004) can be the result of an early biological oxygen release. If O_2 producers did appear as early as before 2.7 or even before 3.7 Ga, what is the cause of the delayed rise of atmospheric oxygen? Several mechanisms have been proposed:

(1)   The net release of atmospheric O_2 by photosynthesis is limited by the burial of organic carbon. Organic carbon that remains at the surface can eventually be oxidized, whether consumed by eukaryotes or directly exposed to atmospheric oxidants. It is thus possible that the rate of carbon burial, and thus the net release of O_2, has been kept at low values by early tectonics and then enhanced as result of the break-up of the Late Archaean Supercontinent after 2.3 Ga (Des Marais et al., 1992).

(2)   The build-up of O_2 can also be frustrated by the release of reducing volcanic gases when it exceeds the biological production of O_2. Kasting et al. (1993) proposed that, due to a progressive oxidation of the mantle, the balance might have favoured the accumulation of O_2 only after 2.3 Ga. However, the mantle redox required to produce such reducing volcanic gases is not consistent with the timing for the

degassing of nitrogen (Libourel et al., 2003; Marty and Dauphas, 2003) and no redox evolution is observed in lavas formed between 3.6 Ga and today (Delano, 2001; Li and Lee, 2004).

(3)   Unstable climate due to coexistence of widespread methanogens and emerging O_2 producers could have prevented the build-up of O_2 until solar luminosity increased to a high-enough value (Selsis, 2002).

(4)   Catling et al. (2001) suggested that methane-producing bacteria must have preceded the rise of O_2 by several hundred million years in order to enhance the escape of hydrogen to space and, consequently, the required irreversible oxidation of the Earth. However, considering the large hydrogen losses that occurred in the earliest history of the Earth (when the H_2-rich atmosphere was submitted to the strong solar XUV radiation, see Figure 6.8), it is unlikely that the hydrogen loss that occurred in the latest period predating the rise of O_2 contributed significantly to the total loss integrated since the formation of the Earth.

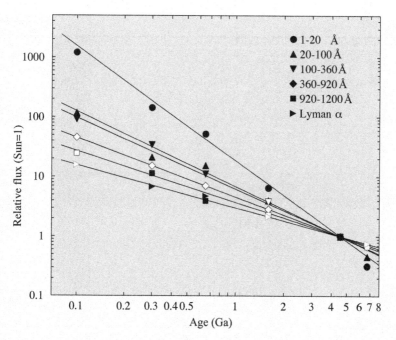

Figure 6.8. XUV emission from Sun-like stars at different ages. This graph represents the Solar-normalized fluxes measured in different wavelength intervals (filled symbols) from 6 Solar-type stars, EK Dra (0.10 Ga), π1 UMa and χ1 Ori (0.30 Ga), κ1 Cet (0.65 Ga), β Com (1.6 Ga), β Hyi (6.7 Ga), and the Sun (4.56 Ga). The corresponding power-law fits are indicated. Empty symbols give the inferred flux where no observation is available. Adapted from Ribas et al. (2005).

According to Kopp et al. (2005), existing geological data and molecular fossils are still consistent with a late emergence of oxygen producers, around 2.3 Ga, that immediately triggered the great oxidation event, while for Melezhik et al. (2005) and Shen et al. (2006), oxygen-rich habitats associated with cyanobacteria undoubtedly existed before 2.7 Ga. The fascinating debate about the origins of oxygenic photosynthesis and its consequences on the whole planetary environment and on the biological evolution, remains extremely active. On this topic, we recommend two inspiring essays (Knoll, 2003; Catling and Claire, 2005).

6.2.1.3. *Evolution of the Sun and climate*
Standard models for the evolution of the Sun give a 37% increase of the luminosity between 4 Ga and present (Figure 6.9). If during this whole period, the Earth had the same atmosphere as today, (an obviously weak assumption) its surface would have remained globally frozen until about 2 Ga. Inversely, geological records prior to 2 Ga indicate that the Earth was warm before 2 Ga, with nearly no traces of glacial deposits except for the

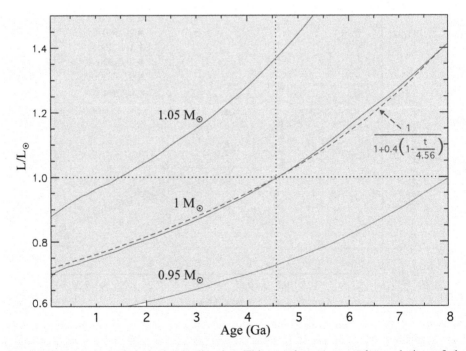

Figure 6.9. Evolution of the solar luminosity. This graph represents the evolution of the bolometric luminosity for 3 different stellar masses and solar metallicity (BCAH98-model-3 from Baraffe et al. 1998). The dashed curve shows a useful approximation for the solar evolution (Gough, 1981).

Pongola (2.9 Ga) (Nhleko, 2004) and Huronian polar glaciations (3 events between 2.45 Ga and 2.32 Ga) (Hilburn et al., 2005) and the Makganyene low-latitude glaciation (Snowball, 2.2 Ga) (Kopp et al., 2005, see part 6.3.2). This apparent contradiction, called the *faint young Sun problem* (Sagan and Mullen, 1972), is no longer a paradox once known that the atmospheric level of greenhouse gases (mainly CO_2) changed during Earth history (see Selsis, 2004, for alternative explanations of the faint young Sun problem). In particular, the atmospheric level of CO_2 is likely to be stabilized over geological periods by the carbonate precipitation (associated with the weathering of silicates) on one hand and the volcanic emission on the other (Walker et al., 1981). This climate regulation, which is assumed to work for Earth-like planets inside the circumstellar habitable zone (Kasting, 1993), can account for most of Earth's climate history. However, it could have been rendered inoperative in two ways. First, under very early tectonics, carbonate-silicate regulation may fail: efficient carbonitization of seafloor (Sleep and Zahnle, 2001b) could lead to surface temperatures below 0 °C while the lack of emerged continents to provide silicate weathering could lead to a build-up of CO_2 and surface temperatures well above 0 °C. Second, the release and consumption of atmospheric gases by the biological activity can overrule the carbonate-silicate regulation. In particular, methanogenesis, known as a very primitive metabolism, could have sustained a warm climate before the rise of oxygen (Pavlov et al., 2000; Kasting, 2005). In the absence of atmospheric O_2, the present production rate of CH_4 by methanogens would result in 100 to 1000 times the present atmospheric abundance of methane (1 PAL $= 1.6 \times 10^{-6}$). To illustrate the efficiency of CH_4 as a greenhouse gas, let us consider that the atmosphere of the Earth at 2.8 Ga had the same composition as today, except for O_2, assumed to be absent (for practical reasons, we assume a 1 bar partial pressure of N_2 to compensate for the lacking 0.2 bars of O_2). The present level of CO_2 (3×10^{-4}) would be too low to sustain a temperature above 0 °C: even with 1 PAL of CH_4, the average surface temperature would be around −10 °C. If we keep the CO_2 level to 1 PAL and we increase the CH_4 to 10, 100 and 1000 PAL, the surface temperature would rise to 0 °C, 5 °C and 15−20 °C, respectively (Pavlov et al., 2000). Moreover, under anoxic conditions, methanogens can colonize most of the habitable environment, while on modern Earth they can grow only under anaerobic conditions, like the ones found in swamps, in hydrothermal systems and in the digestive system of mammals. Even higher levels of atmospheric methane, and higher surface temperatures, are thus likely to have occurred between the emergence of methanogens (at an unknown date) and the great oxidation event (2.3 Ga), although there is no geological evidence to support this hypothesis. At very high levels of CH_4, 1) photochemical hazes start to form like in Titan atmosphere, resulting in an enhanced albedo and 2) CH_4 begins to absorb a significant part of the visible and near infrared solar radiation, producing a stratospheric warming that reduces the surface warming (again like on Titan). For a given level of CO_2, there is thus a limit of the surface temperature that can be reached under a CH_4-rich atmosphere. This limit is however not well determined due to incomplete spectroscopic data of CH_4 and uncertain properties of these hazes, but it could certainly be as high as 30 °C for 1 PAL of CO_2 at 2.8 Ga.

If the emergence of O_2 producers occurred when CH_4 was the main greenhouse gas, climatic consequences would be dramatic: the photochemical lifetime of atmospheric CH_4 drops, anoxic conditions required by methanogens became scarcer yielding a decrease of the production rate of CH_4, and CO_2 consumption increases due to the enhanced biological productivity. Therefore, the beginning of O_2 production is likely to trigger a glaciation or even a snowball[2] event. Under snowball conditions, oxygenic photosynthesis "freezes" while CH_4 and CO_2 can rise again. Under low solar irradiation, such configuration (competing CH_4- and O_2-producers) can induce a steady state characterized by a temperature slightly above 0 °C and a very low abundance of O_2 ($<10^{-5}$). Such steady state is however unstable and can result in repeated glaciations (Selsis, 2000). The collapse of a CH_4-warming is the favoured explanation for the earliest identified snowball (Makganyene glaciation, 2.3–2.2 Ga) (Kopp et al., 2005). The three Huronian glaciations (probably restricted to low-latitudes, Hilburn et al., 2005), which shortly predated the Makganyene event, and the Pongola Glaciation (2.9 Ga) could be caused by such instabilities if oxygen producers were already widespread.

After the great oxidation event, atmospheric levels of methane were limited by the growing oxic environment and by the short photochemical lifetime of CH_4 but could have remained a major greenhouse gas, with a level exceeding 10 PAL (Pavlov et al., 2003). Therefore, the Neoproterozoic Snowball, associated with a second rise of O_2 could have also been triggered by the fall of the CH_4 abundance down to about its present level.

---

[2] Snowball glaciations, characterized by an ice cover down to the equator, are the coldest times recorded in Earth's history. At the beginning of these events, the runaway ice-albedo feedback makes the global mean temperature drop to −50 °C for a few tens of thousands of years (Schrag et al. 2002). This temperature drop is followed by a period of a few million years during which the mean temperature is around −10 °C. Earth recovers from Snowball events thanks to the carbonate-silicate regulation (Hoffman et al., 1998): volcanic gases released in the absence of surface liquid water and carbonate precipitation, allow CO_2 to build-up to a level high enough to warm the surface above 0 °C.

6.2.2. CHEMICAL EVOLUTION OF THE OCEANS

DANIELE L. PINTI

6.2.2.1. *Hadean (4.6–3.8 Ga)*
The chemistry of seawater in the early Hadean was likely controlled by high-temperature water–rock interactions between the hot CO_2–H_2O runaway greenhouse and the basaltic proto-crust (see part 4.2.2.2). Modern analogues of these environments could be mid-oceanic ridges. At mid-oceanic ridges, due to water circulation, rocks partially dissolve thus saturating water in the major constituents of the rock. Similarly, in the early Earth, magma eruption repeatedly brought basalts into contact with the atmospheric water vapour, maintaining its saturation in major elements (Sleep et al., 2001). Today, similar atmosphere-rock reactions are supposed to take place at the surface of Venus (Johnson and Fegley, 2002). When the runaway greenhouse cooled down, possibly before $t_0 + 165$ Ma (see part 4.2.2), condensed seawater started to penetrate deeper in the oceanic basaltic proto-crust, experiencing extensive water–rock interactions. A significant difference with modern seawater is that CO_2 was an abundant volatile in the early runaway atmosphere making early oceans more acidic than today (pH = 5.5, Pinti, 2005). Attack of such solutions on exposed rock was thus more rapid, enriching the solution of dissolved salts (Krauskopf and Bird, 1995).

The first volatile to condense was chlorine, likely outgassed from the Earth's interior in the form of HCl (Holland, 1984; Graedel and Keene, 1996). The amount of chlorine available at the beginning is difficult to evaluate. Holland, (1984) considered that H_2O and Cl were both degassed from the interior of the Earth. Using the Cl/H_2O ratio in the crust, Holland, (1984) calculated an amount of 24.8 g/Kg_{water}, which represents 1.2 times the present value (18.8 g/Kg_{water}). This figure is probably a minimum estimate, and it is expected that higher concentrations of chlorine were available in the primordial oceans. Amounts of chlorine 2 times the present values have been argued by Knauth (1998, 2005), based on mass balance calculations including chlorine from evaporite deposits of marine origin, recycled and deposited on the continental crust during Proterozoic and Phanerozoic times.

The amount of available Na in early seawater was by far enough to combine with Cl to form NaCl. Sodium was removed from the basaltic proto-crust, through pervasively hydrothermal-driven water–rock interactions. As an example, a global layer of basalt of 500-meter thick could supply the total amount of Na actually present in the modern ocean (Sleep and McClure, 2001; Sleep et al., 2001). Continuous genesis and recycling of cooled lava at the surface of the planet could have thus supplied the amount of sodium required for saturating the primordial ocean with a NaCl-dominated brine. Depending on the pressure, solid NaCl may have been in equilibrium with a water–rich high-pressure gas (representing the proto-atmosphere, Pinti, 2005). Temperature and pressure conditions necessary for the transition from two fluids to one fluid plus solid NaCl were 407 °C, at 298 bars (Bischoff and Rosenbauer, 1988).

Other abundant cations were likely K^+, Ca^{2+} and Mg^{2+}. Particularly, Ca^{2+} and Mg^{2+} should have been more concentrated than today due to the higher pH of the primitive oceans, since at low pH they exist together with HCO_3^-. The Ca^{2+}/Mg^{2+} ratio could have been higher than today because it was likely controlled by the reactions with basalts at mid-ocean ridges (Bischoff and Dickson, 1975). Sulphate ion would have been scarce or absent, because sulphur was mainly under its reduced form (Krauskopf and Bird, 1995). Rarer metals whose concentrations at present are kept low by reactions with and such as Ba^{2+}, Sr^{2+} and Mn^{2+} likely played a more important role in Hadean seas (Krauskopf and Bird, 1995; Holland, 2002; Holland, 2003). Halogens could have been also abundant, mainly bromine and iodine, because the absence of a biosphere able to scavenge them from seawater and fix on sediments, as occurs today (Krauskopf and Bird, 1995; Channer et al., 1997). The lower rates of sedimentation expected for a Hadean planet dominated by oceans, with small continental plates and very small or no emerged continent, should have favoured the concentration of halogens in seawater. Finally, in a poorly oxygenated and acidic early ocean, iron was very abundant as soluble ferrous iron (Fe^{2+}) (Holland, 2003, and references therein).

During Hadean, high salinities, together with high temperatures (Knauth and Lowe, 2003), could have limited the hypothetical biological activity in the sea (Knauth, 2005). Complex organic molecules are indeed vulnerable to damage caused by sodium and chlorine in seawater and only organisms adapted to these "drastic" conditions, such as extremophiles, could have developed and survived (Cowen, 2005; Knauth, 2005).

6.2.2.2. *Archaean (3.8–2.5 Ga)*
Major advances in the comprehension of the chemical evolution of the oceans come from two main sources: indirect evidence derived from the mineralogy of sediments of that period (see: Holland, 1984; Holland, 2003, for a detailed analysis) and from the isotopic fractionation of elements sensitive to the redox of their depositional environment (mainly sulphur and nitrogen, Beaumont and Robert, 1999; Farquhar et al., 2000; Farquhar et al., 2001; Pinti et al., 2001; Shen and Buick, 2004). Direct evidence for the chemical composition of seawater is rather scarce in the Archaean sequences and mostly derived from a very few chemical analyses carried out in fluids inclusions into minerals which are considered to have preserved pristine seawater (Channer et al., 1997; de Ronde et al., 1997; Appel et al., 2001; Foriel et al., 2004; Weiershauser and Spooner, 2005).

Indirect evidence of the chemical properties of seawater and of its redox state comes mostly from the study of the few carbonates and Banded Iron Formation (Figure 6.7) (Holland, 1984; Isley, 1995; Holland, 2002; Konkauser et al., 2002). The carbonate minerals in Archaean sediments are mainly calcite, dolomite, aragonite and scarce siderite ($FeCO_3$). This latter mainly occurs in the Banded Iron Formations (Holland, 1984). This means that the oceans were supersaturated in Ca^{2+} and Mg^{2+}, but estimation of their concentrations is difficult to achieve (Holland, 2003). A rough upper limit of 250 for the Ca^{2+}/Fe^{2+} ratio in the Archaean seawater has been derived by Holland (2003) from the absence of significant siderite fraction in the BIF. This value is much lower than the present-day Ca^{2+}/Fe^{2+} ratio in seawater of 9.1×10^6 (Horibe et al., 1974; Gordon et al., 1982). In the modern oxygenated ocean, Fe^{3+} is rapidly precipitated as $Fe(OH)_3$. The higher Fe^{2+} concentrations observed in the Archaean indicates a much lower O_2 content in the atmosphere and in the near-surface ocean. The higher Fe^{2+} and Mn^{2+} contents found in Archaean limestones, compared to their Phanerozoic counterparts (Veizer et al., 1989), support this hypothesis.

Stronger evidences of a poorly oxygenated ocean come from the isotopic fractionation of elements such as sulphur and nitrogen. In the absence of O_2, solar UV interact with SO_2 and generate large mass-independent fractionation (mif) of sulphur isotopes ($\delta^{33}S$) in the reactions products (sulphides and sulphates, Farquhar et al., 2000). The mass-dependent fractionation of the sulphur isotopes ($\delta^{34}S$) in sedimentary sulphides is smaller prior to 2.7 Ga (less than 20%) than in more recent times, when variations of 20% to 40 % are measured. This has been interpreted as the result of a lower concentration of sulphate in the Archaean seawater (0.2 mMol L^{-1}) compared to the present-day (28.7 mMol L^{-1}), which inhibited isotopic fractionation during sulphate reduction (Holland, 2002). In the absence of oxygen, pyrite would not have been oxidized during weathering, reducing drastically the concentrations of sulphates in the Archaean ocean. Nitrogen isotope ratios ($\delta^{15}N$) measured in Archaean (older than 2.7 Ga) organic matter (Beaumont and Robert, 1999; Pinti et al., 2001; Pinti et al., 2003; Ueno et al., 2004), show two distinct populations centred around $\delta^{15}N$ values of -3.6 and $+4.3$, respectively (Shen et al., 2006). Negative nitrogen isotopic values have been interpreted by Beaumont and Robert (1999) as the result of an Archaean nitrogen oceanic cycle different from the modern one, and dominated by nitrogen fixation in a poor-O_2 environment. Alternatively, Pinti et al. (2001) interpreted these values as representative of a marine N cycle dominated by biological cycling of reduced N species by chemolithotrophs at hydrothermal vents, suggesting a chemistry of the ocean largely buffered by interactions with the mantle (see: Shen et al., 2006, for a detailed synthesis).

Direct evidence for the chemical composition of seawater is rather scarce in Archaean sequences. Some studies suggest that pristine fluid inclusions, containing Archaean seawater, could have been preserved in the geological record (Channer et al., 1997; de Ronde et al., 1997; Appel et al., 2001; Foriel et al., 2004; Weiershauser and Spooner, 2005) (Table 6.1). Appel et al. (2001) studied quartz globules preserved in pillow basalts from localized low-strain domains of 3.75 Ga sequences at Isua Greenstone Belt, West Greenland. Fluid-gaseous inclusions in quartz globules contain remnants of two independent fluid/mineral systems including pure CH4 and highly saline aqueous fluids (about 25 wt.% NaCl equivalent), and co-precipitating carbonates (calcite). Appel et al. (2001) interpreted this aqueous system as a relic of sea-floor hydrothermal fluids (Table 6.1). However, the salinity of modern venting solutions is close of that in modern seawater (3.5 wt% equivalent NaCl), deviating by 40–200% from the average only at specific vents (0.4–7 wt% equivalent NaCl, von Damm et al., 1995). The high salinities measured in these inclusions (10 times the modern hydrothermal systems) may have been produced by Archaean seawater, with salinity 3 times the present value, modified by high temperature hydrothermal reactions with basalts (Appel et al., 2001). Alternatively, phase separation could have been produced during prograde metamorphism of the rocks, leaving an enriched saline residual in the inclusions. Isua sequences are indeed highly metamorphosed, at amphibolite facies ($T460 - 480\,C$ and 4 kb), and it cannot be excluded that phase separation could have affected the fluid content of these inclusions.

High salinities have also been observed in primary fluid inclusions from intra-pillow quartz from the North Pole Dresser Formation (3.490 Ga), Pilbara craton, Western Australia (Foriel et al., 2004). These rocks never experienced high-T metamorphism ($T \leq 200\,°C$). The chemistry of the fluid inclusions, particularly the Cl/Br ratios, point to the occurrence of at least three fluids: a saline fluid (Cl/Br = 631 and a 12 wt% equivalent NaCl); a Fe-rich fluid (Cl/Br = 350); and a Ba-rich fluid (Cl/Br = 350). The saline fluid is likely seawater (Foriel et al., 2004). The Cl/Br ratio of 630 is indeed very close to that of modern seawater (Cl//Br = 647). The amount of chlorine, 4 times the present value, is explained by intense evaporation of seawater. The Fe-rich and Ba-rich fluids have Cl/Br ratios close to bulk Earth value (420), possibly indicating mantle buffering (Channer et al., 1997; de Ronde et al., 1997) and a hydrothermal origin for these fluids. This seems to be corroborated by the occurrence of nitrogen and argon having a pure pristine mantle signature in intra-pillow cherts from the same formation (Pinti et al., 2003; Shen et al., 2006). The Dresser formation sequence at North Pole has been interpreted as an estuarine littoral sedimentary deposit with a stratified body of deeper water enriched of primary hydrothermal barium and silica emanating from white smokers (Nijman et al., 1998; Van Kranendonk et al., 2001). Therefore, it appears logic to find a mixing between evaporated seawater and hydrothermal fluids, in these inclusions.

TABLE 6.1

Chemistry of Archaean, Paleoproterozoic seawater and hydrothermal fluids, together with that of Na–Ca–Cl brines from Precambrian Shields

Location	Age Ma	Stratigraphy	Type of inclusion/sample name	Cl⁻ mmol/L	Br⁻ mmol/L	SO_4^{2-} mmol/L	Na⁺ mmol/L	K⁺ mmol/L	Mg^{2+} mmol/L	Ca^{2+} mmol/L	Sr^{2+} mmol/L	Cl/Br mmol/L	Type of fluids
North Pole, W. Australia	3490	Warrawoona; Dresser Fmt	Metal-depleted[1] (Pi0l-21)	2284	3.8	n.d.	n.d.	49.1	n.d.	525	3.7	596	Seawater
North Pole, W. Australia	3490	Warrawoona; Dresser Fmt	Metal-depleted[1] (Pi02-39/1)	1772	2.7	n.d.	n.d.	44.4	nd	385	2.4	642	Seawater
North Pole, W. Australia	3490	Warrawoona; Dresser Fmt	Metal-depleted[1] (Pi02-39/2)	2580	4.2	n.d.	n.d.	78.0	n.d.	679	4.3	607	Seawater
North Pole, W. Australia	3490	Warrawoona; Dresser Fmt	Metal-depleted[1] (Pi02-39/3)	2065	3.4	n.d.	n.d.	52.1	n.d.	486	3.0	652	Seawater
North Pole, W. Australia	3490	Warrawoona; Dresser Fmt	Iron-rich[1]	2580	4.2	n.d.	n.d.	78.0	n.d.	679	4.3	402	Hydrothermal
North Pole, W. Australia	3490	Warrawoona; Dresser Fmt	Barium-rich[1]	1599	4.0	n.d.	n.d.	603.9	n.d.	437	3.5	370	Hydrothermal
Ironstone Pods, RAS	3230	Fig Tree Group	Seawater end-member[2]	920	2.3.	2.3	789	18.9	50.9	232	4.5	409	Seawater
Ironstone Pods, RAS	3230	Fig Tree Group	Hydrothermal end-member[2]	730	2.6		822	21.5		43	0.2	282	Hydrothermal
Ironstone Pods, RAS	3230	Fig Tree Group	Seawater end-member[2]	758	2.4	n.d.	n.d.	n.d.	n.d.	n.d.	n.d.	313	Seawater
Griqualand West, RAS	2220	Transvaal; Ongeluk Fmt	Quartz (Bosch Aar)	33538	262.0	185	6446	605.0	164	12258	316.0	128	Hydrothermal
Griqualand West, RAS	2220	Transvaal; Ongeluk Fmt	Quartz (Bosch Aar)	11559	95.0	46	2178	201.0	45	4160	99.0	122	Hydrothermal
Griqualand West, RAS	2220	Transvaal; Ongeluk Fmt (Bovenongeluk)	Qtz-II	1675	16.4	42	283	60.0	n.d.	n.d.	n.d.	102	Seawater

TABLE 6.1
Continued

Location	Age Ma	Stratigraphy	Type of inclusion/ sample name	Cl⁻ mmol/L	Br⁻ mmol/L	SO_4^{2-} mmol/L	Na⁺ mmol/L	K⁺ mmol/L	Mg^{2+} mmol/L	Ca^{2+} mmol/L	Sr^{2+} mmol/L	Cl/Br mmol/L	Type of fluids
Griqualand West, RAS	2220	Transvaal; Ongeluk Fmt (Bovenongeluk)	Qtz-II	4190	38.9	87	869	160.0	n.d.	n.d.	n.d.	108	Seawater
Griqualand West, RAS	2220	Transvaal; Ongeluk Fmt (Bovenongeluk)	Qtz-I	8084	68.0	87	1333	88.0	n.d.	n.d.	n.d.	119	Seawater
Griqualand West, RAS	2220	Transvaal; Ongeluk Fmt (Bovenongeluk)	Qtz-I	14716	108.1	50	4532	476.0	n.d.	n.d.	n.d.	136	Seawater
Griqualand West, RAS	2220	Transvaal; Ongeluk Fmt (Bovenongeluk)	Qtz-I	753	6.1	47	211	10.0	n.d.	n.d.	n.d.	123	Seawater
Griqualand West, RAS	2220	Transvaal; Ongeluk Fmt (Bovenongeluk)	Qtz-I	5429	38.6	35	1098	120.0	n.d.	n.d.	n.d.	141	Seawater
Griqualand West, RAS	2220	Transvaal; Ongeluk Fmt (Bovenongeluk)	Qtz-I	1513	10.9	34	343	13.0	n.d.	n.d.	n.d.	139	Seawater
Precambrian Shields			Ca–Na–Cl Brines[4]	1520	6.2	2.89	468.78	2.5	28.19	498	4.7	244	Brine
Precambrian Shields			Na–Ca–Cl Brines[5]	383	1.0	1.01	277.14	2.6	4.61	65	0.5	379	Brine
Modern seawater			Na–Cl-dominated	556	0.9	28.7	477	10.1	54.2	11	0.1	647	Seawater

Note.
1. Average have been weighted by the number of inclusions analyzed. The average values are slightly different from those calculated by Foriel et al. (2004).
2. As calculated by de Ronde et al. (1997).
3. Average of 27 analyses from Charmer et al. (1997).
4. Average from 26 analyses of brines from Finland, Sweden and Canada, as reported by Frape it al. (2003).
5. Average from 16 analyses of brines from Finland, Sweden, Canada, UK and West Europe as reported by Frape et al. (2003).

However, the fact that seawater could have evolved in a closed basin does not help in determining its initial saline content.

De Ronde et al. (1997) studied the fluid chemistry of what they believed to be Archaean (3.2 Ga) seafloor hydrothermal vents. Quartz crystals found in iron oxide structures (Ironstone Pods) from the 3.5–3.2 Ga Barberton greenstone belt, South Africa, contain fluid inclusions, which may derive from a mixing between a $NaCl–CaCl_2$ dominated seawater and a $CaCl_2–FeCl_2$ hydrothermal fluid. The Na/Cl ratio in this presumed Archaean seawater is the same as the present-day ocean (0.858), but the total amount of sodium and chlorine is 1.6 times the present day value (Table 6.1). The concentrations of Ca^{2+}, Sr^{2+} are 22 and 50 times higher than the present-day ocean, possibly, because the more acidic ocean did not allow them to precipitate as carbonate, as today. The concentration of is lower than in modern seawater because sulphur was largely in reduced form (Krauskopf and Bird, 1995; Holland, 2003). Finally, the Cl/Br ratio of 400 is very close to the bulk Earth value (420) and of those observed at mid-oceanic ridges, suggesting that halogens were buffered by mantle. Lowe and Byerly (2003) have recently contested the Archaean age of these deposits, showing that the Ironstone Pods are largely composed of goethite (a thermally unstable hydrated iron oxide mineral), deriving from quaternary dissolution of the Archaean siderite. The debate that followed is casting doubts on the validity of these data, which were considered the most compelling evidence of the chemistry of the Archaean seawater.

Very recently, Weiershäuser and Spooner (2005) reported the chemistry of primary fluid inclusions in quartz-filled vesicles and interstitial and drainage cavity quartz from pillowed flows in the area of Ben Nevis, in the 2.7 Ga Abitibi Greenstone Belt, Ontario, Canada. They measured salinities in what they assumed to be Archaean seawater, ranging from 4.8 to 11.4 wt% equivalent NaCl with a small subset of inclusions yielding salinities between 20 and 25 wt% equivalent NaCl. These results point out again the possible higher salinity of the Archaean ocean compared to modern seawater. Three main hypothesis can account for high salinities in fluid inclusions: (1) Younger high saline fluids of unknown origin were trapped and erroneously interpreted as primary, modified high-salinity seawater; (2) phase separation occurred, leaving a highly saline brine as the liquid phase (which got trapped in the inclusions), and a low-salinity vapour phase, which escaped the system before trapping could occur; and (3) a pristine Archaean high-salinity seawater was trapped (Weiershauser and Spooner, 2005).

The first hypothesis seems not plausible. There are compelling mineralogical evidence that the structures, where the fluid inclusions reside, are primary structures. They are mostly undisturbed by post-deposition metamorphism (de Ronde et al., 1997; Foriel et al., 2004; Weiershauser and Spooner, 2005), with the possible exception of inclusions from Isua (Appel et al., 2001). Weiershäuser and Spooner (2005) argued against phase separation, because there is no evidence of liquid and vapour-dominated inclusions in the same fluid inclusion assemblages, which will be the rule for samples affected by boiling (Roedder, 1984). Although assuming these fluid inclusions as pristine seawater, several questions can be raised. Do these inclusions give a real representation of the early global ocean? Alternatively, do the high salinities recorded in these inclusions reflect local conditions, such as close basins, where seawater was affected by high degree of evaporation? This latter hypothesis, for example, seems to be the likely explanation for the high salinities observed in the North Pole aqueous inclusions (Foriel et al., 2004).

Salinity of Archaean fluids could have been strongly modified by inter-actions with the volcanic pile at seafloor, prior to be definitively trapped in rocks. Several of these inclusions contain Na−Ca−Cl dominated fluids, which are characteristics of deep continental water circulating or having interacted with the crystalline basement (Frape et al., 2003). Ca−Na−Cl and Na−Ca−Cl brines, with salinities 10 times the modern seawater, have been found in most Precambrian shields, in Finland, Sweden, Russia and Canada (Fritz and Frape, 1987; Frape et al., 2003, and references therein). In (Figure 6.10), we reported the concentrations of dissolved species, normalized to those found in modern seawater, for Archaean seawater (de Ronde et al., 1997), Palaeoproterozoic seawater (Gutzmer et al., 2003) and the average composition of Ca−Na−Cl and Na−Ca−Cl-dominated brines (calculated from compiled literature data from Frape et al., 2003). Except for a lower K^+ and Mg^{2+} amount in Ca−Na−Cl brines, the variations in dissolved species and the dilution-enrichment factors of these brines are very close to those estimated for the Archaean seawater. Deep-brines from Precambrian shields have been interpreted, based on their H, O, Cl and Sr isotopic ratios, as derived from intense evaporation of modern-like seawater. Prolonged water−rock interactions are thus responsible for their variations in the dis-solved species (Frape et al., 2003). The resemblance with Archaean seawater may suggest that early oceans had chemistry and Total Dissolved Salinity (TDS) practically identical to modern seawater. Most of the variations in the amounts of dissolved species (Figure 6.10) could be thus derived from sea-floor water−rock interactions or contamination by hydrothermal fluids and may not reflect a different chemistry of the early oceans. Further studies on this issue are needed to answer to this hypothesis.

It is only in Proterozoic terrains that fluid inclusions with a chemistry similar to that of modern seawater is finally found. Palaeoproterozoic aqueous inclusions found in 2.22 Ga pods and short veins of quartz within basaltic andesites of the Ongeluk Formation, Transvaal Supergroup, South Africa, show the presence of Ca-rich high saline fluids, resulting from wa-ter−rock interactions with the host volcanic rocks together with seawater having a NaCl concentration equivalent to that of modern seawater (~4 wt% equivalent NaCl, Gutzmer et al., 2003). This could indicate that the

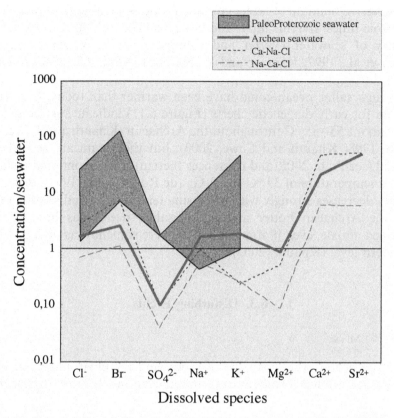

Figure 6.10. Concentration of dissolved anions and cations in Archaean and Paleoproterozoic seawater (data from de Ronde et al., 1997 and Gutzmer et al., 2001, respectively) normalized to the concentrations in modern seawater. Seawater-normalized average concentrations of Ca−Na−Cl and Na−Ca−Cl-dominated brines flowing in crystalline Precambrian shields of Finland, Sweden, Russia and Canada are also reported for comparison (data from Frape et al., 2003).

chemistry of seawater was starting to evolve towards the present-day con- ditions, just after the 2.5 Ga transition. However, Knauth (2005) suggested that the first great lowering of oceanic salinity probably occurred in latest Precambrian (ca. 1 Ga ago), when enormous amounts of salt and brine were sequestered in giant Neoproterozoic evaporite basins.

   Gutzmer et al. (2003) measured Cl/Br values in 2.22 Ga inclusions and found that they are similar to those predicted for Archaean seawater (bulk Earth value), but significantly below to those of the present-day seawater. Gutzmer et al. (2003) suggested that ambient seawater in continental shelf environments at ca. 2.22 Ga was still buffered by vent fluids. This conclusion seems to be common in most of the Archaean studies: the chemistry of the Archaean ocean was buffered by the mantle, through emission of dissolved chemical species and volatiles from mid-oceanic vents and high-temperature pervasive water−rock interactions throughout the oceanic crust. Only with the progressive build up of larger continents and of an abundant biosphere in Proterozoic times, riverine input and biological controls start to regulate the chemistry of seawater (Veizer and Compston, 1976; Veizer et al., 1989; Channer et al., 1997; Kamber and Webb, 2001; Pinti, 2005, and references therein).

This high saline ocean could have been warmer than today. Oxygen isotope data for early diagenetic cherts (Figure 6.11) indicate surface temperatures of about 55–85 °C throughout the Archaean (Knauth and Lowe, 1978; Knauth, 1998; Knauth and Lowe, 2003), but these data are actually controversial (see Pinti, 2005, and references therein). Data from fluid inclusions suggest a temperature of 39 °C at 3.2 Ga (de Ronde et al., 1997). Because O_2 solubility decreases strongly with increasing temperature and salinity (Weiss, 1970), the Archaean hotter and highly saline ocean has been probably maintained anoxic, even if atmospheric O_2 were somehow as high as 70% of the modern level (Knauth, 2005).

6.3. Disturbing Events

PHILIPPE CLAEYS

Among the brutal events capable of modifying the environment of life on Earth surface are meteoritic impacts and glaciations. If most Archaean and

Figure 6.11. 3.445 Ga old chert from the Hooggenoeg formation in Barberton greenstone belt (South Africa). This chert is considered as formed due to hydrothermal activity in an environment possibly similar to modern mi-ocean ridges. Chemical analyse of such rocks provides information on water composition at the time of their formation. (Photo H. Martin).

Proterozoic impact craters disappeared due to erosion and plate tectonics, their traces are preserved in sedimentary records, as spherule layers. Similarly, glaciation periods strongly affected the environment and modified exogenous mechanisms such that their existence and influence have also been imprinted in the sedimentary record.

6.3.1. SPHERULE LAYERS: THE RECORD OF ARCHAEAN AND PROTEROZOIC IMPACT EVENTS.

Among the ~175 impact structures known on Earth, only a small fraction is Proterozoic in age, but this fraction includes two giants: Sudbury in Canada (~250 km in diameter and 1.850 Ga old) and Vredefort in South Africa (estimated to reach 300 km in diameter and dated at 2.023 Ga). So far, no crater of Archaean age has been identified, and due to the limited amount of preserved lithologies of this age, it is rather unlikely that one will ever be found. It is undeniable that also during this period, asteroids and/or comets were colliding with Earth. The ejecta layers recorded in ancient sedimentary sequences constitute the only remaining trace of these early impacts. These layers are generally composed of particles ejected during the cratering process, transported for hundreds to thousands of km and deposited in sedimentary sequences. When an extraterrestrial body collides with the Earth, a huge amount of energy is released leading to the vaporization and melting of both the projectile and the upper parts of the target-rock. The cratering process is accompanied by the ejection of target-rock debris as solid, vapour and molten phases. These different impact products are recorded in sedimentary sequences as distinct ejecta layers containing shocked minerals, spherules made of glass (or its alteration product) and crystallites, Ni-rich spinels, and/or geochemical anomalies (such as the famous positive anomaly in Iridium and in other Platinum group elements).

The study of Phanerozoic ejecta, especially at the Cretaceous-Tertiary (KT) boundary, demonstrates that rounded millimetre-sized spherules constitute the type of ejecta by far easiest to recognize in the field. The spherules originate by the melting or vaporization of the target-rock and of part of the projectile. They condense or quench in flight, inside or outside the upper atmosphere, before landing back at distal sites from the impact point. The spherules are usually rounded but other aerodynamically shaped morphologies such as elongated, teardrops or dumbbells are also frequent. Smit et al. (1992) and Smit (1999) have discussed in detail the characteristics and geographic distribution of the different types of spherules occurring in the KT boundary layer. Two kinds of spherules are classically distinguished: (a) spherules containing crystallites of clinopyroxene, spinels and olivine and called microkrystites (Glass and Burns, 1988); they most likely condensed out of the vapour plume generated by the impact; (b) spherules entirely made of glass and called microtektites, which result from the melting of the upper part of the target-rock and subsequent rapid quenching in flight. Another didactic example exists in the Late Eocene with two closely spaced layers of microkrystites and microtektites respectively produced by the Popigai and Chesapeake Bay impacts (see Montanari and Koeberl, 2000, for review).

The current record of spherule layers of Archaean and Proterozoic ages is presented in Table 6.2. Ten ejecta layers crop out in Western Australia and South Africa; they concentrated in two time-windows between 2.65 to 2.50 Ga and 3.47 to 3.24 Ga. Few ejecta layers are known outside these two "impact windows". Between 2.13 and 1.84 Ga, Chadwick et al. (2001) identified a single unit rich in impact spherules, reaching maybe 1 m thick, in dolomitic sediments of the Ketilidian orogenesis in West Greenland (Figure 6.12). Recently, the probable ejecta layer produced by the Sudbury crater, was recognized in several sections of the uppermost part of the Gunflint Formation near the US–Canadian border in Minnesota and Ontario (Addison et al., 2005). Dated between 1.88 and 1.84 Ga, this ~50 cm thick ejecta unit contains shocked quartz and impact spherules. The younger (Late Proterozoic, 0.59 Ga) ejecta produced by the Acraman crater in Australia is described by Wallace et al. (1990). So far, only the ejecta layers found in some of the least metamorphosed and tectonised sedimentary units of the Barberton Greenstone Belt in South Africa (Lowe et al., 2003, for review) and of the Hamersley Basin in Western Australia (Simonson, 2003, for review) allow drawing some interesting hypotheses relevant for the ancient record of impact events.

Proterozoic and Archaean spherule layers are most likely to be preserved when deposited below the zone of surface wave influence, in rather deep-water environments of the continental shelf. In such low-energy environments, they avoid reworking and the absence of bioturbation ensures that they remain undisturbed. The known ejecta layers commonly occur in beds coarser than the surrounding units, interpreted as due to high energy depositional episodes, perhaps related to tsunamis or massive sediment transports triggered by the impact (Hassler et al., 2000; Hassler and Simonson, 2001; Simonson and Glass, 2004). Because they are free of burrowing and/or reworking by organisms, the Archaean and Proterozoic ejecta units provide a clear view of their depositional process and sorting. It is likely that both spherule-types are represented in some of these ancient layers.

Today K-feldspar, sericite, chlorite, quartz and carbonate have replaced the original spherule composition (Lowe et al., 2003; Simonson, 2003). The highly resistant dendritic spinels probably represent the only primary crystals preserved in the early to middle Archaean spherules of the Barberton Greenstone Belt (Byerly and Lowe, 1994). At the KT boundary, dendritic spinels formed either by condensation out of the impact induced vapour

TABLE 6.2

Main spherule layers recorded in Archaean and Proterozoic terrains, Modified after Simonson and Glass (2004)

Layer	Location	Formation	Age (Ga)	Lateral extend (km)	Layer thickness (cm)	Accumulation spherules (cm)	Spherule size (mm)	Ir (ppb)
Acraman	South Australia, Adelaide Geosyncline, Officer Basin	Bunyeroo, Rodda beds	0.59	700	0–40	Traces	1.0	2.0
Sudbury	Ontario Minnesota	Gunflint-Rove boundary Biwabik-Virginia boundary	1.84–1.88	260	25–70	?	?	n.d
n.n	Southern Greenland	Vallen Group, Graensesø	2.13–1.85	26*	100	20	1.4	<DL
n.n	Western Australia Hamersley basin	Dales Gorge Member of Brockman	2.48	135	30	6.0	2.1	19.9
n.n	Western Australia Hamersley basin	Wittenoom	2.54	330	10–100	3.5	1.1	0.43
n.n	South Africa Griqualand West basin	Reivilo	2.56 ?	25	2–20	1.8	1.3	n.d
n.n	Western Australia Hamersley basin	Carawine**	2.63 ?	75	940–2470	30.0	2.3	1.54
n.n	Western Australia Hamersley basin	Jeerinah**	2.63	160	0.3–280	7.0	2.0	11.4
n.n	South Africa Griqualand West basin	Monteville**	2.60–2.63	230	55	5.0	1.6	6.4
S4	South Africa, Barberton belt	Mapepe	3.24	Single outcrop	15	7.5	1.6	450
S3	South Africa, Barberton belt	Mapepe and Unlundi	3.24	30	15–200	10.0	4.0	725
S2	South Africa, Barberton belt	Mapepe	3.26	12	20–310	10	2.5	3.9
S1	South Africa, Barberton belt	Hooggenoeg	3.47	25	10–35	5.0	1.0	3.0
	Pilbara craton Western Australia	Apex Basalt**		1	110	7.5	0.8	n.d

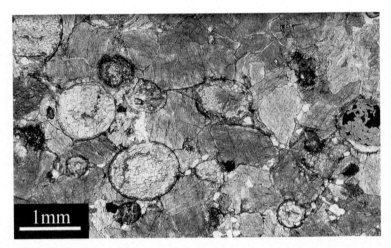

Figure 6.12. Polarized light microphotograph of spherules in a dolomitic matrix. These spherules come from the Ketilidian orogenesis (2.1–1.8 Ga) in Greenland. (Photo Ph. Claeys).

cloud (Kyte and Bostwick, 1995) or by the fusion and oxidation of the impacting projectile during its passage through the atmosphere (Robin et al., 1992). Although being of indisputable impact origin, the Barberton spinels differ in composition from their Phanerozoic counterparts; they are enriched in Cr, depleted in Mg and have higher Ni/Fe ratio. Their significantly lower Fe^{3+}/Fe_{total} ratio is consistent with their formation in the Archaean oxygen-poor atmosphere (Byerly and Lowe, 1994).

The ejecta layers exposed in the sedimentary sequences of the Barberton Group (South Africa), and of the Hamersley Basin (Western Australia) present some major differences with the ejecta units recognized in the Phanerozoic. These ancient layers contain thicker spherule accumulations and their composition seems to be more mafic than their Phanerozoic equivalent. Simonson and Harnik (2000) estimate that the spherule concentration is 10 to 100 times greater than that measured in the Phanerozoic ejecta. Most layers are a few centimetres thick but some can reach 30 cm (Table 6.2). In the most recent layers, pure spherule accumulations are barely thicker than 0.1 cm, except for the KT boundary and that only at proximal sites to the Chicxulub crater (Smit, 1999). The ejecta layer found in the 2.6 Ga Wittenoom formation of the Hamersley basin (Australia) maintains a consistent thickness of several cm over more than 20,000 km^2 (Simonson, 1992; Simonson and Harnik, 2000), while other layers probably extend over several hundreds of km (Simonson et al., 1999). In Western Australia, the Jeerinah ejecta layer of the Hamersley Basin is correlated with the Carawine horizon in the Oakover river region of the Pilbara Craton to the East and possibly with the layer occurring in the Monteville Formation of South Africa. If such correlations are reliable, they imply that the area of spherule distribution was greater than 32,000 km^2 (Simonson et al., 1999; Rasmussen and Koeberl, 2004; Simonson and Glass, 2004). The possibility of a global distribution of these ejecta debris certainly deserves to be considered. Based on the Phanerozoic record, a crater of ~100 km in diameter is probably capable of spreading ejecta worldwide. The location of their source craters is of course unknown, but based on the spherule accumulation and consistency over important distance, Simonson et al. (199) interpreted them as rather distal (deposited > 5 crater radii) ejecta layers.

The size of these ancient spherules ranges between 1 and 4 millimetres. Meteoritic contamination of these layers is generally quite significant (Lowe et al., 2003). The reported Iridium concentrations range from a few ppb (equivalent to most KT boundary sites) up to values of 725 ppb for one of the Barberton layers (Table 6.2). Chromium isotopes (^{53}Cr/^{52}Cr) indicate that the projectile responsible for the formation of the Carawine and Jeerinah spherules likely corresponded to an ordinary chondrite, while the S4, S3, and S2 layers were probably formed by the fall of carbonaceous chondrites (Shukolyukov et al., 2000; Shukolyukov et al., 2002; Kyte et al., 2003).

Based on texture and overall composition, the Archaean and Proterozoic spherules were probably originally basaltic. Due to secondary replacement, it is difficult to estimate their primary composition. However, it appears that in many spherules, the replacement occurred with (some) preservation of the original texture. The K-feldspar crystals commonly display a typical elongated and fibrous lath-shaped morphology that resembles that formed by the rapid cooling of basaltic glasses (Simonson and Harnik, 2000). In other spherules now containing calcite, quartz or chlorite pseudomorphs, quenched and devitrification textures can still be recognized along with botroydal shapes and flow-bands (Simonson, 2003). Phanerozoic spherules are most commonly replaced by smectite and their chemistry rather reflects the composition of the continental crust. So far, only one sample of the Jeerinah layer in Western Australia contains shocked quartz grains (Rasmussen and Koeberl, 2004). The mafic character of the Archaean and Proterozoic spherules, combined with the scarcity of shocked quartz has been interpreted as the result of impacts on the oceanic crust (Lowe et al., 2003; Simonson, 2003). Oceanic impacts are further substantiated by the occurrence of many spherules in high-energy sedimentary units that could be caused by a major disturbance of the water masses during deposition (Hassler and Simonson, 2001; Lowe et al., 2003; Simonson et al., 2004). The dominance of oceanic impacts agrees with the suggestion that before 2.5 Ga the volume of oceanic crust was more important than the present-day 60% (McCulloch and Bennett, 1994). In the Phanerozoic oceanic impacts are rare, mafic ejecta is exceptional, and so far no crater is known to have excavated the oceanic crust (Dypvik et al., 2003). The apparently common cratering of the oceanic crust in the Archaean and Proterozoic may be explained either by very large projectiles or by much shallower oceans than in the Phanerozoic, or both.

It thus seems likely that collisions the size of the KT event, or even significantly bigger, took place during two distinct intervals between 3.47 to 3.24 Ga and 2.65 to 2.5 Ga. The thickness of the ejecta, – if indeed they are distal layers –, the rather large dimension of spherule, and mass balance estimations based on the concentration of Ir and extraterrestrial Cr, – assuming the global distribution of the ejecta –, all these evidences indicate that the projectiles responsible for the formation the S2, S3, S4 Barberton impact layers varied between 20 and 50 km in diameter (Lowe et al., 2003). These sizes imply the formation of craters between 400 and 1000 km in diameter, quite bigger than Chicxulub, Sudbury or Vredefort. Simonson et al. (1999) also advocates that large projectiles, KT equivalent or more, formed some of the younger impact layers identified in Western Australia.

The clustering of spherule layers between 3.47 to 3.24 Ga and 2.65 to 2.5 Ga might be due to the coincidental preservation of two exceptional periods marked by a higher delivery rate of large impactors. It could also be interpreted as evidence that the whole Archaean and Early Proterozoic period was marked by a high flux of projectiles on Earth. The latter is compatible with the dating of lunar spherules that documents a progressive decrease in the collision frequency, after the end of the Late Heavy Bombardment 3.8 Ga ago (Culler et al., 2000, see discussion in Claeys, chapter 4.5). Glikson (2001) speculated that the Barberton dense spherule record coincides with a possible spike of impacts on the Moon around 3.2 Ga. These impacts could have influenced the evolution of early life, affected the oceans/atmosphere system or even modified the plate tectonic regime. Large oceanic impact could evaporate part of the ocean, and/or the associated disturbance could contribute to the mixing of a strongly stratified ocean increasing the exchange of nutriment-rich waters (Lowe et al., 2003). Glikson (1999; 2001) has speculated that they also lead to crustal deformation, rifting, fracturing of the crust, massive volcanism and uplift. The on-going searches for ejecta deposits at other stratigraphic levels in the Archaean and the Proterozoic should provide answers to some of these questions.

6.3.2. THE PALEOPROTEROZOIC GLACIATIONS, THE FIRST SNOWBALL EARTH

The first evidence for possible mid-latitude glaciations occurs in the diamictite of the Pongola Supergroup in South Africa, which is dated at about 2.9 Ga (Young et al., 1998; Nhleko, 2004) unfortunately few studies have been devoted to them. Between 2.45 and 2.22 Ga, three episodes of glaciations are recorded in the Huronian Supergroup of Canada (Hilburn et al., 2005). From oldest to youngest, diamictite deposits are present in the Ramsey

Lake, the Bruce and the Gowganda Formations. The precise ages of these events remain poorly constrained. The Nipissing diabase, dated at 2.22 Ga, crosscuts the entire Huronian sequence. Therefore, the Gowganda diamictite must be older than 2.2 Ga, as it is covered by km-thick sediments belonging to the Lorrain, Gordon Lake and Bar River Formations (Kopp et al., 2005). Based on the existing paleomagnetic data, the latitudinal extension of the three Huronian glacial deposits cannot be defined (Kopp et al., 2005). At this point, it is thus difficult to compare them with the Late Proterozoic glaciations, which probably covered the whole planet leading to the concept of Snowball Earth (Hoffman et al., 1998). Moreover, the Huronian glacial sediments do not seem to be covered by warm water carbonates (the so-called cap carbonates) (Kopp et al., 2005), which in the Late Proterozoic are indicative of an abrupt transition to much warmer climates.

Recent isotopic dates and new correlations indicate that the glacial deposits occurring in the Transvaal Supergroup of South Africa probably postdate the Huronian diamictites (Kopp et al., 2005). The Bosheok diamictite cropping out in the eastern part of the Transvaal region probably correlates with the Makganyene diamictite deposited in the Griqualand region, to the west of the Transvaal Basin. A Re-Os age of 2.3 Ga has been established for the Timeball Hill formation that is found below the Boshoek diamictite (Hannah et al., 2004). The Makganyene glacial deposits are interfingered with the Ongeluk marine basalt. The latter correlates with its subaerial equivalent the Hekpoort volcanics, which contains zircons, dated at 2.2 Ga (Cornell et al., 1996). Paleomagnetic data acquired on the Ongeluk basalts imply that the Makganyene glacial unit was deposited at tropical latitude, around $11° \pm 5°$ (Evans et al., 1997). Based on this stratigraphy and low latitude occurrence, Kopp et al., (2005) advocated that the Makganyene event represents the oldest evidence for global glaciation, in other words, the first case of a Snowball Earth in the history of the planet.

The origin of oxygen producing cyanobacteria remains a major topic of discussion (see part 7.1). Several organic Biomarkers, traces of microfossils – considered by some authors as controversial –, the presence of stromatolites and isotopic fractionation, which origin has been debated, have widely been used as indicators of oxygen production by organisms since 2.8 Ga or even as early as 3.7 Ga (see Brocks et al., 1999; Des Marais, 2000; Catling et al., 2001; Brasier et al., 2002; Brocks et al., 2003a; Rosing and Frei, 2004). However, the significant mass independent fractionation of sulphur isotopes detected only in Archaean and Palaeoproterozoic sulphides supports an atmospheric oxygenation taking place only 2.5–2.2 Ga ago (Farquhar et al., 2002; Farquhar and Wing, 2003). The presence of only a few percent of oxygen in the lower atmosphere would hamper the penetration of the light of wavelength under 200 nm responsible for the photolysis of SO_2 to S, which results in the fractionation of S isotopes (Farquhar et al., 2001).

The presence of thick red beds, lateritic paleosols and major Mn deposits indicative of oxygen conditions occurs immediately below or above the Makganyene glacial deposits. This succession let Knopp et al., (2005) to propose that O_2-producing cyanobacteria appeared just before the Makganyene glaciation, between 2.3 and 2.2 Ga. The active oxygenic photosynthesis would have induced the rapid collapse of the pre-existing methane greenhouse leading to a Snowball event on a time scale of ~1 Ma (Kopp et al., 2005).

References

Addison, W. D., Brumpton, G. R., Vallini, D. A., McNaughton, N. J., Davis, D. W., Kissin, S. A., Fralick, P. W. and Hammond, A. L.: 2005, *Geology* **33**(3), 193–196.

Albarède, F.: 1998, *Tectonophysics* **296**, 1–14.

Albarède, F.: 2005, in M. Gargaud, P. Claeys and H. Martin (eds.), *Des atomes aux planètes habitables*, Presses Universitaires de Bordeaux, Bordeaux, pp. 79–102.

Anbar, A. D. and Knoll, A. H.: 2002, *Science* **297**, 1137–1141.

Appel, P. W. U., Rollinson, H. R. and Touret, J. L. R.: 2001, *Precambrian Res.* **112**, 27–49.

Baraffe, I., Chabrier, G., Allard, F. and Hauschildt, P. H.: 1998, *Astron. Astrophys.* **337**, 403.

Beaumont, V. and Robert, F.: 1999, *Precambrian Res.* **96**(1–2), 63–82.

Bekker, A., Holland, H. D., Wang, P.-L., Rumble III, D., Stein, H. J., Hannah, J. L., Coetzee, L. L. and Beukes, N. J.: 2004, *Nature* **427**, 117–120.

Berner, R. A., Beerling, D. J., Dudley, R., Robinson, J. M. and Wildman, R. A. Jr: 2003, *Ann. Rev. Earth Planetary Sci.* **31**, 105–134.

Bischoff, J. L. and Dickson, F. W.: 1975, *Earth Planetary Sci. Lett.* **25**, 385–397.

Bischoff, J. L. and Rosenbauer, R. J.: 1988, *Geochim. Cosmochim. Acta* **52**, 2121–212.

Bourdon, E., Eissen, J.-P., Gutscher, M.-A., Monzier, M., Hall, M. L. and Cotten, J.: 2003, *Earth Planetary Sci. Lett.* **205**(3–4), 123–138.

Bourrouilh, R.: 2001, in M. Gargaud, D. Despois and J.-P. Parisot (eds.), *Evolution parallèles de la Terr et de la Vie. L'environnement de la Terre primitive*, Presses Universitaires de Bordeaux, Bordeaux, pp. 287–320.

Brasier, M. D., Green, O. R., Jephcoat, A. P., Kleppe, A. K., Van Kranendonk, M. J., Lindsay, J. F., Steele, A. and Grassineau, N. V.: 2002, *Nature* **416**, 76–81.

Brocks, J. J., Buick, R., Logan, G. A. and Summons, R. E.: 2003a, *Geochemica et Cosmochemica Acta* **67**, 4289–4319.

Brocks, J. J., Logan, G. A., Buick, R. and Summons, R. E.: 1999, *Science* **285**, 1033–1036.

Brocks, J. J., Buick, R., Summons, R. E. and Logan, G. A.: 2003, *Geochimica et Cosmochimica Acta* **67**(22), 4321–4335.

Brown, G. C.: 1985, in N. Snelling (ed.), *The Chronology of the Geological Record*, Memoir - Geological Society of London, London, pp. 326–334.

Buick, R.: 1992, *Nature* **255**, 74–77.

Byerly, G. R. and Lowe, D. R.: 1994, *Geochemica et Cosmochemica Acta* **58**(16), 3469–3486.

Canfield, D. E. and Teske, A.: 1996, *Nature* **382**, 127–132.

Catling, D. C. and Claire, M. W.: 2005, *Earth Planetary Sci. Lett.* **237**, 1–20.

Catling, D. C., Zahnle, K. J. and McKay, C. P.: 2001, *Science* **293**, 839–843.

Chadwick, B., Claeys, P. and Simonson, B.: 2001, *J. Geol. Soc. London* **158**, 331–340.

Channer, D. M. D. R., de Ronde, C. E. J. and Spooner, E. T. C.: 1997, *Earth Planetary Sci. Lett.* **150**, 325–335.

Commeyras, A., Taillades, J., Collet, H., Boiteau, L., Vandenabeele-Trambouze, O., Pascal, R., Rousset, A., Garrel, L., Rossi, J., Biron, J., Lagrille, O., Plasson, R., Souaid, E., Danger, G., Selsis, F., Dobrijévic, M. and Martin, H.: 2004, *Origins Life Evol. Biosphere* **34**, 35–55.

Condie, K. C.: 1989. *Plate Tectonics and Crustal Evolution*, Pergamon, Oxford, 476 pp.

Condie, K. C.: 1998, *Earth Planetary Sci. Lett.* **163**(1–4), 97–108.

Cornell, D. H., Schütte, S. S. and Elington, B. L.: 1996, *Precambrian Res.* **79**, 101–123.

Cowen, R.: 2005, The History of Life, 4th edn. Blackwell Scientific, 324 pp.

Culler, T. S., Becker, T. A., Muller, R. A. and Renne, P. R.: 2000, *Science* **287**, 1785–1788.

de Ronde, C. E. J., Channer, D. M. d., Faure, K., Bray, C. J. and Spooner, T. C.: 1997, *Geochimica and Cosmochimica Acta* **61**(19), 4025–4042.

Delano, J. W.: 2001, *Origins Life Evol. Biosphere* **31**(4–5), 311–341.

Des Marais, D. J.: 2000, *Science* **289**, 1703–1705.

Des Marais, D. J., Strauss, H., Summons, R. E. and Hayes, J. M.: 1992, *Nature* **359**, 605–609.

Dypvik, H., Burchell, M. J. and Claeys, P.: 2003, in H. Dypvik, M. J. Burchell and P. Claeys (eds.), *Cratering in Marine Environments and on Ice*, Springer, Berlin, pp. 1–19.

Beukes, N. J. and Kirschvink, J. L.: 1997, *Nature* **386**, 262–266.

Farquhar, J., Bao, H. M. and Thiemens, M.: 2000, *Science* **289**(5480), 756–758.

Farquhar, J., Savarino, J., Airieau, S. and Thiemens, M. H.: 2001, *J. Geophys. Res. Planets* **106**(E12), 32829–32839.

Farquhar, J. and Wing, B. A.: 2003, *Earth Planetary Sci. Lett.* **213**, 1–13.

Farquhar, J., Wing, B. A., McKeegan, K. D., Harris, J. W., Cartigny, P. and Thiemens, M. H.: 2002, *Science* **298**, 2369–2372.

Foriel, J., Philippot, P., Rey, P., Somogyi, A., Banks, D. and Menez, B.: 2004, *Earth Planetary Sci. Lett.* **228**, 451–463.

Frape, S. K., Blyth, A., Blomqvist, R. and McNutt, R. H.: 2003, in J. I. Drever (ed.), *Deep fluids in the continents: II Crystalline rocks. The Oceans and Marine Geochemistry. Treatise of Geochemistry*, Elsevier-Pergamon, Oxford, pp. 541–580.

Fritz, P. and Frape, S.K. (eds.), 1987. *Saline Water and Gases in Crystalline Rocks. GAC Special Paper, 33*, Geological Association of Canada, Ottawa, 259 pp.

Gladman, B., Dones, L., Levison, H. F. and Burns, J. A.: 2005, *Astrobiology* **5**(4), 483–496.

Glass, B. P. and Burns, C. A.: 1988, *Proc. Lunar Planetary Sci. Conf.*, 455–458.

Glikson, A. Y.: 1999, *Geology* **27**(5), 387–390.

Glikson, A. Y.: 2001, *J. Geodynamics* **32**(1–2), 205–229.

Gomes, R., Levison, H. F., Tsiganis, K. and Morbidelli, A.: 2005, *Nature* **435**, 466–469.

Gordon, R. M., Martin, J. H. and Knauer, G. A.: 1982, *Nature* **299**, 611–612.

Gorman, B. E., Pearce, T. H. and Birkett, T. C.: 1978, *Precambrian Res.* **6**, 23–41.

Gough, D. O.: 1981, *Solar Phys.* **74**, 21–34.

Graedel, T. E. and Keene, W. C.: 1996, *Pure Appl. Chem.* **68**, 1689–1697.

Gutscher, M.-A., Maury, F., Eissen, J.-P. and Bourdon, E.: 2000, *Geology* **28**(6), 535–538.

Gutzmer, J., Banks, D., Lüders, V., Hoefs, J., Beukes, N. J. and von Bezing, K. L.: 2003, *Chem. Geol.* **201**, 37–53.

Gutzmer, J., Pack, A., Luders, V., Wilkinson, J.-J., Beukes, N. J. and van Niekerk, H. S.: 2001, *Contrib. Mineral. Petrol.* **142**, 27–42.

Hanna, J. L., Bekker, A., Stein, H. J. J., Markey, R. J. and Holland, H. D.: 2004, *Earth Planetary Sci. Lett.* **225**, 43–52.

Hassler, S. W., Robey, H. F. and Simonson, B.: 2000, *Sedimentary Geol.* **135**, 283–294.

Hassler, S. W. and Simonson, B. M.: 2001, *J. Geol.* 109.

Hilburn, I. A. L. J., Kirschvink, J. L. E. T., Tada, R., Hamano, Y. and Yamamoto, S.: 2005, *Earth Planetary Sci. Lett.* **232**(3–4), 315–332.

Hoffman, P. F., Kaufman, A. J., Halverson, G. P. and Schrag, D. P.: 1998, *Science* **281**, 1342–1346.

Holland, H. D.: 1984. *The Chemical Evolution of the Atmosphere and Oceans*, Princeton Series in Geochemistry. Princeton University Press, Princeton, 582 pp.

Holland, H. D. 1994, in S. Bengston (ed.), *Early Life on Earth*. Columbia University Press, New York, pp. 237–244.

Holland, H. D.: 1999, *The Geochemical News* **100**, 20–22.

Holland, H. D.: 2002, *Geochimica et Cosmochimica Acta* **66**(21), 3811–3826.

Holland, H. D.: 2003, in H. D. Holland and K. K. Turekian (eds.), *The Oceans and Marine Geochemistry. Treatise of Geochemistry*, Elsevier-Pergamon, Oxford, pp. 583–625.

Horibe, Y., Endo, K. and Tsubota, H.: 1974, *Earth Planetary Sci. Lett.* **23**(1), 136–140.

Isley, A. E.: 1995, *J. Geol.* **103**, 169–185.

Johnson, N. M. and Fegley, B. Jr.: 2002, *Adv. Space Res.* **29**, 2333–241.

Kamber, B. S. and Webb, G. E.: 2001, *Geochimica et Cosmochimica Acta* **65**, 2509–2525.

Karhu, J. A. and Holland, H. D.: 1996, *Geology* **24**, 867–870.

Kasting, J. F.: 1993, *Earth's Early Atmosphere* **259**, 920–926.

Kasting, J. F.: 2005, *Precambrian Res.* **137**, 119–129.

Kasting, J. F., Eggler, D. H. and Raeburn, S. P.: 1993, *J. Geol.* **101**, 245–257.

Knauth, L. P.: 1998, *Nature* **395**, 554–555.

Knauth, L. P.: 2005, *Palaeogeogr. Palaeoclimatol. Palaeoecol.* **219**(1–2), 53–69.

Knauth, L. P. and Lowe, D. R.: 1978, *J. Geol.* **41**, 209–222.

Knauth, L. P. and Lowe, D. R.: 2003, *Geol. Soc. Am. Bull.* **115**(5), 566–580.

Knoll, A. H.: 2003, *Deobiology* **1**, 3–14.

Konkauser, K. O., Hamade, T., Morris, R. C., Ferris, G. F., Southam, G. and Canfield, D. E.: 2002, *Geology* **30**, 1079–1082.

Kopp, R. E., Kirschvink, J. L., Hilburn, I. A. and Nash, C. Z.: 2005, *Proc. Nat. Acad. Sci.* **102**, 11131–11136.

Krauskopf, K. B. and Bird, D. K.: 1995. *Introduction to Geochemistry*, McGraw-Hill, New York, 647 pp.

Kress, M. E. and McKay, C. P.: 2004, *Icarus* **168**(2), 475–483.

Kyte, F. T., Shukolyukov, A., Lugmair, G. W., Lowe, D. R. and Byerly, G. R.: 2003, *Geology* **31**, 283–286.

Kyte, F. T. and Bostwick, J. A.: 1995, *Earth Planetary Sci. Lett.* **132**(1–4), 113–127.

Li, Z.-X. A. and Lee, C.-T. A.: 2004, *Earth Planetary Sci. Lett.* **228**, 483–493.

Libourel, G., Marty, B. and Humbert, F.: 2003, *Geochimica et Cosmochimica Acta* **67**, 4123–4135.

Lowe, D. R. and Byerly, G. R.: 2003, *Geology* **31**(10), 909–912.

Lowe, D. R., Byerly, G. R., Kyte, F. T., Shukolyukov, A., Asaro, F. and Krull, A.: 2003, *Astrobiology* **3**(1), 7–48.

Maher, K. A. and Stevenson, D. J.: 1988, *Nature* **331**, 612–614.

Martin, H.: 1986, *Geology* **14**, 753–756.

Martin, H.: 1999, *Lithos* **46**(3), 411–429.

Martin, H. and Moyen, J.-F.: 2002, *Geology* **30**(4), 319–322.

Martin, H., Smithies, R. H., Rapp, R., Moyen, J.-F. and Champion, D.: 2005, *Lithos* **79**(1–2), 1–24.

Marty, B. and Dauphas, N.: 2003, *Earth Planetary Sci. Lett.* **206**, 397–410.

McCulloch, M. T. and Bennet, V. C.: 1993, *Lithos* **30**, 237–255.

McCulloch, M. T. and Bennett, V. C.: 1994, *Geochimica and Cosmochimica Acta* **58**, 4717–4738.

Melezhik, V. A., Fallick, A. E., Hanski, E. J., Kump, L. R., Lepland, A., Prav, A. R. and Strauss, H.: 2005, *Geol. Soc. Am. Today* **15**(11), 4–11.

Mojzsis, S. J., Coath, C. D., Greenwood, J. P., McKeegan, K. D. and Harrison, T. M.: 2003, *Geochimica et Cosmochimica Acta* **67**, 1635–1658.

Montanari, A. and Koeberl, C.: 2000. *Impact Stratigraphy Lecture Notes in Earth Sciences, 93*, Springer Verlag, Berlin, 364 pp.

Moyen, J.-F., Jayananda, M., Nédelec, A., Martin, H., Mahabaleswar, B. and Auvray, B.: 2003, *J. Geol. Soc. India* **62**, 753–758.

Moyen, J.-F., Martin, H. and Jayananda, M.: 1997, *Compte Rendus de l'Académie des Sciences de Paris* **325**, 659–664.

Navarro-Gonzalez, R., McKay, C. P. and Nna Mvondo, D.: 2001, *Nature* **412**, 61–64.

Nhleko, N.: 2004, *The Pongola Supergroup in Swaziland*. Rand Afrikaans University.

Nijman, W., de Bruijne, K. C. H. and Valkering, M. E.: 1998, *Precambrian Res.* **88**(1–4), 25–52.

Nisbet, E. G., Cheadle, M. J., Arndt, N. T. and Bickle, M. J.: 1993, *Lithos* **30**, 291–307.

Nisbet, E. G. and Fowler, C. M. R.: 2003, in W. H. Schlesinger (ed.), *Biogeochemistry. Treatise of Geochemistry*, Elsevier-Pergamon, Oxford, pp. 1–61.

Nisbet, E. G. and Sleep, N. H.: 2001, *Nature* **409**, 1083–1091.

Oberbeck, V. and Fogelman, G.: 1989, *Nature* **339**, 434.

Pavlov, A. A., Hurtgen, M. T., Kasting, J. F. and Arthur, M. A.: 2003, Geology **31**, 87–90.

Pavlov, A. A. and Kasting, J. F.: 2002, *Astrobiology* **2**(1), 27–41.

Pavlov, A. A., Kasting, J. F., Brown, L. L., Rages, K. A. and Freedman, R.: 2000, *J. Geophys. Res.* **105**, 11981–11990.

Peltier, W. R., Butler, S. and Solheim, L. P.: 1997, in D. J. Grossley (ed.), *Earth's* Deep Interior, Gordon and Breach, Amsterdam, pp. 405–430.

Pinti, D. L.: 2005, in M. Gargaud, B. Barbier, H. Martin and J. Reisse (eds.), *Lectures in Astrobiology. Advances in Astrobiology and Biogeophysics*, Springer-Verlag, Berlin, pp. 83–107.

Pinti, D. L., Hashizume, K. and Matsuda, J.: 2001, *Geochimica et Cosmochimica Acta* **65**(14), 2301–2315.

Pinti, D. L., Hashizume, K., Philippot, P., Foriel, J. and Rey, P.: 2003, Geochimica et Cosmochimica Acta, A287.

Rapp, R. P., Shimizu, N. and Norman, M. D.: 2003, *Nature* **425**, 605–609.

Rapp, R. P., Shimizu, N., Norman, M. D. and Applegate, G. S.: 1999, *Chem. Geol.* **160**, 335–356.

Rasmussen, B. and Koeberl, C.: 2004, *Geology* **32**, 1029–1032.

Ribas, I., Guinan, E. F., Gudel, M. and Audard, M.: 2005, *The Astrophis J.* **622**(1), 680–694.

Robin, E., Bonté, P., Froget, L., Jéhanno, C. and Rocchia, R.: 1992, *Earth Planetary Sci. Lett.* **108**, 181–190.

Roedder, E.: 1984. *Fluid Inclusions*, Rewiews in Mineralogy 12., Mineralogical Society of America.

Rosing, M. T.: 1999, *Science* **283**, 674–676.

Rosing, M. T. and Frei, R.: 2004, *Earth Planetary Sci. Lett.* **217**, 237–244.

Rouxel, O. J., Bekker, A. and Edwards, K. J.: 2005, *Science* **307**, 1088–1091.

Rye, R. and Holland, H. D.: 1998, *Am. J. Sci.* **298**, 621–672.

Sagan, C. and Mullen, G.: 1972, *Science* **177**, 52–56.

Samaniego, P., Martin, H., Robin, C. and Monzier, M.: 2002, *Geology* **30**(11), 967–970.

Samaniego, P., Martin, H., Robin, C., Monzier, M. and Cotten, J.: 2005, *J. Petrol.* **46**, 2225–2252.

Selsis, F.: 2000, Evolution of the atmosphere of terrestrial planets. From early Earth atmosphere to extrasolar planets. Ph. D. Thesis, University of Bordeaux.

Selsis, F.: 2002, Occurrence and detectability of O_2-rich atmosphere in circumstellar "habitable zones". ASP 269: The Evolving Sun and its Influence on Planetary Environments.

Selsis, F.: 2004. *The Prebiotic Atmosphere of the Earth.*, *Astrobiology, Future Perspective,* Kluwer, Astrophysics and Space Science Library.

Schrag, D. P., Berner, R. A., Hoffman, P. F. and Halverson, G. P.: 2002. *Geochem. Geophys. Geosyst.*, **3**(6), doi: 10.1029/2001GC000219.

Shen, Y. and Buick, R.: 2004, *Earth-Sci. Rev.* **64**(3−4), 243−272.

Shen, Y., Pinti, D. L. and Hahsizume, K.: 2006, in K. Benn, J.-C. Mareschal and K. Condie (eds.), *Archean Geodynamics and Environments. AGU Geophysical Monograph*, American Geophysical Union, Washington, DC, Vol 164, pp. 305−320.

Shirey, S. B. and Hanson, G. N.: 1984, *Nature* **310**, 222−224.

Shukolyukov, A., Castillo, P. and Simonson, B. W. L. G., 2002. Chromium in Late Archean spherule layers from Hamersley basin, Western Australia; isotopic evidence for extraterrestrial component, Lunar and Planetary Science Conference, Houston, Texas, pp. 1369 (abstract).

Shukolyukov, A., Kyte, F. T., Lugmair, G. W., Lowe, D. R. and Byerly, G. R.: 2000, in I. Gilmour and C. Koeberl (eds.), *Impacts and the Early Earth*, Springer, Berlin, pp. 99−115.

Simonson, B.: 2003, *Astrobiology* **3**(1), 49−65.

Simonson, B., Hassler, S. W., Smit, J. and Summer, D.: 2004, How Many Late Archean Impacts are Recorded in the Hamersley Basin of Western Australia, Lunar and Planetary Science, Houston Texas, pp. CD-ROM 1718 [Abstract].

Simonson, B. M.: 1992, *Geol. Soc. Am. Bull.* **104**, 829−839.

Simonson, B. M. and Glass, B. P.: 2004, *Ann. Rev. Earth Planet Sci.* **32**, 329−361.

Simonson, B. M. and Harnik, P.: 2000, *Geology* **28**(11), 975−978.

Simonson, B. M., Hassler, S. W. and Beukes, N. J.: 1999, Late Archean impact spherule layer in South Africa that may correlate with a Western Australian layer. Geological Society of America, Special paper 339.

Sleep, B. E. and McClure, P. D.: 2001, *J. Contam. Hydrol.* **50**(1−2), 21−40.

Sleep, N. H., Meibom, A., Fridriksson, T., Coleman, R. G. and Bird, D. K.: 2004, *Proc. Nat. Acad. Sci.* **101**(35), 12818−12823.

Sleep, N. H. and Zahnle, K.: 2001a, *J. Geophys. Res. Planets* **106**(E1), 1373−1399.

Sleep, N. H., Zahnle, K. and Neuhoff, P. S.: 2001, *Proc. Nat. Acad. Sci. USA* **98**(7), 3666−3672.

Sleep, N. H. and Zahnle, K. J.: 2001b, *J. Geophys. Res.* **106**, 1373−1400.

Smit, J.: 1999, *Ann. Rev. Earth Planetary Sci.* **27**, 75−113.

Smit, J., Alvarez, W., Montanari, A., Swinburne, N., Van Kempen, T. M., Klaver, G. T. and Lustenhouwer, W. J.: 1992, *Proc. Lunar Planetary Sci.* **22**, 87−100.

Smithies, R. H.: 2000, *Earth Planetary Sci. Lett.* **182**, 115−125.

Stein, M. and Hofmann, A. W.: 1994, *Nature* **372**, 63−68.

Svetov, S. A., Svetova, A. I. and Huhma, H.: 2001, *Geochem. Int.* **39**, 24−38.

Tian, F., Toon, O. B., Pavlov, A. A. and De Sterck, H.: 2005, *Science* **308**, 1014−1017.

Ueno, Y., Yoshioka, H., Maruyama, S. and Isozaki, Y.: 2004, *Geochimica et Cosmochimica Acta* **68**(3), 573−589.

Van Kranendonk, M. J., Hickman, A. H., Williams, I. R. and Nijman, W.: 2001, Archaean geology of the East Pilbara granite-greenstone terrane, Western Australia − A field guide, Geological Survey of Western Australia, Perth.

Veizer, J. and Compston, W.: 1976, *Geochimica et Cosmochimica Acta* **40**, 905−914.

Veizer, J., Hoefs, J., Ridler, R. H., Jensen, L. S. and Lowe, D. R.: 1989, *Geochimica et Cosmochimica Acta* **53**, 845–857.
von Damm, K. L., Oosting, S. E., Kozlowski, R., Buttermore, L. G., Colodner, D. C., Edmonds, H. N., Edmond, J. M. and Grebmeir, J. M.: 1995, *Nature* **375**, 47–50.
Walker, J. C. G., Hays, P. B. and Kasting, J. F.: 1981, *J. Geophys. Res.* **86**, 9776–9782.
Wallace, M. W., Gostin, V. A. and Keays, R. R.: 1990, *Geology* **18**(2), 132–135.
Weiershauser, L. and Spooner, E. T. C.: 2005, *Precambrian Res.* **138**(1–2), 89–123.
Weiss, R. F.: 1970, *Deep-Sea Res.* **17**, 721–735.
Wells, L. E., Armstrong, J. C. and Gonzalez, G.: 2003, *Icarus* **162**, 38–46.
Young, G. M., von Brunn, V., Gold, D. J. C. and Minter, W. E. L.: 1998, *J. Geol.* **106**, 523–538.
Zahnle, K. and Sleep, N. H.: 2002, in C. M. R. Fowler, C. J. Ebinger and C. J. Hawkesworth (eds.), *The Early Earth: Physical, Chemical and Biological Development*, Geological Society, London, pp. 231–257.
Zegers, T. E., de Wit, M. J., Dann, J. and White, S. H.: 1998, *Terra Nova* **10**, 250–259.

Earth, Moon, and Planets (2006) 98: 247–290
DOI 10.1007/s11038-006-9091-9

7. Ancient Fossil Record and Early Evolution (ca. 3.8 to 0.5 Ga)

PURIFICACIÓN LÓPEZ-GARCÍA and DAVID MOREIRA

Unité d'Ecologie, Systématique et Evolution, Université Paris-Sud, Orsay, France
(E-mail: puri.lopez@ese.u-psud.fr)

EMMANUEL DOUZERY

Institut des Sciences de l'Evolution, Université Montpellier II, Montpellier, France
(E-mail: douzery@isem.univ-montp2.fr)

PATRICK FORTERRE

Institut de Génétique et Microbiologie, Université Paris-Sud, Orsay, France
(E-mail: forterre@igmors.u-psud.fr)

MARK VAN ZUILEN

Equipe Géobiosphère Actuelle et Primitive, Institut de Physique du Globe, Paris, France
(E-mail: vanzuilen@ipgp.jussieu.fr)

PHILIPPE CLAEYS

Department of Geology, Vrije Universiteit, Brussels, Belgium
(E-mail: phclaeys@vub.ac.be)

DANIEL PRIEUR

Université Bretagne Occidentale, Brest, France
(E-mail: daniel.prieur@univ-brest.fr)

(Received 1 February 2006; Accepted 4 April 2006)

Abstract. Once life appeared, it evolved and diversified. From primitive living entities, an evolutionary path of unknown duration, likely paralleled by the extinction of unsuccessful attempts, led to a last common ancestor that was endowed with the basic properties of all cells. From it, cellular organisms derived in a relative order, chronology and manner that are not yet completely settled. Early life evolution was accompanied by metabolic diversification, i.e. by the development of carbon and energy metabolic pathways that differed from the first, not yet clearly identified, metabolic strategies used. When did the different evolutionary transitions take place? The answer is difficult, since hot controversies have been raised in recent years concerning the reliability of the oldest life traces, regardless of their morphological, isotopic or organic nature, and there are also many competing hypotheses for the evolution of the eukaryotic cell. As a result, there is a need to delimit hypotheses from solid facts and to apply a critical analysis of contrasting data. Hopefully, methodological improvement and the increase of data, including fossil signatures and genomic information, will help reconstructing a better picture of life evolution in early times as well as to, perhaps, date some of the major evolutionary transitions. There are already some certitudes. Modern eukaryotes evolved after bacteria, since their mitochondria derived from ancient bacterial endosymbionts. Once prokaryotes and unicellular eukaryotes had colonized terrestrial ecosystems for millions of years, the first pluricellular animals appeared and radiated, thus inaugurating the Cambrian. The following sections constitute a collection of independent articles providing a general overview of these aspects.

Keywords: Biomarkers, Cambrian explosion, early evolution, microfossil, origin of eukaryotes

7.1. The First Traces of Life

MARK VAN ZUILEN

Approximately a century ago the fossil record, based predominantly on macroscopic morphologic evidence, could only be traced to the beginning of the Cambrian (544 Ma). The record of life therefore comprised only about 12% of the total history of the Earth (4500 Ma). However, the abrupt appearance of complex organisms suggested that more primitive forms of life must have occurred before the Cambrian. Indeed, as more rock formations of Precambrian age were studied, many microfossils and even macrofossils of primitive life forms were found (Schopf, 2000). The understanding of life through time was greatly improved with the development of new types of tracers, or 'biosignatures', which include isotopic, mineralogic and chemical indicators. In addition some indirect geochemical evidence has been used to invoke the presence of life; e.g. the first appearance of oxygenic photosynthesizing bacteria is believed to have preceded the rise of oxygen in the atmosphere at ca. 2.3 Ga ago. Due to these new developments, the start of the evolution of life was slowly pushed back in time (Schopf and Klein, 1992; Knoll, 1999). Beyond the Proterozoic–Archean boundary at 2.5 Ga, however, two fundamental changes do occur that greatly challenge the search for traces of early life.

Firstly, a change in geologic processes due to which a simple interpretation of paleo-environmental conditions is difficult. The Hadean and Archean were dominated by high temperature ultramafic volcanism (e.g. komatiites) and chemical sediment deposition (e.g. Banded Iron Formations – BIF – and cherts). The inferred high geothermal gradient (see Chapter 6.1: evolution in geological mechanisms: the 2.5 Ga transition) and predominantly anoxic surface conditions (see Chapter 6.3: atmospheric and ocean physiochemical evolution) must have shaped the nature and habitat of early forms of life. Hydrothermal settings such as are found today at mid-ocean ridges were common and chemoautotrophic life would have been the most dominant life form on the early Earth (Nisbet and Sleep, 2001). The metabolism of these organisms depends on reduced chemical species (e.g. CH_4, H_2, S, H_2S, Fe^{2+}) that are released during alteration of ocean floor volcanic rocks. The search for traces of early life should therefore be directed to those specific environments (hydrothermal vents, subaerial hotspring deposits, chilled margins of pillow basalts) that were likely to have harbored these primitive groups of organisms.

Secondly, a change in degree of preservation of the rock record due to which conventional paleontological tools become highly ineffective. In the progressively metamorphosed rock record of the Early Archean all currently reported types of biosignatures have been found to be ambiguous. Possible microfossil structures have lost most of their original morphology, organic compounds including molecular biomarkers have turned into kerogen or crystalline graphite (of uncertain origin), and isotope signatures have been blurred by exchange reactions and hydrothermal processes. Furthermore, several abiologic metamorphic reactions have been identified that can produce kerogen or graphite, and specific abiologic processes have been described that can generate complex structures that resemble microfossils. Macroscopic fossil evidence in the form of stromatolites has been controversial as well, since several abiologic processes were identified that resulted in similar degrees of structural complexity. In summary, several processes associated with metamorphism (strain, deformation, hydrothermal fluid circulation, metasomatic mineral deposition, thermal degradation) have made the search for traces of early life a great challenge, and has led to several ongoing controversies.

Currently it is difficult to declare with certainty what the oldest trace of life is, and importantly what its nature and habitat were. Life can be traced unambiguously to approximately 2.7 Ga ago, based on well-described morphological microfossils and especially molecular fossils. Beyond this point many claims for biologic processes have been made, and all of them have to some degree been drawn into question (Fig.7.1). Some of the intriguing but controversial early Archean traces include (1) isotopically light graphite inclusions in older than 3.8 Ga rocks from Akilia island and the Isua Supracrustal Belt in southwest Greenland, (2) kerogenous microstructures, stromatolites and diverse stable isotope ratio anomalies in 3.5 Ga cherts from the Pilbara Granitoid–Greenstone Belt in Western Australia, (3) kerogenous microstructures, stromatolites, and diverse stable isotope ratio anomalies in cherts, as well as microscopic tubes in altered pillow basalts from the 3.4–3.2 Ga Barberton Greenstone Belt in South Africa. The purpose of this chapter is to show the problems associated with the search for the earliest traces of life. Paragraph 7.1.1 describes some commonly used tools (biomarkers, microfossil morphology, isotope ratios), and their limitations with regard to the Archean rock record. Paragraph 7.1.2 is an account of some important field examples of the earliest traces of life. This is by no means a rigorous discussion of all work that has been done on describing the early biosphere on Earth. For a classical account of all fossil evidence of the Archean, the reader is referred to 'The Earth's earliest biosphere' (Schopf, 1983).

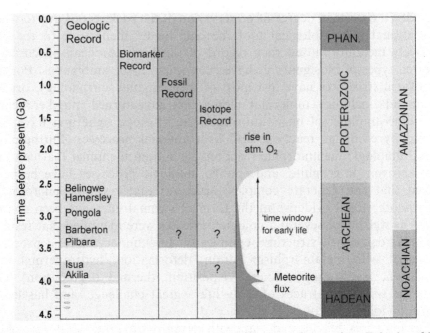

Figure 7.1. A simplified overview of the record of life on Earth over time. Episodes of high meteorite flux to the Earth during the Hadean ceased after ca. 3.9 Ga. Life did not exist or was frequently destroyed during this time (Chyba, 1993). The rise in atmospheric oxygen at ca. 2.3 Ga (Bekker et al., 2004) was caused by oxygenic photosynthesizing bacteria, which are relatively highly evolved organisms. These two events therefore define the 'time window' for the origin and evolution of early life on Earth. Metamorphic alteration of rocks older than 2.7 Ga has caused ambiguity (indicated by question marks) in the interpretation of microfossil and carbon isotopic evidence for life.

7.1.1 THE TRACERS

7.1.1.1 *Morphological fossils*

Macroscopic fossils of Archean age are scarce. Apart from occurrences in the Neoproterozoic of preserved multicellular life (see, for instance Schopf and Klein, 1992; Xiao et al., 1998; Chen et al., 2004), the only macroscopic evidence of Precambrian life comes from stromatolites (Grotzinger and Knoll, 1999). These are laminated, accretionary structures which are commonly regarded to have formed by the sediment-binding or direct carbonate precipitating activities of microbial mats or biofilms composed of photosynthesizing and associated bacteria. Fossil stromatolites only rarely contain individual microfossils. The fine microfabric that is observed in extant stromatolites (Stolz et al., 2001) has been destroyed by diagenesis and metamorphism, leaving only the overall macroscopic appearance of Archean stromatolites as an indicator for biogenicity. The biologic processes that control growth of stromatolites is currently the subject of intense study (see, for instance Reid et al., 2000; Bosak et al., 2004) and different mathematical models for stromatolite surface growth have been proposed to either argue for (Batchelor et al., 2004) or against (Grotzinger and Rothman, 1996) a biologic control. Several stromatolite structures have been documented in greenstone belts from South Africa and Western Australia that are older than 3.2 Ga. The biologic origin of these structures is the subject of debate, since they only meet several but not all of the criteria for biogenicity (Buick et al., 1981). Abiologic explanations include evaporitic precipitation, soft-sediment deformation, or silicious sinter formation around hot springs (Lowe, 1994).

Most microfossils in silicified Proterozoic microbial mats have relatively simple coccoid or filamentous morphologies and possess a limited number of attributes available for taxonomic characterization. Many characteristics of cell cultures can be modified during post-mortem degradation. Elevated temperature, pressure and strain can cause structures to flatten and ultimately loose their original three-dimensional shape (Schopf and Klein, 1992). These problems have led to many misinterpretations, and to classifications such as 'pseudofossils', 'non-fossils', or 'dubiofossils'. In fact undeformed microfossil shapes in moderately metamorphosed rocks are rather suspicious.

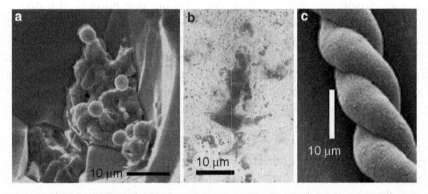

Figure 7.2. Potential problems in microfossil recognition. (a) Endolithic coccoids within a crack in an Isua BIF sample. The coccoids are embedded in extracellular polymeric substances, and although they are partially fossilized they can be recognized as post-metamorphic contamination. (Reprinted from Precambrian Research, Vol. 126, Westall, F. and Folk, R. L., Exogenous carbonaceous microstructures in Early Archaean cherts and BIFs from the Isua Greenstone Belt: implications for the search for life in ancient rocks, pages 313–330, copyright (2003), with permission from Elsevier). (b) Carbonaceous microstructure from the Apex Chert, Pilbara called it the 'ballerina', clearly showing a general problem of morphology; certain 'characteristic' shapes can easily be produced by metamorphic processes. (Reprinted by permission from Macmillan Publishers Ltd: Nature, Brasier M., Green O.R., Jephcoat A. P., Kleppe A., van Kranendonk M. J., Lindsay J. F., Steele A., and Grassineau N. (2002), Questioning the evidence for Earth's oldest fossils, Vol. 416, 76–81, copyright 2002). (c) Microstructures produced by abiologic processes in the laboratory. Such structures can absorb organics, and resemble true Archean microfossils (From Garcia-Ruiz J. M., Hyde S. T., Carnerup A. M., Van Kranendonk M. J., and Welham N. J. (2003) Self: assembled silica-carbonate structures and detection of ancient microfossils. Science 302, 1194–1197).

For instance, spherical objects in a 3.8 Ga metachert from the Isua Supra-crustal Belt were interpreted as microfossils (Pflug and Jaeschke-Boyer, 1979). Yet, the rock itself was shown to have experienced intense strain, which should have caused any spherical shape to deform into ellipsoids or more likely rod-shaped objects (Appel et al., 2003). These spheres are therefore clearly epigenetic, and indeed have been interpreted as limonite-stained fluid inclusions, cavities, or post-metamorphic endolithic contamination (Westall and Folk, 2003) (Figure 7.2a). In order to rule out simple shapes such as fluid inclusions or cavities, it has been suggested that putative microfossils should be of organic character. Carbonaceous struc-tures can be recognized using *in-situ* Laser-Raman Spectroscopy (Ku-dryavtsev et al., 2001; Schopf et al., 2002; Schopf and Kudryavtsev, 2005). However, the Raman spectrum of a putative microfossil can also be derived from abiologic forms of carbon such as graphitic coatings of fluid inclusions (Pasteris and Wopenka, 2002, 2003). Laser-Raman spectroscopy therefore is a necessary but inconclusive analysis for microfossil identification. In addi-tion to simple shapes such as fluid inclusions and cavities, abiologic processes have been recognized that can produce complex microscopic shapes that are capable of absorbing simple abiogenic organic compounds (Figures 7.2b, 7.2c). When metamorphosed such structures display the morphology and the Raman spectrum of a typical microfossil (Brasier et al., 2002; Garcia-Ruiz et al., 2003). In summary, both stromatolites and microfossils are difficult to interpret in the metamorphosed early Archean rock record. For this reason additional chemical, mineralogical, and isotopic tracers have been developed to provide further insight in this part of Earth history.

7.1.1.2 *Molecular fossils*

Molecular fossils, or biomarkers, are derived from characteristic cellular macromolecules, such as membrane lipids. Bacterial and thermal degrada-tion will destroy most biologic material, but some biolipids are transformed into highly resistant geolipids that still carry enough information to identify the original biologic source. Such geolipids have been used extensively to trace life in the geologic rock record. When exposed to higher degrees of thermal alteration, however, these compounds will slowly alter into insol-uble macromolecular kerogen (Durand, 1980). As only very small amounts of geolipids can be extracted from metamorphosed material, contamination issues become the biggest hurdle for unambiguous biomarker research of Archean rock samples. Brocks et al. (1999) stressed the importance to identify anthropogenic contamination (e.g. petroleum products from drilling activities), and different forms of post-Archean contamination (including local subsurface biological activity, groundwater containing biolipids, and most importantly migrated petroleum from another source rock that carries geolipids), before a biomarker is recognized as both indigenous to and syngenetic with the host rock. Currently, small amounts of such unambiguous biomarkers, representative of e.g. cyanobacteria, have been found in the 2.6 Ga Marra Mamba Formation of the Hamersley Group, and the 2.715 Ga Maddina Formation of the Fortesque Group, which both occur in Western Australia (Brocks et al., 1999). Most Archean rocks older than that have experienced more severe metamorphic alteration (lower-greenschist and up), and currently the biomarker record has only been traced to 2.7 Ga (Figure 7.1). New techniques, however, such as hydropyrolytic degradation (Brocks et al., 2003) provide promising venues for obtaining higher yields of extractable geolipids from Archean kerogen. It is therefore certainly possible that the biomarker record will be further extended into deep Archean time.

7.1.1.3 Isotope ratios

The carbon isotope ratio $\delta^{13}C$ is expressed as $\delta^{13}C = ([(^{13}C/^{12}C)_{sample}/(^{13}C/^{12}C)_{ST}]-1)*1000$ in per mil (‰), relative to a standard (ST = Vienna Pee Dee Belemnite, VPDB). Carbon isotope ratios have been used extensively to trace back life over the geological record (Hayes et al., 1983; Schidlowski, 1988, 2001). The two main reservoirs of carbon in sediments are carbonates with an average $\delta^{13}C$ of 0‰, and the remains of biologic material with an average $\delta^{13}C$ value of −25‰. This characteristic difference in isotope ratio between the two carbon reservoirs has been observed in many organic-rich sediments of different ages and is due to a kinetic isotope effect associated with irreversible enzyme-controlled metabolic pathways of autotrophic organisms (most of them photosynthetic). Carbonaceous material in Archean cherts has a low average $\delta^{13}C$ value of ca. −35 to −30 ‰ (Hayes et al., 1983; Ueno et al., 2004). If it is assumed that mantle-derived CO_2 at that time had a $\delta^{13}C$ similar to that of today (−5 ‰, Des Marais and Morre, 1984), CO_2-fixation by organisms should have produced a carbon isotope fractionation close to −30‰ relative to the source. Photosynthesizing organisms and methanogens are capable of producing this degree of carbon isotope fractionation (House et al., 2003). Unfortunately in moderately to highly metamorphosed rocks these initial isotopic ratios can be lost. Processes that can cause changes include isotope exchange with carbonates or CO_2-rich fluids (Schidlowski et al., 1979; Robert, 1988; Kitchen and Valley, 1995) and devolatilization reactions during metamorphism (Hayes et al., 1983). These processes shift the $\delta^{13}C$ of sedimentary biological material to higher values, making it isotopically indistinguishable from e.g. graphite that forms abiologically during metamorphic processes (van Zuilen et al., 2002, 2003).

The sulfur isotope ratio $\delta^{34}S$ is expressed as $\delta^{34}S = ([(^{34}S/^{32}S) \text{ sample}/(^{34}S/^{32}S)_{ST}]-1)*1000$ in per mil (‰), relative to a standard (ST = Canyon Diablo troilite, CDT). Sulfate-reducing bacteria preferentially reduce the light isotope, leading to isotopically depleted sulfides with a range in $\delta^{34}S$ between -10 to $-40\%_0$ (Canfield and Raiswell, 1999). In general a small range in $\delta^{34}S$ is observed for igneous rocks of about $0\%_0 \pm 5\%_0$. Therefore it has been suggested by many workers that the significantly low $\delta^{34}S$ values of sedimentary sulfide deposits provide a record of sulfur reducing bacteria over time (Ohmoto et al., 1993; Rasmussen, 2000; Shen et al., 2001). In rocks older than approximately 2.7 Ga most $\delta^{34}S$ values of sedimentary sulfides cluster around mantle sulfur isotope values (0 $\%_0$ with a range of ca. $10\%_0$). Hydrothermal fluid circulation can cause inorganic sulfate reduction and potentially cause isotope effects that fall in the observed range. The search for unambiguous biologic $\delta^{34}S$ signatures is further complicated by metamorphic overprinting. Especially in early Archean deposits it becomes difficult to establish a syngenetic origin of sulfide deposits.

The nitrogen isotope ratio $\delta^{15}N$ is expressed as $\delta^{15}N = ([(^{15}N/^{14}N)_{sample}/(^{15}N/^{14}N)_{ST}]-1)*1000$, in per mil $(\%_0)$ relative to a standard (ST = nitrogen-air standard, Nier, 1950). Beaumont and Robert (1999) have shown that the $\delta^{15}N$ of kerogen in Archean metasediments is several per mil $(\%_0)$ lower than that found in the modern biosphere (ca. $+5\%_0$), which they suggest is due to the absence of nitrifying and denitrifying bacteria, and the presence of nitrogen fixing bacteria in the mildly reducing Archean oceans. Specific low nitrogen isotope compositions have been observed in certain Archean hydrothermal settings (Pinti et al., 2001). Such isotopic ratios may be indicative of chemo-autotrophic bacteria, which occur in deep sea hydrothermal vent communities and derive NH_3 directly from hydrothermal fluids. However, van Zuilen et al. (2005) have observed low $\delta^{15}N$ in graphites from the 3.8 Ga Isua Supracrustal Belt, and argued that mantle type nitrogen $(-5\%_0)$ could have been incorporated during secondary metasomatic processes. Nitrogen that is lost from biogenic material by devolatilization during metamorphism, can be incorporated in clay minerals in the form of NH_4^+ were it substitutes for K^+. When these clay minerals recrystallize at high-grade metamorphism this ammonium ion is retained in the resulting mica (e.g. tobelite). It has therefore been suggested that NH_4^+ concentration in micas and the associated $\delta^{15}N$, could act as a potential indirect biomarker in metasedimentary rocks (Papineau et al., 2005 and references therein). However, metamorphism could drive the $\delta^{15}N$ of residual nitrogen in rocks to higher values (Bebout and Fogel, 1992), making it difficult to interpret $\delta^{15}N$ as a biosignature.

The iron isotope ratio $\delta^{56}Fe$ is expressed as $\delta^{56}Fe = ([(^{56}Fe/^{54}Fe)_{sample}/(^{56}Fe/^{54}Fe)_{ST}]-1)*1000$ in per mil $(\%_0)$ relative to a standard (IRMM-014 reference material). Igneous rocks worldwide have a near-constant $\delta^{56}Fe$ of zero $\%_0$, and geologic processes such as melting/crystallization or weathering do not cause significant isotope fractionation (Johnson et al., 2004b). The isotopic composition of iron is affected by biological processes and can potentially be used to trace life in ancient environments (Johnson et al.,

2004b and references therein). For instance Fe(II)-oxidizing anoxygenic photosynthesizers can cause a shift in δ^{56}Fe of +1.5‰ (Croal et al., 2004). Shifts towards negative δ^{56}Fe are observed for Fe(III)-reducing bacteria (Johnson et al., 2005). A range of δ^{56}Fe values (between ca. −1 and +1‰) has been observed in Precambrian BIFs (Johnson et al., 2003; Dauphas et al., 2004; Yamaguchi et al., 2005); negative values in the more reduced mineral phases (pyrite, siderite) and positive values in oxidized mineral phases (magnetite, hematite). It has been suggested before that BIFs are the direct (Konhauser et al., 2002) or indirect (Beukes, 2004) result of photosynthetic activity. The observed positive shifts in Fe-isotope ratio could be a further confirmation of these hypotheses. However, there are abiologic processes by which BIFs can form. For instance direct photodissociation of ocean surface water by UV-radiation could lead to oxidation and precipitation of BIF (Braterman et al., 1983). It is not known to what extent iron isotopes are fractionationed during such a process, and therefore it remains difficult to use BIFs and their Fe-isotope ratio as a biosignature. Furthermore, there are abiologic redox processes that can cause a shift to positive δ^{56}Fe (Bullen et al., 2001). The analysis of δ^{56}Fe is a relatively new field of research (Anbar, 2004) many aspects of iron isotope fractionation still remain to be studied. The use of δ^{56}Fe therefore remains a promising tool for tracing early Archean life.

7.1.2 EXAMPLES FROM THE FIELD

7.1.2.1 Before 3.8 Ga: Akilia Island, West Greenland

A highly metamorphosed quartz–pyroxene rock on the southwestern tip of Akilia Island has for long been the center of attention regarding the oldest traces of life on Earth. This five-meter wide outcrop (Figure 7.3) was interpreted as a BIF and was found to contain graphite inclusions within apatite crystals (Mojzsis et al., 1996). The low δ^{13}C of these graphite inclusions suggested a biologic source material that had retained its original carbon isotope signature. This claim has since been the center of controversy, as it was argued that the protolith of this rock was not a BIF, but instead a highly metasomatized ultramafic rock which does not represent a marine depositional setting and would not be able to harbor traces of ancient life (Fedo and Whitehouse, 2002a). Since then, geochemical data has been presented to either argue for or against a sedimentary origin (Fedo and Whitehouse, 2002b; Friend et al., 2002; Mojzsis and Harrison, 2002b; Palin, 2002; Mojzsis et al., 2003; Whitehouse et al., 2005). Recently, iron isotope systematics and trace element ratios have been used to establish more firmly a sedimentary origin. Dauphas et al. (2004) observed δ^{56}Fe up to +1.1‰ in the fine-grained part of this quartz–pyroxene rock, which is in line with values observed in

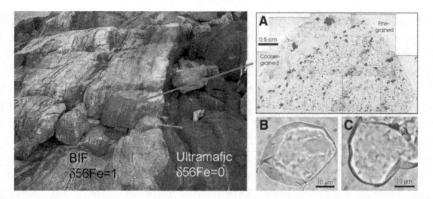

Figure 7.3. (*Left*) Overview of Akilia outcrop, (A) Petrographic thin section of sample G91-26, in which traces of life were found (Lepland et al., 2005). The dashed line shows the contact between fine- and coarser-grained layers. Apatite crystals are common in the fine-grained part, but inclusions of biologically derived graphite are extremely rare. Two of such inclusion-free apatite crystals are shown below (B, C) (Figure 7.3B is reprinted from Lepland A., van Zulien M. A., Arrhenius A., Whitehouse M., and Fedo C. M. (2005) Questioning the evidence for Earth's earliest lif - Akilia revisited. Geology, Vol. 33, 77–79).

younger BIFs (Figure 7.3). It is important to establish that such a positive δ^{56}Fe is not the result of metasomatic alteration of an original igneous protolith. For instance altered mid-ocean ridge basalts (MORB) have positive δ^{56}Fe values (Rouxel et al., 2003). However, such positive δ^{56}Fe correlates with a depletion in Fe concentration. Loss of Fe would be evident in the comparison of the ratios of Fe to an immobile trace element (e.g. Ti, Nb, Hf). If a similar loss of Fe occurred as a result of metasomatic alteration of the quartz–pyroxene rock on Akilia Island, it would be evident in the comparison of the Fe/Ti ratio between this rock and the surrounding igneous rocks. A high Fe/Ti ratio in the quartz–pyroxene rock indicates that Fe was not preferentially lost. On the contrary, the high Fe/Ti ratio resembles those of BIFs found in Isua (Dymek and Klein, 1988).

Apart from the discussion regarding the protolith, the age of this rock is still debated (Whitehouse et al., 1999; Mojzsis and Harrison, 2002a), and the claim for the oldest trace of life on Earth is still strongly contested. As was discussed the positive δ^{56}Fe is in itself not an unambiguous biosignature, since an abiologic process such as photodissociation of ocean water by UV-radiation could have caused iron isotope fractionation. In addition, the graphite inclusions in apatite, claimed to be the remnants of microorganisms, are extremely rare (Lepland et al., 2005; Mojzsis et al., 2005; Nutman and Friend, 2006). Furthermore, it is contested whether the apatite crystals themselves are as old as the surrounding rock matrix (Mojzsis et al., 1999; Sano et al., 1999) casting doubt on their syngenetic character.

7.1.2.2 *3.8 Ga: The Isua Supracrustal Belt, West Greenland*

Early work on the 3.8 Ga old Isua Supracrustal Belt (ISB) in southern West Greenland showed evidence for a marine depositional setting; BIFs, metacherts, pillow lava structures, carbonates, and felsic metasediments in which graded bedding is locally preserved. The occurrences of siderite and dolomite, occasionally interlayered with quartzite in the ISB, appears similar to marine platform deposits that are found throughout the Precambrian and the Phanerozoic, and in early studies this field appearance led to the interpretation of a shallow marine, subtidal depositional environment (Dimroth, 1982). The ISB has a complex metamorphic history; evidence has been reported for multiple episodes of early Archean deformation and metamorphism (Nutman et al., 1996). These events were responsible for amphibolite-facies metamorphism, reaching temperatures between 500°C and 600°C and pressures to 5–5.5 kbar (Boak and Dymek, 1982). Biologic remains in these sedimentary sequences would therefore have been converted to crystalline graphite. It has been suggested in several studies that graphite contained in the ISB could be biogenic in origin (Schidlowski et al., 1979; Hayes et al., 1983; Mojzsis et al., 1996). The wide range of carbon isotope ratios ($\delta^{13}C$ range from -25 to -6%) of graphite in carbonate rich rocks has been interpreted to reflect post-depositional isotopic equilibration of graphitizing organic matter with co-existing carbonates. More recent work has shown inconsistencies in this interpretation (van Zuilen et al., 2002, 2003). Protoliths of several carbonate-rich rocks in Isua have been reinterpreted (Rose et al., 1996; Rosing et al., 1996) as secondary metasomatic and not as sedimentary in origin. This fundamental reinterpretation of the protolith would rule out a biogenic origin of graphite in metasomatic rocks. Graphite was found in large quantities in such metasomatic carbonate-rich rocks, whereas no distinguishable graphite particles were found in sedimentary BIF and metacherts. In these latter samples a very low concentration of reduced carbon was measured (less than 100 ppm), that could be combusted at relatively low temperature (450 °C). Since graphite typically combusts around 700–800 °C and all syngenetic organic material in Isua should have turned into graphite during metamorphic events, it can be concluded that the small amounts of isotopically light reduced carbon in these samples are mainly derived from post-metamorphic (and thus much younger, non-indigenous) organic material. In contrast, the metasomatic carbonate veins within mafic country rocks contain graphite-siderite-magnetite assemblages (Figure 7.4), suggesting that graphite and magnetite in these rocks are the products of partial thermal disproportionation of the carbonate. The siderite ($FeCO_3$) disproportionation reaction, yielding graphite and magnetite ($6FeCO_3 \rightarrow 2Fe_3O_4 + 5CO_2 + C$) has been studied in detail at metamorphic P, T, fO_2-conditions (French, 1971), and has been suggested earlier as a

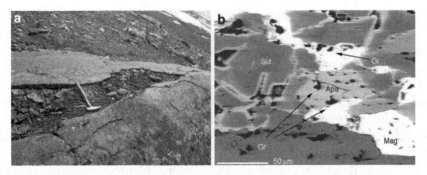

Figure 7.4. a Outcrop of a metacarbonate vein within mafic country rock, eastern part of the ISB. (b) SEM-BSE image of a metacarbonate thinsection. Mineral phases Sid: MgMn–siderite; Apa: apatite; Mag: magnetite; Gr: graphite (Reprinted by permission from Macmillan Publishers Ltd: Nature, van Zulien M. A., Lepland A., and Arrhenius, G., (2002), Reassessing the evidence for the earliest traces of life, vol. 418, pages 627–630).

possible mechanism for graphite formation in the amphibolite facies (T ca. 550 °C; P ca. 5 kBar) ISB (Perry and Ahmad, 1977).

Micrometer-size graphite inclusions with a pronounced light $\delta^{13}C$ value (weighted mean $-30 \pm 3\%_o$; ion microprobe data) were reported to occur in apatite crystals from the ISB (Mojzsis et al., 1996). The graphite was thought to have escaped isotope exchange with the associated carbonates due to armoring by the host apatite. This claim was based on a rock sample that at the time was believed to represent a sedimentary BIF. More recent petrographic analysis has revealed that it contains MgMn–siderite–magnetite–graphite associations and is compositionally akin to Isua metacarbonates (Lepland et al., 2002; van Zuilen et al., 2002). Furthermore, the REE pattern of these graphite-bearing apatites is distinctly different from apatites occurring in sedimentary rocks (Lepland et al., 2002). As is shown in Figure 7.4b graphite is not restricted to apatite, but occurs as inclusions in most other phases too. The petrographic and geochemical evidence strongly suggests that this graphite is produced epigenetically through thermal disproportionation of ferrous carbonate during one or several thermal events later than 3.8 Ga. The isotopic systematics of the process responsible for formation of isotopically light graphite (weighted mean $-30 \pm 3\%_o$) enclosed in apatite crystals remains to be studied, but petrographic evidence clearly excludes a primary biogenic origin.

Several other isotopically light graphitic globules have been reported from the Isua region (Rosing, 1999; Ueno et al., 2002). The most intriguing are those that occur in graded beds from the western part of Isua (Rosing, 1999; Rosing and Frei, 2003). This rock outcrop is characterized by significant graphite content, lack of Fe-bearing carbonate, and graphite $\delta^{13}C$ values that are significantly lower than the Fe-carbonate derived graphite described above. A biologic origin can therefore not be excluded, and further research is necessary to confirm this claim.

7.1.2.3 3.5 Ga: Pilbara, Western Australia

Some of the oldest traces of life on Earth have been found in chert horizons (predominantly the Dresser Formation at North Pole Dome, the Apex chert, and the Strelley Pool chert) occurring in the lower Warrawoona Group, eastern Pilbara Craton, Western Australia (Van Kranendonk and Pirajno, 2004), and include stromatolites (Allwood et al., 2006; Walter et al., 1980; Awramik et al., 1983; Lowe, 1983; Hoffman et al., 1999; Van Kranendonk et al., 2003), microfossils of photosynthesizing bacteria (Schopf, 1983, 1993; Schopf et al., 2002), sulfur isotopic evidence of sulfate-reducers (Shen et al., 2001), carbon isotopic evidence for autotrophic life (Hayes et al., 1983; Ueno and Isozaki, 2001; Ueno et al., 2004), and nitrogen isotopic evidence of chemoautotrophic life (Pinti et al., 2001). The validity of these claims strongly depends on the geological context, and especially on the process of formation of the chert horizons in which most of these traces of life are found. The origin of several chert units in the Warrawoona Group has been controversial. The bedded chert–barite horizons in the Dresser Formation were originally interpreted as evaporitic and clastic deposits that were silicified by later hydrothermal circulation (Buick and Dunlop, 1990). In contrast, several lines of evidence suggest that such chert–barite horizons were formed as exhalite deposits associated with hydrothermal seafloor alteration (Van Kranendonk and Pirajno, 2004). This is particularly evident from the swarms of chert–barite veins (Figure 7.5a) that terminate into shallow evaporitic bedded chert–barite units. Such episodes of hydrothermal activity are thought to be associated with caldera formation, as is inferred from chert veins that developed in active growth faults (Nijman et al., 1999). The Apex

Figure 7.5. (a) Swarms of hydrothermal chert feeder dikes within the Dresser Formation at North Pole Dome, Pilbara, Western Australia. (b) Outcrop of the Apex Chert near Chinaman Creek, Pilbara, Western Australia. Microfossil structures described by Schopf (1993) and Schopf et al. (2002) occur in a hydrothermal feeder dike, not in the actual exhalative seafloor portion of the Apex Chert.

Chert (lower Warrawoona Group) probably formed by a similar process of syngenetic hydrothermal activity, since many feeder dikes have been observed to terminate into it (Figure 7.5b). Recently a similar syngenetic process has been suggested for the origin of black chert dikes and veins within the Strelley Pool chert (Lindsay et al., 2005), although post-depositional hydrothermal activity could have played a role as well. In summary, high temperature ocean floor hydrothermal processes were of key importance for the emplacement of many cherts, and by implication could have been fully or partially responsible for the observed 'biosignatures'.

Most importantly it has been suggested that serpentinization by circulating CO_2-rich fluids at depth in the ocean floor basaltic crust could have prompted hydrocarbon formation by Fischer–Tropsch (FT) type reactions (Figure 7.6a). Modern analogues for this process are mid ocean ridge hydrothermal systems, where abiologic hydrocarbon formation has been observed (Charlou et al., 1998; Holm and Charlou, 2001). FT reactions indeed seem to produce hydrocarbons with a $\delta^{13}C$ range that is similar to biologic material (McCollom and Seewald, 2006). Lindsay et al. (2005) observed low $\delta^{13}C$ carbonaceous clumps and wisps only within a specific depth range of the chert feeder dikes that terminate into the Strelley Pool Chert. They suggest that this depth range corresponds to the optimal conditions for

---

Abiologic formation of organic compounds:

A) Hydrothermal alteration of ultramafic rocks leads to serpentinization and production of H_2, via the general, non-stoichoimetric reaction scheme (Fo_{88}= 88 Forsterite, 12 Fayalite):

1) Olivine (Fo_{88}) + H_2O = Serpentine + Brucite + Magnetite + H_2

Under these conditions dissolved CO_2 can be reduced to CH_4 and hydrocarbons.
This process can be described as a Fischer-Tropsch synthesis, where certain mineral phases (e.g. chromite, magnetite, awaruite) act as catalysts:

2) $12 \, CO_2 + 37 \, H_2 = C_{12}H_{26} + 24 \, H_2O$

B) Siderite decomposition in the presence of water vapor (300°C) can generate a variety of organic compounds (McCollom, 2003). A general (non-stoichoimetric) reaction of this kind could take place during partial decomposition of iron carbonates in a hydrothermal system (Garcia-Ruiz et al., 2003) or could lead to graphite formation during moderate to high grade metamorphism (van Zuilen et al., 2002):

3) $FeCO_3 + H_2O = Fe_3O_4 + CO_2 + CO + H_2 +$ organic compounds

---

Figure 7.6. Abiologic formation of organic compounds. (a) FT-type reactions associated with serpentinization produce methane and low molecular weight organics that could have been a source of energy for lithoautotrophic bacteria. Alternatively, high-molecular weight organics and kerogen could have been produced directly by FT-type reactions, providing an entirely abiologic explanation of carbonaceous structures within hydrothermal feeder dikes. (b) Thermal decomposition of iron carbonates in the presence of water vapor can lead to the formation of a complex mixture of hydrocarbons.

FT reactions (P ca. 500 kBar, T ca. 300 °C). It must be stressed, though, that FT reactions appear to be only efficient in the vapor phase, and are exceedingly more sluggish in hydrothermal systems (McCollom and Seewald, 2001). Ueno et al. (2004) observed low $\delta^{13}C$ carbonaceous structures in the chert feeder dike swarms of the Dresser Formation and noted the absence of an effective catalyst mineral phase and a relatively low hydrothermal temperature (ca. 100–200 °C). They therefore suggest that chemolithoautotrophic organisms may actually have been present in these feeder dikes. The notion that hydrothermal processes could at least in part have produced carbonaceous material in cherts of the Warrawoona Group has led to some important reinterpretations of previously recognized microfossil life. Carbonaceous microstructures resembling fossilized bacteria were reported from the 3.5-Ga-old Apex chert (Schopf, 1993; Schopf et al., 2002). At the time this chert was thought to represent a shallow marine depositional setting in which photosynthetic bacteria could thrive. However, it was subsequently shown by Brasier et al. (2002) that the samples studied by Schopf (1993) actually represent one of the hydrothermal feeder dikes that terminates in a bedded chert horizon (Figure 7.5b). It is highly unlikely that photosynthetic bacteria occurred in such a sub-seafloor hydrothermal setting. Instead, Brasier et al. (2002) argued for the abiologic origin of the observed carbonaceous particles by FT reaction (Figure 7.6). Alternatively, the carbonaceous structures in this feeder dike may have formed by partial decomposition of iron carbonates (as suggested by Garcia-Ruiz et al., 2003, see Figure 7.6b) or may be actually the remnants of chemoautotrophic organisms, in analogy to the suggestion made by Ueno et al. (2004). Many aspects of hydrothermal systems and associated conditions for chemoautotrophic life remain to be studied. As Lindsay et al. (2005) suggest, the abiotic organic output of such systems may overwhelm the signatures of primitive life that are present, and therefore make it the most difficult environments in which to recognize a record of the early biosphere.

Hydrothermal processes would also have caused the emplacement of secondary mineral phases. For instance pyrite with low $\delta^{34}S$ and of putative biological origin could therefore have had a hydrothermal origin. Shen et al. (2001), however, provide compelling evidence that rules out a hydrothermal origin; pyrite grains in bedded chert–barite horizons of the Dresser Formation were found to occur along the original crystal phases of primary gypsum (Buick and Dunlop, 1990). Since gypsum is unstable above 60 °C it implies that these sulfide crystals were emplaced before hydrothermal conversion to barite took place.

7.1.2.4 *3.4–3.2 Ga: Barberton, South Africa*
Traces of life have been found in chert deposits occurring in the Onverwacht Group of the 3.4–3.2 Ga Barberton Greenstone Belt, South Africa (de Wit

and Hart, 1993; Lowe and Byerly, 1999), and include stromatolites (Byerly et al., 1986), microfossils (Walsh, 1992; Westall et al., 2001), sulfur isotopic evidence for sulfate-reducers (Ohmoto et al., 1993), carbon isotopic evidence for autotrophic life (Hayes et al., 1983; Robert, 1988), and stratigraphically constrained carbonaceous structures (Tice et al., 2004; Tice and Lowe, 2004). In addition, (Furnes et al., 2004) described micrometer scale mineralized tubes within the chilled margins of pillow lava structures from basaltic units within the Onverwacht Group.

As was concluded for the traces of life in Pilbara, the validity of these claims strongly depends on the geological context. The Onverwacht Group represents a predominantly thoileiitic and komatiitic volcanic sequence in the lower part of the Swaziland supergroup, that has experienced regional metamorphic alteration as late as 2.7 Ga, when temperatures between 200 °C and 320 °C were reached (Tice et al., 2004). Carbonaceous Chert beds often occur as top layers on volcanic formations (capping cherts) throughout the Onverwacht Group (Figure 7.7a).

Knauth and Lowe (1978) argued that most cherts in the Overwacht Group formed from low-energy diagenetic replacement of preexisting sedimentary and pyroclastic deposits. In the absence of silica-precipitating organisms in the early Archean ocean, silica concentration would be high. Under such circumstances the diffusive flux of silica is directed from ocean water to the sediments (Siever, 1994), leading to silicification of the sediments below (Lowe and Byerly, 1999). In addition to silicified sediments, many underlying volcanic units have been silicified as well. Lowe and Byerly (1986) suggested that these silicified parts of volcanic units represent flow-top alteration zones. Such zones of shallow marine/subaerial alteration formed during intervals of volcanic quiescence. Regional subsidence led to deposition of volcaniclastic material and local growth of stromatolites in marginal evaporitic environ-

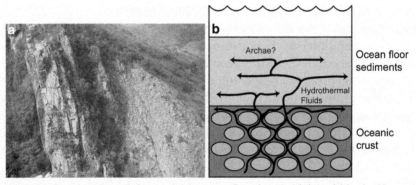

Figure 7.7. (a) Capping chert (left) overlying ocean floor basalt (right) within the Hooggenoeg Formation. (b) Schematic of a convective seawater circulation model (background drawing based on Paris et al., 1985). Hot fluids circulate the basaltic ocean floor and cause serpentinization, and precipitate silica when cooled.

ments. Synsedimentary silicification of these deposits then produced impermeable carbonaceous capping cherts on top of these komatiite flows. In another model Paris et al. (1985) have suggested that silica addition to both volcanic units and overlying sediments is the result of convective seawater circulation that acted directly on the oceanic crust and overlying sediments. Higher heat flux and greater availability of Mg,Fe-rich silicates (ultramafic/mafic basalts) in the early Archean caused effective serpentinization of oceanic crust that produced large quantities of silica-enriched hydrothermal fluids. Lateral migration of such fluids caused silicification of the upper ocean floor basalts and overlying sediments (Figure 7.7b). In addition to this relatively low-temperature hydrothermal circulation model, it has been suggested that some ferruginous chert deposits in Barberton could represent exhalites from local high-temperature hydrothermal vents. This model is based on the occurrence of ironstone pods in close association with BIFs and ferruginous cherts in the region (de Wit et al., 1982; de Ronde and Ebbesen, 1996). The origin of these ironstone pods, however, is still controversial. For instance, it has been suggested that these ironstone bodies are of Quaternary age, and are therefore irrelevant to interpretations of Archean hydrothermal events (Lowe and Byerly, 2003). Alternative explanations for the origin of these iron stone pods, however, have been suggested (Lowe et al., 2003).

As was discussed for Pilbara, serpentinization by circulating CO_2-rich fluids at depth in the ocean floor basaltic crust could have prompted hydrocarbon formation by Fischer–Tropsch (FT) type reactions. However, organics should then be found at a specific depth range within chert feeder dikes that cross-cut the underlying ultramafic/mafic volcanic sequence. Although such carbonaceous veining has been observed, it is clearly not as abundant as in Pilbara. Instead, there is a clear division between the regionally traceable carbonaceous cherts and underlying organics-free volcanic alteration zones. A simple comparison with carbonaceous cherts from Pilbara can therefore not be made, and careful studies are required to resolve the possible biologic origin of the observed organic structures. For example Tice and Lowe (2004) described a stratigraphic sequence within the Buck Reef Chert in the lower Kromberg Formation, and found that carbonaceous debris were restricted to specific shallow marine siderite-dominated successions. The apparent confinement to the photic zone, and the absence of a locally oxidized environment (such as is normally inferred from magnetite or hematite rich BIF's, Beukes, 2004), led them to conclude that anoxygenic photosynthetic life was present at this time. Such organisms use H_2, H_2S or Fe^{2+} as an electron donor for their metabolism. These elements would be readily available in an environment that was dominated by ocean floor hydrothermal alteration processes. The observations by Tice et al. (2004) and by Furnes et al. (2004) suggest that early Archean life may have been intimately linked to ocean floor hydrothermal alteration processes.

7.1.3 THE CHALLENGES AHEAD

From the field examples discussed above, it can be concluded that studies of early life in the incomplete and strongly metamorphosed Archean rock record face specific challenges. There is a strong need for careful description of geological context, identification of secondary metamorphic processes, and detailed structural, isotopic and chemical description of microstructures that are indigenous to and syngenetic with the rock formations. One improvement for this field of research is the increased availability of representative, relatively fresh Archean rock samples. Recently several scientific drilling projects have been initiated in Archean terrains, with the specific purpose to study traces of early life. Such drill cores represent samples that have been protected from weathering processes, and are less prone to biologic surface contamination. On the other hand, this type of sampling introduces new problems and challenges. Organic contamination can be avoided by working with well-characterized drilling mud or water. Discrimination between indigenous carbonaceous structures and extant deep biosphere can be achieved by on-site biologic monitoring of drilling mud. Another improvement for this field of research is the wide variety of new isotopic and chemical techniques that can be applied to small rock samples. Laser Raman spectroscopy is now used routinely to determine the organic character (Schopf and Kudryavtsev, 2005) and degree of structure (Beyssac et al., 2002; Tice et al., 2004) of carbonaceous material. It enables the recognition of indigenous metamorphosed structures, and excludes fluid inclusions or post-metamorphic contamination. *In situ* isotopic analysis of putative microfossils is made possible using ion probe techniques. For instance carbon isotope ratios have been determined of individual carbonaceous structures within Archean rock samples (House et al., 2000; Ueno and Isozaki, 2001; Ueno et al., 2002). Detailed sub-micron chemical analysis and chemical mapping of microstructures is achieved using a Nano-SIMS (Robert et al., 2005) and detailed sub-micron chemical characterization of fluid inclusions has been achieved for Archean chert samples (Foriel et al., 2004). These *in situ* techniques make it possible to directly link microfossil morphology to both chemistry and isotopic characteristics, greatly improving the discrimination between biologic and abiologic processes. Finally, our understanding of Archean surface processes is greatly expanded with the use of multicollection ICP-MS. This technique enables the precise determination of the natural isotopic variation of e.g. transition metals and other biologically significant elements (Johnson et al., 2004a).

7.2. Microbial and Metabolic Diversification

DANIELL PRIEUR

Organisms living today on Earth use chemicals and/or light to gain the energy required for cellular functions, and they carry out biosynthetic processes converting carbon dioxide or already existing organic compounds. Considered as a whole, the prokaryotes (microscopic unicellular organisms whose genetic material is not separated from cytoplasm by a membrane) possess all types of metabolic pathways known so far. Their study might serve as models for establishing a scenario.

7.2.1 HOW DO CONTEMPORARY CELLS GAIN THEIR ENERGY?

The variety of metabolisms used by prokaryotes has been presented in details by Madigan et al. (2003) in a text book which inspired this paragraph.

7.2.1.1 *Chemotrophic metabolisms*
Chemotrophic organisms gain their energy from oxidation–reduction chemical reactions which obligatorily involve inorganic and/or organic electron donors and acceptors. When molecular oxygen (O_2) is present, it can play as an electron acceptor; then processes are called aerobic respirations. When molecular oxygen is lacking, other molecules (organic or inorganic) play as electron acceptors for processes called anaerobic respirations. In both cases, electrons are transported by electron carriers whose number and type depend on the donor and acceptor involved. During this process, a proton motive force is established across the cytoplasmic membrane, allowing ATP synthesis through an oxidative phosphorylation. When molecular oxygen and other electron acceptors are missing, certain organic molecules can be degraded through an energy generating process called fermentation, in which electron acceptors are provided by intermediate organic compounds deriving from the degradation of the initial organic compound.

7.2.1.2 *Phototrophic metabolisms*
Photosynthesis can be defined as a conversion of light energy into chemical energy: it is one of the most important biological processes on Earth. Organisms carrying out photosynthesis are called phototrophs. Most of them are usually autotrophs, and utilize carbon dioxide as carbon source. Photosynthesis depends on light sensitive pigments present in all phototrophs.

Photosynthesis is the sum of two series of reactions: light reactions and dark reactions. During light reactions, light energy is converted into chemical energy. When excited by light, some pigment molecules are converted into a strong electron donor with a very electronegative reduction potential. Electrons are then released, and transported by different ways, according to the organisms concerned. This electron flow leads to ATP synthesis.

During dark reactions, chemical energy is utilized to reduce carbon dioxide into organic compounds. Electrons required for carbon dioxide reduction come from the reduced form NADPH (for nicotinamide-adenine dinucleotide-phosphate) of the coenzyme $NADP^+$ (for nicotinamide-adenine dinucleotide-phosphate), previously reduced by electrons whose origin may vary according to the type of photosynthesis concerned.

They are two types of photosynthetic processes. Terrestrial and aquatic green plants, algae, and some prokaryotes of the Bacteria domain (Cyanobacteria) carry out oxygenic photosynthesis. In this case, electrons required for reduction of NADP + into NADPH are generated by photolysis of water, with the reaction:

$$H_2O \rightarrow 2H^+ + 2e^- + 1/2O_2.$$

Molecular oxygen released during this reaction gives its name to this particular photosynthesis.

The second type of photosynthesis is carried out exclusively by other prokaryotes from the Bacteria domain, called the green bacteria and the purple bacteria. These prokaryotes harbor photo-sensitive pigments (bacteriochlorophylls) that slightly differ from the pigments harbored by oxygenic phototrophs. Transport of electrons released by the light-excited pigments, and following phosphorylation is also different from those encountered during oxygenic photosynthesis. But, in the case of autotrophic growth, purple bacteria use electrons given by reduced molecules present in the environment such as hydrogen sulfide (H_2S) or various organic molecules for reduction of NADP + into NADPH. This photosynthesis that does not produce molecular oxygen is called anoxygenic photosynthesis.

7.2.1.3 *Electron carriers and ATP synthesis*

During an oxidation–reduction reaction within a cell, electrons are transferred from the donor to the acceptor, by one or more intermediates located in the cytoplasmic membrane and called electron carriers. These carriers are oriented within the membrane in such a way that electrons are transported along this chain, while protons are extruded outside the cell. The result is the formation of a proton gradient and an electrochemical potential across the membrane, with the inside of the cytoplasm alkaline and electrically negative, and the outside of the membrane acidic and electrically positive. This energized state of the membrane is called the proton motive force. This proton motive force is then used to synthesize ATP (adenosine

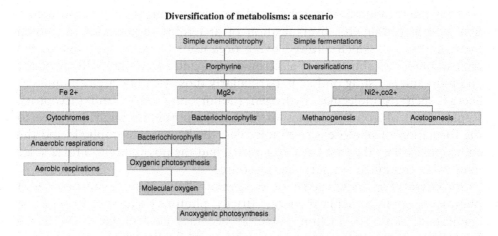

Figure 7.8. Diversification of metabolisms, a tentative scenario.

tri-phosphate), a molecule with high energy phosphate bonds, which represents the most frequent form of energy conservation in the cell. The enzyme that catalyses this reaction is an ATP synthase, or ATPase, which functions as a proton channel from outside to inside the cell, and synthesizes ATP during a process called oxidative phosphorylation.

7.2.2 WHAT WERE THE MOST PROBABLE MILESTONES?

This section is inspired by the hypothesis proposed by Madigan et al. (2003), and summarized in Figure 7.8.

Many actual prokaryotes (both Bacteria and Archaea) utilize molecular hydrogen (and many other inorganic and organic electron donors) as an electron donor, in combination with a variety of electron acceptors such as nitrate, sulfate, ferric iron, and molecular oxygen. Electrons originated from molecular hydrogen are transported towards the final electron acceptor using a variety of specific electron carriers, which number along a particular electron transport chain is rather depending on the difference of reduction potentials between the electron donor and the electron acceptor. Among these electron carriers are the cytochromes. Cytochromes are proteins with iron-containing porphyrin rings that carry out oxidation and reduction through loss or gain of one electron by the iron atom in the porphyrin ring.

Interestingly, other complex molecules involved in energy generating mechanisms are also porphyrin-based. They are particularly chlorophylls and bacteriochlorophylls involved in photosynthesis, and coenzyme F430 involved in methanogenesis, but instead of iron, they include magnesium and nickel, respectively.

Thus the formation of porphyrin-like molecules might have constituted a first step preceding the diversification of anaerobic respirations (including methanogenesis), which allowed living organisms to take advantage of the almost infinite combination of electron donors and acceptors (both inorganic and inorganic) existing on Earth, particularly those yielding large amounts of energy such as molecular hydrogen (donor) and ferric iron (acceptor). Replacement of iron by magnesium within the porphyrin might have led to the formation of photosensitive molecules: the bacteriochlorophylls and the chlorophylls. For the first time, organisms had the possibility to utilize solar light as an unlimited primary energy source.

Photosynthesis (anoxygenic or oxygenic) allows the transformation of photon energy into chemical energy. Briefly, photons excite specific sensitive compounds whose reduction potential becomes electronegative with the excitation, which generates an electron flow, and finally ATP synthesis. Then these phototrophic organisms have to use their energy for synthesizing macromolecules. Some of them can utilize already formed organic compounds from their environment, but many are autotrophs and utilize carbon dioxide as a carbon source, via various biochemical pathways.

Transformation of carbon dioxide into organic carbons requires not only energy, but also a reducing power. In case of molecular hydrogen-oxidizing organisms, hydrogen can reduce carbon dioxide via the electron carriers NADH or NADPH. For some anoxygenic photosynthetic Bacteria (Green sulfur Bacteria and Heliobacteria) the primary electron acceptor is sufficiently electronegative for carbon dioxide reduction through the reverse citric acid cycle or the hydroxypropionate cycle. When an electron donor with a less favorable reduction potential (sulfide, ammonium, etc) is involved, organisms concerned must have recourse to an inverse electron flow. This is the case for purple bacteria (anoxygenic phototrophs), which uptake their reducing power from the environments, using electron donors such as hydrogen sulfide, ferrous iron or organic compounds.

A second, but major step, in the evolution of energy generating metabolisms, was the evolution of oxygenic photosynthesis, carried out by ancestors of the Cyanobacteria. These organisms achieved a mechanism through which the reducing power required for carbon dioxide fixation came from photolysis of water, producing electrons, protons and molecular oxygen as a by-product. This production of molecular oxygen, probably hidden at its beginning by the presence of large amounts of reducing substances and/or its immediate consumption by evolving aerobic organisms, would provide the living organisms the best electron acceptor, in terms of energy generation, and lead to the explosion of aerobic respiration.

Finally, the transfer of energy generating mechanisms invented by prokaryotes (oxygenic photosynthesis and aerobic respiration) to primitive eukaryotes through endosymbiosis, gave the best energetic processes to lineages that evolved successfully into multicellular complex organisms. One must note that even if today photosynthetic driven life seems to dominate on earth, chemotrophy (organo-but also lithotrophy) is widespread (Karl et al., 1980; Cavanaugh, 1983; Parkes et al., 2000) and are the driving forces of many ecosystems such as deep-sea hydrothermal vents or deep marine sediments (Karl et al., 1980; Cavanaugh, 1983; Parkes et al., 2000).

This tentative scenario might appear feasible but would require supports that will be impossible to obtain. Particularly, one cannot say if the scenario started before or after LUCA (see part 5.7), and how it could have been coupled with the early genome evolution.

The most evident proof of this tentative succession of metabolisms is the increase of molecular oxygen in the atmosphere, following the oxygenic photosynthesis carried out by Cyanobacteria or their ancestors. But one can remind that today, when an aquarium is well balanced (chemically and biologically speaking), nitrite (which results from aerobic ammonium oxidation) is immediately oxidized into nitrate by specific micro-organisms, and is almost or totally undetectable in the aquarium environment. Similarly, the first production of molecular oxygen could have been masked by immediate chemical oxidations, then by the first aerobic respirations, before oxygen production overcame oxygen utilization, and finally reached the equilibrium of the actual atmosphere. For these reasons, the discussion about the first record of Cyanobacteria (or oxygenic photosynthetic organisms) and the oxygenation of the atmosphere is not finished yet.

Also, one must consider the size of micro-organisms and their habitats. A one micrometer microbe is actually depending on physico-chemical conditions which exist in the surrounding micrometers: 1 mm for a micro-organism is equivalent to 1 km for a man. A remote analysis of Earth, and particularly its atmosphere, could not detect the anaerobic life which most probably occurred first and still exist.

7.3. The Origin of Eukaryotes

Purificación López García, David Moreira, Patrick Forterre and Emmanuel J. P. Douzery

The terms 'prokaryote' and 'eukaryote' were introduced with their modern meaning by the microbiologists R. Stanier and C.B. van Niel in 1962 (Stanier and Van Niel, 1962; Sapp, 2005). They reflect the two major structural patterns in cellular organization. Eukaryotes, either unicellular (e.g. micro-algae, dinoflagellates, amoebas) or multicellular (e.g. plants, animals, fungi), are characterized by three major features that are missing in prokaryotes: a well-developed cytoskeleton of actin filaments and tubulin microtubules, membrane-bounded organelles (mitochondria, where respiration takes place

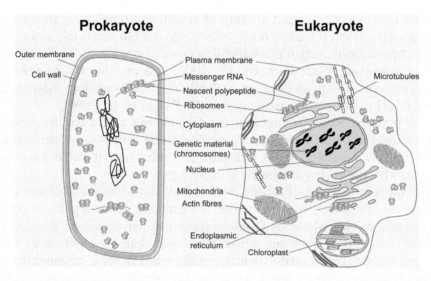

Figure 7.9. Schematic organisation of prokaryotic and eukaryotic cells.

and, in photosynthetic eukaryotes, chloroplasts, where photosynthesis occurs) and, most importantly, a nucleus, a region surrounded by a double membrane that contains the genetic material (Figure 7.9). A few years later, the development of molecular phylogeny, which infers evolutionary relationships among organisms from their conserved molecules, challenged this structural dichotomy. The comparison of sequences of RNAs from the small ribosomal subunit (SSU-rRNA) of different organisms revealed three, instead of two, major phylogenetic groups. Eukaryotes were one of them, but prokaryotes appeared to be divided in two groups that were as far from one another as they were from eukaryotes. They were initially called 'Eubacteria' and 'Archaebacteria' (Woese and Fox, 1977), and later re-baptized in the so-called domains Bacteria and Archaea (Woese et al., 1990). Molecular biology and biochemistry studies subsequently revealed many fundamental differences between them (Zillig, 1991). Despite the major differences between non-eukaryotic organisms, eukaryotes constitute both, phylogenetically and structurally a distinctive set of life forms. When, and particularly, how they originated is a matter of vivid controversy, as we will try to briefly summarize in the following.

7.3.1 DIFFERENT HYPOTHESES FOR THE ORIGIN OF EUKARYOTES

Due to its simpler cell organization, most researchers believe that some kind of prokaryotic ancestor gave rise to eukaryotes, an idea already implicit in the work of the German evolutionist E. Haeckel (1866). However, the discovery that prokaryotes are profoundly divided in two groups phylogenetically distinct complicated this view. Furthermore, the first gene comparisons, later corroborated by the analysis of complete genome sequences, uncovered a paradox: eukaryotic genes related to DNA replication, transcription and translation – the basic informational core – resembled archaeal genes, whereas genes involved in energy and carbon metabolism resembled their bacterial counterparts (Rivera et al., 1998). How could this mixed heritage in eukaryotes be explained? A variety of competing models have been put forward for reviews, see (López-García and Moreira, 1999; Martin et al., 2001) (Figure 7.10).

7.3.1.1 *Autogenous models*
The most widely accepted proposal is an elaboration of the traditional prokaryote-to-eukaryote transition, whereby the emergence of the nucleus and most of the other eukaryotic features occurred by complexification of ancestral structures that appeared in a single prokaryotic lineage (Cavalier-Smith, 1987). Since the first attempts to find the origin of the tree of life placed the root along the bacterial branch, implying the sisterhood of archaea and eukaryotes (Gogarten et al., 1989; Iwabe et al., 1989) (see Figure 5.10.A in Chapter 5.7), the prokaryotic lineage from which eukaryotes emerged would be archaeal-like (Woese et al., 1990; Brown and Doolittle, 1997). In this model, the eukaryotic bacterial-like genes would have been imported from mitochondria and chloroplasts to the nucleus, since it has been unambiguously demonstrated that both organelles evolved from bacterial endosymbionts: mitochondria derive from alphaproteobacteria, and chloroplasts from photosynthetic cyanobacteria (Gray and Doolittle, 1982). In a variant, the 'you are what you eat' model, those bacterial genes would come also from bacterial preys (Doolittle, 1998). The nucleus would have formed by invagination of the cell membrane at the same time and by the same mechanism that the endoplasmic reticulum, an internal membrane system for macromolecule transport (Jekely, 2003) (Figure 7.10).

7.3.1.2 *Chimeric models*
To explain the mixed composition of eukaryotic genomes, various hypotheses propose that eukaryotes resulted from the union of two prokaryotic lines, one archaeal and one bacterial (see Figure 5.10.B in Chapter 5.7). Some proposals are relatively simple and suggest a direct fusion or some kind of unspecified symbiosis of one archaeon and one bacterium (Zillig, 1991), either by an engulfment of one archaeon by a bacterium (Lake and Rivera, 1994; Gupta and Golding, 1996) or by a hypothetical descendant of an RNA world (Sogin, 1991) (Figure 7.10). Symbiosis-based models are the more comprehensive of the chimeric proposals. Symbiosis (*living-together*) between different organisms does not imply merely the sum of the parts but may lead to the creation of novel functions and properties as a consequence of gene

Autogenous model

The nucleus evolved in an archaeal-like lineage. Mitochondria derive from a bacterial endosymbiont.

Alphaproteobacterium

The model of Zillig 1991

Eukaryotes derive from a fusion or unspecified symbiosis between one archaeon and one bacterium

The models of Lake & Rivera 1994 Gupta & Golding 1996

One bacterium engulfs a crenarchaeote ('eocyte'), a member of the Archaea

The model of Sogin 1991

An RNA-based proto-eukaryote engulfs an archaeon

The model of Searcy 1992

Eukaryotes derive from a sulphur-based symbiosis between a bacterium that became the mitochondrion and a *Thermoplasma*-like archaeon

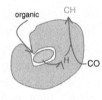

Hydrogen hypothesis
Martin & Müller 1998

Eukaryotes emerged from a hydrogen-based symbiosis between a fermentative alphaproteobacterium that became the mitochondrion and a methanogenic archaeon (Euryarchaeota)

Serial Endosymbiosis Model
Margulis 1993

A *Thermoplasma*-like archaeon established a symbiosis with spirochetes, acquiring motility. Later on, mitochondrial ancestors becames endosymbionts

Alphaproteobacterium

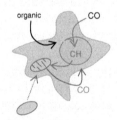

Syntrophy hypothesis
Moreira & López-García 1998

Eukaryotes emerged from a hydrogen-based symbiosis between a H₂-producing myxobacterium and a methanogenic archaeon. Mitochondria derived from methanotrophic alphaproteobacteria

A eukaryotic-like cenancestor

A nucleus-bearing eukaryotic-like organism would be ancestral. It acquired mitochondria later in evolution.

Alphaproteobacterium

A viral origin

The eukaryotic nucleus would be the result of the infection of a proto-eukaryotic lineage by a complex virus

Figure 7.10. Different models for the origin of eukaryotes.

redundancy and increased evolutionary rate (Kirschner and Gerhart, 1998; Margulis and Fester, 1993). Excellent examples are mitochondria and chloroplasts, once bacterial endosymbionts, which have positively contributed to the make-up of contemporary eukaryotes, attesting to the incontestable importance of symbiosis in eukaryotic evolution. Under these models, a long-term symbiotic relationship between one archaeon and one bacterium could have ended up in a novel entity, a eukaryote.

Detailed symbiotic proposals for the origin of eukaryotes, those that explicit a selective advantage for the involved partners, can be classified in two main classes. A first group of models states that eukaryotes stemmed from a symbiosis established between the ancestor of mitochondria (a bacterium) and one archaeon, more precisely one euryarchaeote (Euryarchaeota and Crenarchaeota are the two major branches of Archaea) (Figure 7.10). One hypothesis proposed a symbiosis between a wall-less euryarchaeote (a *Thermoplasma*-like species) able to reduce S_0 to H_2S and a bacterium oxidizing H_2S to S_0. To increase the efficiency of the exchange of sulfur species, the bacterium (the future mitochondrion) became an endosymbiont of the archaeon (Searcy, 1992). The hydrogen hypothesis states that the ancestor of mitochondria, one alphaproteobacterium able to ferment organics liberating H_2, established a symbiosis with a methanogenic archaeon (one euryarchaeote) using H_2 to reduce CO_2 to CH_4. As in the previous case, the bacterium became an endosymbiont of the archaeon and, with time, the mitochondrion (Martin and Muller, 1998). A second group of models envisages that the bacterium involved in the initial eukaryogenetic symbiosis was different from the mitochondrial ancestor, and that mitochondria derived from a second symbiotic event (Figure 7.10). The Serial Endosymbiosis Theory (SET) thus states that a first symbiosis was established between a *Thermoplasma*-like archaeon and spirochetes, which would have conferred motility to the ensemble, becoming eukaryotic flagella. The consortium evolved to form a proto-eukaryotic cell that then acquired the mitochondrial endosymbiont (Margulis, 1981). The syntrophy hypothesis proposes that eukaryotes arose from a symbiosis based on interspecies hydrogen transfer between an ancestral myxobacterium (gliding bacteria with complex developmental cycles belonging to the deltaproteobacteria) and a methanogenic archaeon. The ancestor of mitochondria would be a versatile alphaproteobacterium able, among others, to oxidize the CH_4 produced by the archaeon (Moreira and López-García, 1998; López-García and Moreira, 1999). Whereas in the above-mentioned models the eukaryotic nucleus was formed *de novo* in the cytoplasm of the archaeal host, in the syntrophy hypothesis the nucleus would be a relic of the archaeal partner.

Since the different hypotheses predict that the eukaryotic nucleus should contain genes coming from various specific prokaryotic groups, comparative genomic analysis of the rapidly increasing number of prokaryotic and eukaryotic complete genome sequences should allow in a next future to corroborate or refute some of these hypotheses.

7.3.1.3 *Other models*

Although most authors favor a prokaryote-to-eukaryote transition, various scientists have argued that prokaryotes are reduced descendants of more complex, eukaryotic-like ancestors already endowed with a nucleus (Bisset, 1973; Reanney, 1974; Poole et al., 1998; Forterre and Philippe, 1999). In this case, the origin of eukaryotes would be in essence the origin of the last common ancestor to extant organisms (see Figure 5.10.D in chapter 5.7) which, later in evolution, acquired mitochondria. The recent discovery of a membranous envelope containing the genetic material and part of the cytoplasm in the Planctomycetales together with their apparent early-branching position in the bacterial tree has led some authors to suggest that they could be intermediates between a eukaryotic ancestor and the rest of bacteria (Fuerst and Webb, 1991; Brochier and Philippe, 2002). However, the analysis of the genome sequence of the planctomycete *Rhodopirellula baltica* does not reveal any particular similarity with eukaryotes (Glockner et al., 2003). Finally, a recent set of models proposes that the eukaryotic nucleus derived from complex viruses related to Poxviruses (Bell, 2001; Takemura, 2001; Villarreal, 2005) (Figure 7.10). Several features of the Poxviruses cell cycle are reminiscent of the eukaryotic nucleus biology. In its original version, the authors of the viral eukaryogenesis theory suggested that the virus at the origin of the nucleus infected a wall-less methanogenic archaeon. Later on, it was proposed that the host was a more primitive cell (even possibly an RNA cell) (Forterre, 2005).

7.3.2 THE LAST COMMON ANCESTOR OF CONTEMPORARY EUKARYOTES

Be as it may, there is one certitude that is relevant for chronological aspects: modern eukaryotes emerged only after prokaryotes had appeared and diversified. This is the only explanation to the fact that all known eukaryotes have or have had mitochondria, which themselves derive from already quite modified bacteria, the alphaproteobacteria. Eukaryotes lacking apparent mitochondria exist and, in addition, they branched at the base of the eukaryotic tree in initial phylogenetic analyses. For some time, they were thought to be primitive eukaryotes that preceded the mitochondrial acquisition (Sogin, 1991). However, subsequent studies refuted this view. First, several genes of undeniable mitochondrial origin were found in the genomes of these mitochondrial-lacking eukaryotes, suggesting that they harbored these organelles once but lost them, in many cases because of a radical adaptation to a parasitic lifestyle (Simpson and Roger, 2002). Second, the improvement of evolutionary models in phylogenetic analyses together with the incorporation of many more eukaryotic sequences showed that those lineages had been misplaced to the base of the tree due to methodological artifacts (Simpson and Roger, 2002; Baldauf, 2003). Today, although the eukaryotic tree is not fully resolved, it is widely accepted that the last common ancestor of extant eukaryotes possessed mitochondria. Most scientists also think that eukaryotes suffered a radiation, i.e. they diversified in a very short time span, which would explain the difficulties to determine the relative order of emergence of the major eukaryotic lineages (Philippe et al., 2000). The cause of that sudden radiation is unclear. For some authors it was the acquisition of mitochondria which, providing O_2 respiration, granted a great selective advantage to colonize a variety of new ecological niches (Philippe et al., 2000). In conclusion, eukaryotes *sensu stricto* – i.e. possessing a nucleus, a well-developed cytoskeleton and organelles – evolved after and derived, at least partly, from prokaryotes. But when?

7.3.3 WHEN DID EUKARYOTES APPEAR AND DIVERSIFY?

7.3.3.1 *Fossil record*
There is little agreement as to when the first eukaryotic traces appeared in the fossil record. The oldest traces claimed to be of eukaryotic origin are biochemical: sterane compounds found in 2.7 Ga old kerogenes (Brocks et al., 1999). Steranes are fossil lipids derived from sterols, typically synthesized by eukaryotes. However, this finding is highly controversial because (i) a later contamination of this material by eukaryotes is possible, and (ii) many bacterial groups also synthesize sterols including, at least, methanotrophic bacteria, cyanobacteria, myxobacteria, actinobacteria and planctomycetes (Ourisson et al., 1987; Pearson et al., 2003). The oldest morphological eukaryotic fossils have been claimed to correspond to the coiled, spaghetti-like *Grypania spiralis* (~2 Ga ago) because of their large size (Han and Runnegar, 1992). However, size alone is not a definitive criterion as prokaryotic and eukaryotic sizes overlap (Javaux et al., 2003). The eukaryotic status of *Grypania* is highly discussed, as it may rather simply correspond to large cyanobacterial filaments (Cavalier-Smith, 2002). Surface ornamentation appears to be a more defining criterion since unicellular eukaryotes often display different scales and other surface structures that confer them an idiosyncratic aspect, while prokaryotes lack decoration in individual cells. The oldest decorated fossils, acritarchs, date from 1.5 Ga ago and were identified in the Mesoproterozoic Roper Group (Northern Australia) (Javaux et al., 2001). It is however difficult to relate this fossil group to any of the extant eukaryotic lineages (Javaux et al., 2003). The oldest fossils that have been assigned to a modern eukaryotic lineage correspond to *Bangiomorpha pubescens*, claimed to belong to red algae (Butterfield et al., 1990), and dated

from 723 Ma to 1.267 Ga ago, yet certainly closer to the latter bound (Butterfield, 2001). In summary, although there is reasonable morphological evidence suggesting that eukaryotes have developed by 1.5 Ga ago, their traces are sparse in rocks older than 1 Ga; Cavalier-Smith (2002) has even proposed that eukaryotes appeared in the fossil record only relatively recently, ~850 Ma ago, and that all older eukaryotic-like fossils corresponded to extinct groups of morphologically complex bacteria.

7.3.3.2 *Molecular dating*

Due to the scarcity of the eukaryote fossil record, the chronology of their origin and diversification has always been difficult to establish (Knoll, 2003). Comparative analysis of genomic data – homologous DNA, RNA, and protein sequences sampled from eukaryotic genomes – provided an alternative approach to reconstruct the history of life on Earth by (inter)connecting geological, paleontological, and biological information (Benner et al., 2002). Biomacromolecular sequences retain information about past history, and their degree of divergence among organisms has been correlated to the time of separation from their last recent common ancestor: the greater the number of differences between genomes or proteomes, the deeper the age of the split between the corresponding species. This hypothesis of a molecular clock ticking in biomolecules, i.e. the relative constancy of molecular evolutionary rate over time (Zuckerkandl and Pauling, 1965), is presented and discussed in Chapter 2.4 (*Biological chronometers: the molecular clocks*).

Several molecular dating studies attempted to evaluate divergence times of the major lineages of eukaryotes, with special focus on plants, animals, and fungi (Table 7.1). A striking concern about these independent multigene studies is the large range of estimates proposed for eukaryotic divergence times. For example, the dichotomy between animals and fungi is supposed to have occurred between 1.513 Ga ago (beginning of the Mesoproterozoic: Hedges et al., 2004) and 984 Ma ago (beginning of the Neoproterozoic; Douzery et al., 2004). Deeper divergence times corresponding to the split between unicellular (protists) and multicellular eukaryotes also display a wide range of estimates, from 1.545 to 2.309 Ga ago (Table 7.1). It would be desirable to directly compare all these results, but the different dating approaches (global vs. relaxed molecular clocks), genes and proteins (variable number of sites sampled from nuclear and plastid compartments), paleontological calibrations (from one to six, with or without incorporation of fossil record uncertainty), and taxon samplings make this difficult. However, more reliable estimates are expected to be inferred from larger data sets (typically more than 100 markers in order to reduce stochastic errors), under relaxed molecular clocks that are not hampered by the detection of constant rate sequences, and with the aid of several independent primary calibrations. In this context, the age of the dichotomy between

TABLE 7.1.
Molecular dating of the divergence of the major groups of eukaryotes compiled from several references

Splits	Ages (Ma)	Markers	Clocks	Calibrations	References
Animals + Fungi	984 (SD ± 65)	129 proteins 30,399 aa	Relaxed (rate auto correlation)	6 constraints (P)	Douzery et al. (2004)
Animals + Fungi	1513 (SE ± 66)	69–92 proteins 31,362 aa	Global/Relaxed	1 constraint (P) 3 points (S)	Hedges et al. (2004)
Plants–Animals–Fungi	1215–1272 (–)	64 proteins 25,000 aa	Global	6 points (P)	Feng et al. (1997)
Plants–Animals–Fungi	1576 (SD ± 88)	38 proteins 22,888 aa	Global	1 point (P)	Wang et al. (1999)
Plants–Animals–Fungi	1392 (SE ± 256)	11 proteins 3310 aa	Global	1 point (P)	Nei et al. (2001)
Plants/[Animals + Fungi]	1085 (SD ± 79)	129 proteins 30,399 aa	Relaxed (rate auto correlation)	6 constraints (P)	Douzery et al. (2004)
Plants/[Animals + Fungi]	1609 (SE ± 60)	99–143 proteins 60,274 aa	Global/Relaxed	1 constraint (P) 3 points (S)	Hedges et al. (2004)
Photosynthetic eukaryotes/Animals	>1558 (1531–1602)	6 plastid genes 7111 nt	Relaxed (penalized)	6 constraints (P)	Yoon et al. (2004)
Protists/Plants–Animals–Fungi	1545 (–)	64 proteins 25,000 aa	Global	6 points (P)	Feng et al. (1997)
Protists/Plants–Animals–Fungi	1717 (SE ± 349)	11 proteins 3310 aa	Global	1 point (P)	Nei et al. (2001)

TABLE 7.1.
Continued

Splits	Ages (Ma)	Markers	Clocks	Calibrations	References
Giardia/Other eukaryotes	2230 (SE ± 120)	17 proteins	Global	3 points (S)	Hedges et al. (2001)
Giardia/Plants–Animals–Fungi	2309 (SE ± 194)	28–32 proteins 11,251 aa	Global/Relaxed	1 constraint (P) 3 points (S)	Hedges et al. (2004)

Estimates of divergence ages and their uncertainties (when available) are given in million years ago. The number of markers, the type of molecular clock methodology, and the number of fossil calibrations used for the dating are also given. Calibrations might be primary (P, derived from fossils) or secondary (S, derived from molecules), and incorporate paleontological uncertainty (constraints) or not (points). Abbreviations – nt: nucleotide sites; aa: amino acid sites; SD: standard deviation; SE: standard error.

plants (i.e., photosynthetic eukaryotes) and animals + fungi would be closer to 1.085 Ga ago (Douzery et al., 2004) than to 1.609 Ga ago (Hedges et al., 2004). This suggests that primary plastids resulting from the endo-symbiosis of a free-living cyanobacterium appeared at the end of the Mesoproterozoic some 1.1 Ga ago.

The postulated anoxic and sulfidic redox status of oceans until 1 Ga might have limited the rise of photosynthetic eukaryotes (Anbar and Knoll, 2002), whereas cytoskeletal and ecological prerequisites for their diversification were already established some 1.5 Ga ago (Javaux et al., 2001). The more oxygenic environments of the Neoproterozoic (1 Ga to 540 Ma) could possibly have triggered the diversification of the major eukaryotic lineages, culminating with the seemingly abrupt appearance of animals in the Cambrian explosion (Conway Morris, 2000).

7.4. The Neoproterozoic–Cambrian Transition (~1000 to 542 Ma)

PHILIPPE CLAEYS

The end of the Proterozoic (Neoproterozoic ~1000 to 542 Ma) corresponds to a period of major global changes most likely initiated by the break-up of the supercontinent Rodania around 750 Ma ago (Kah and Bartley, 2001). Three widespread and severe glaciation events occur during the Neoproterozoic: the Sturtian (~710 to 725 Ma), the Marinoan (~635 to 600 Ma), and Gaskiers (~580 Ma) (Figure 7.11). These events are identified based on isotopic profiles ($^{13}C/^{12}C$ and $^{87}Sr/^{86}Sr$) and the repetitive accumulation of thick packages of glacial sediments, recognized worldwide. These tillites (or diamictites) are commonly covered by distinctive cap carbonates, which, curiously, almost certainly precipitated inorganically under warm-water conditions. The Sturtian and Marinoan events were probably the most extreme glaciations recorded on Earth. Evans (2000) provided paleomagnetic evidence for the presence of glaciers at sea level within 10 ° from the Neoproterozoic equator. Hoffman et al. (1998a) proposed that the Sturtian and Marinoan were global glaciations covering the entire planet, commonly referred to as the *Snowball Earth* hypothesis. The thick ice cover (up to 1 km) implies a drastic reduction in photosynthesis and a collapse of biological productivity that best explains the intensity of the negative carbon isotopic anomalies (up to -14% in surface ocean $\partial^{13}C$) recorded in carbonates bracketing glacial sediments (Hoffman et al., 1998a). The presence of continental ice inhibited silicate weathering, another CO_2 sink, leading to its accumulation in the atmosphere. Outgassing by subaerial volcanoes contributed further to the increase of CO_2 in the atmosphere throughout the icehouse period. The glacial conditions ended abruptly when greenhouse gas concentrations

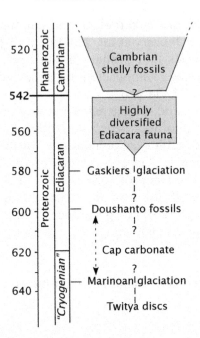

Fig. 7.11. Schematic stratigraphy of the Proterozoic–Cambrian transition.

became high enough (~0.12 bars of CO_2 in the atmosphere) to overcome the albedo effect, causing rapid melting of the ice and subsequent precipitation of warm-water carbonates (Hoffman et al., 1998a). On such a fully glaciated Earth, extraterrestrial material would accumulate on and within the ice. Based on Ir flux measured at the base of the cap carbonates, the duration of the Marinoan glaciation is estimated around 12 Ma (Bodiselitsch et al., 2005). The Snowball hypothesis is currently subject to lively debates (see Jenkins and Scotese, 1998; Christie-Blick et al., 1999; and replies by Hoffman et al., 1998b; Hoffman and Schrag, 1999 for example) focusing on the initiation and termination of the glaciations, as well as the global extent and average thickness of the ice. Some authors consider that (large) parts of the ocean must have remained ice-free, forming a sanctuary for marine organisms where photosynthesis could continue. This alternative view is called the "Slushball Earth" hypothesis (Hyde et al., 2000; Crowley et al., 2001).

Because of the absence of skeletonized fossils, Proterozoic lithostrati-graphic units are often difficult to correlate precisely and remain subdivided in broad periods defined essentially on the basis of the chronometric ages obtained by isotopic dating of specific but sporadic layers. Recently, a new stratigraphic period: the Ediacaran (Knoll et al., 2004), defined in analogy with its Phanerozoic counterparts, has been approved by the International Union of Geosciences (IUGS) to represent the most recent part of the Pro-terozoic. The top of the Ediacaran corresponds to the well-defined base of the

Cambrian, dated at 542 Ma (Gradstein et al., 2004). Its base is placed in a distinct carbonate layer that overlies sediment deposited by the Marinoan glaciation in the Flinders Ranges of South Australia (Figure 7.11). The base of the Ediacaran is not precisely dated. It is younger than an U-Pb date of 635.5 ± 1.2 Ma measured on zircons from within the glacial diamictites of Namibia and older than a Pb–Pb date of 599 ± 4 Ma obtained on post-glacial phosphorites in China (Barfod et al., 2002; Hoffman et al., 2004; Knoll et al., 2004). In term of event stratigraphy, the Ediacaran is bounded above by the rapid diversification of shelly organisms and below by the Marinoan ice age of global extent (Knoll et al., 2004).

The Ediacaran period is characterized by the presence of the traces of soft-body organisms fossilized as impression on sandstone beds or less commonly on ash-layers (Figure 7.12). Such fossils lacking a mineralized shell (or skeleton) differ significantly from their Phanerozoic counterparts. Named after the remarkable collection recognized in South Australia, almost 60 years ago, the Ediacara fauna occurs worldwide (30 localities on 5 continents, Narbonne, 1998). At most localities, the preserved fossils attest of highly diversified and sophisticated organisms displaying a great variety of

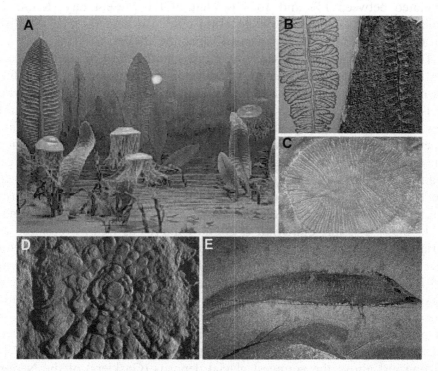

Figure 7.12. Ediacaran fauna. (A) reconstruction of the Ediacaran environment; (B) Chania (Cnidaria?); (C) Dickensonia (worm? cnidaria?); (D) Mawsonites (medusa); (E) Pikaia, Burgess Pass (first known chordate).

shapes and sizes. This complex fauna forms a biological transition between the modern shelly organisms of the Phanerozoic and the essentially microbial communities of the Proterozoic, dominated by prokaryotes and microbial eukaryotes including algae. Although, the first Eukaryotes probably appeared between 2.7 and 1.8 Ga ago, they do not seem to have been abundant or diversified until the end of the Proterozoic (Knoll, 2003). Javaux et al. (2004) reported the presence of a moderate diversity of probable eukaryotic remains, among a fauna rich in protistan microfossils in carbonaceous shales dated between 1.5 and 1.4 Ga from Australia.

Either Snowball or Slushball, both scenarios seem capable to strongly influence the evolution of life. Ediacara fauna diversified within a few million years after the last Neoproterozoic glacial event (Narbonne, 2005). The less severe glaciation at the end of the Ordovician caused one of the major mass extinctions of the Phanerozoic (Sheehan et al., 1996). Although not clearly linked to extinction, these Neoproterozoic glaciations followed by greenhouse conditions, must have been harsh selection factors, possibly triggering in their aftermath the radiation of Ediacaran organisms. An issue that remains unclear is the role of the rising oxygen concentrations (Knoll, 2003). The level of oxygen necessary for the development of large-scale metazoans is estimated between 1% and 10% of that of the present day (Knoll and Holland, 1994). Post-glacial oxygenation may also have favored the diversification of the Ediacara organisms (Narbonne, 2005).

The characteristic Ediacara biota clearly marks the first appearance on Earth of large, complex and highly diversified communities. These fossil assemblages contain radial and bilateral organisms constituting, perhaps the root-stock of the Cambrian radiation, possible life-forms belonging to other, now extinct, eukaryotic phyla or kingdoms, and/or what appear as "failed experiments" in animal evolution (see Narbonne, 2005 for a detailed review). The stratigraphic distribution of the Ediacara organisms is rather well constrained. The Twitya formation of northwestern Canada contains, just below the Marinoan tillites, a poorly diversified assemblage of "Twitya discs" considered perhaps as the oldest known Ediacara-type fossils (Hofmann et al., 1990). Possible bilaterian eggs and embryos as well as fossilized cnidarian may be present in the Doushantuo formation in China (Xiao et al., 1998), which is dated by Pb–Pb and Lu–Hf on phosphates between 599 and 584 Ma (Barfod et al., 2002). These still enigmatic fossils clearly predate the major radiation of Ediacara organisms. Conway Morris (1998) recognizes rare Ediacara survivors among the Cambrian Burgess shale fauna. Nevertheless, the typical, highly diversified Ediacara fauna is restricted to a well-constrained stratigraphic interval between ~575 and 542 Ma (Figure 7.11); starting just above the youngest glacial deposits (Gaskiers) of the Neoproterozoic and extending to the very base of the Cambrian (Narbonne, 2005).

The typical Ediacara fossils are commonly a few cm to 10 cm in size but some giant forms reach more than a meter in length. They display a great range of shapes such as disks, fronds or segmented morphologies, somewhat reminiscent to those of modern organisms (Narbonne, 2005). Other forms are highly unusual and completely unique to the Ediacaran fauna. Ediacara fossils lived on soft and muddy seafloor and are best preserved when rapidly buried by a coarse sedimentary event such as the deposition of turbidites, or ash layers. In the best-preserved sections, the abundance of organisms is comparable to that found in equivalent modern seafloor communities (Narbonne, 2005).

Most of the Ediacara organisms cluster in 3 assemblages (see Narbonne, 2005 for a detailed discussion): (1) the Avalon assemblage (575–560 Ma), characterized by apparently more primitive and bizarre shapes, found in deep-water settings, (2) the White Sea assemblage (560–542 Ma), more diverse, living in shallow-water and composed of segmented, disk, and front morphologies, with some bilaterian organisms capable of mobility, (3) the Nama assemblage (549–542 Ma), also of shallow-water but marked by the presence of some early calcified metazoans. The relationships between Ediacara biota and modern organisms were rather controversial during the mid-1980s and early 1990's. Early work based on morphologies had advocated similarities with the jellyfishes. For Seilacher (1992), Ediacara fauna differ drastically from modern organisms and represent an extinct Kingdom of life, which he called Vendobionta. Today, there seems to be an agreement that ancestors of radial phyla such as Porifera and Cnidaria dominated the Ediacara fauna (Narbonne, 2005). McCaffrey et al. (1994) reported the presence of sponge biomarkers in Neoproterozoic hydrocarbon deposits. Other body fossils with evident bilateral symmetry and segmentation point to possible ancestors of the phyla Arthropoda, Mollusca, Annelida and Echinodermata (Budd and Jensen, 2000). A fraction of Ediacara taxa appears unrelated to modern organisms and may represent extinct phyla (Knoll, 2003).

The transition between the Ediacara fauna and the Cambrian shelly fossils is not clearly understood. The apparently abrupt disappearance of the Ediacara fossils, just below the base of the Cambrian could be linked to a major anoxic event (Kimura and Watanabe, 2001) or to the rise of widespread predation among organisms (Bengtson and Yue, 1992). The subsequent "Cambrian Explosion of life" reflects again a major diversification of biosphere and, because of the appearance of a hard shell a much better preservation of the fossil remains. The small shelly fossils either made of carbonates or phosphates are widespread in the very basal Cambrian, preceding the rich and highly diversified fauna such as that preserved later in the Burgess shales (Conway Morris, 1998) or other fossiliferous beds. All present day phyla, along with a few enigmatic taxa, are already present at this level of the Cambrian. No new phylum will emerge in the next 500 Ma.

References

Allwood, A. C., Walter, M. R., Kamber, B. S., Marshall, C. P. and Burch, I. W.: 2006, *Nature* **441**, 714–718.

Anbar, A. D.: 2004, *Earth Planet. Sci. Lett.* **217**, 223–236.

Anbar, A. D. and Knoll, A. H.: 2002, *Science* **297**, 1137–1142.

Appel, P. W. U., Moorbath, S. and Myers, J. S.: 2003, *Precambrian Res.* **126**, 309–312.

Awramik, S. M., Schopf, J. W. and Walter, M. R.: 1983, *Precambrian Res.* **20**, 357–374.

Baldauf, S. L.: 2003, *Science* **300**, 1703–1706.

Barfod, G. H., Albarede, F., Knoll, A. H., Xiao, S., Telouk, P., Frei, R. and Baker, J.: 2002, *Earth Planet. Sci. Lett.* **201**, 203–212.

Beaumont, V. and Robert, F.: 1999, *Precambrian Res.* **96**, 63–82.

Bebout, G. E. and Fogel, M. L.: 1992, *Geochim. Cosmochim. Acta* **56**, 2839–2849.

Bekker, A., Holland, H. D., Wang, P.-L., Rumble III, D., Stein, H. J., Hannah, J. L., Coetzee, L. L. and Beukes, N. J. 2004, *Nature*, 117–120.

Bell, P. J.: 2001, *J. Mol. Evol.* **53**, 251–256.

Bengtson, S. and Yue, Z.: 1992, *Science* **257**, 367–369.

Benner, S. A., Caraco, M. D., Thomson, J. M. and Gaucher, E. A.: 2002, *Science* **296**, 864–868.

Beukes, N. J.: 2004, *Nature* **431**, 522–523.

Beyssac, O., Goffé, B., Chopin, C. and Rouzaud, J. N.: 2002, *J. Metamor. Geol.* **2002**(20), 859–871.

Bisset, K. A.: 1973, *Nature* **241**, 45.

Boak, J. L. and Dymek, R. F.: 1982, *Earth Planet. Sci. Lett.* **59**, 155–176.

Bodiselitsch, B., Koeberl, C., Master, S. and Reimold, W. U.: 2005, *Science* **308**, 239–242.

Bosak, T., Souza-Egipsy, V., Corsetti, F. A. and Newman, D. K.: 2004, *Geology* **32**, 781–784.

Brasier, M., Green, O. R., Jephcoat, A. P., Kleppe, A., Van Kranendonk, M. J., Lindsay, J. F., Steele, A. and Grassineau, N. V.: 2002, *Nature* **416**, 76–81.

Brasier, M., Green, O. R., Lindsay, J. F., McLoughlin, N., Steele, A. and Stoakes, C.: 2005, *Precambrian Res.* **140**, 55–102.

Brochier, C. and Philippe, H.: 2002, *Nature* **417**, 244.

Braterman, P. S., Cairns-Smith, A. G. and Sloper, R. W.: 1983, *Nature* **303**, 163–164.

Brocks, J. J., Logan, G. A., Buick, R. and Summons, R. E.: 1999, *Science* **285**, 1025–1027.

Brocks, J. J., Love, G. D., Snape, C. E., Logan, G. A., Summons, R. E. and Buick, R.: 2003, *Geochim. Cosmochim. Acta* **67**, 1521–1530.

Brown, J. R. and Doolittle, W. F.: 1997, *Microbiol. Mol. Biol. Rev.* **61**, 456–502.

Budd, G. E. and Jensen, S.: 2000, *Biol. Rev.* **75**, 253–295.

Buick, R. and Dunlop, J. S. R.: 1990, *Sedimentology* **37**, 247–277.

Buick, R., Dunlop, J. S. R. and Groves, D. I.: 1981, *Alcheringa* **5**, 161–181.

Bullen, T. D., White, A. F., Childs, C. W., Vivit, D. V. and Schulz, M. S.: 2001, *Geology* **29**, 699–702.

Butterfield, N. J., Knoll, A. H. and Swett, K.: 1990, *Science* **250**, 104–107.

Butterfield, N. J.: 2001, *Precambrian Res.* **111**, 235–256.

Batchelor, M. T., Burne, R. V., Henry, B. I. and Jackson, M. J.: 2004, *Physica A – Stat. Mech. Appl.* **337**, 319–326.

Byerly, G. R., Lowe, D. R. and Walsh, M.: 1986, *Nature* **319**, 489–491.

Canfield, D. E. and Raiswell, R.: 1999, *Am. J. Sci.* **299**, 697–723.

Cavanaugh, C.: 1983, *Nature* **302**, 340–341.

Cavalier-Smith, T.: 1987, *Ann. N. Y. Acad. Sci.* **503**, 17–54.

Cavalier-Smith, T.: 2002, *Int. J. Syst. Evol. Microbiol.* **52**, 7–76.

Charlou, J. L., Fouquet, Y., Bougault, H., Donval, J. P., Etoubleau, J., Jean-Baptiste, P., Dapoigny, A., Appriou, P. and Rona, P. A.: 1998, *Geochim. Cosmochim. Acta* **62**, 2323–2333.

Chen, J.-Y., Bottjer, D. J., Oliveri, P., Dornbos, S. Q., Gao, F., Ruffins, S., Chi, H., Li, C. -W. and Davidson, E. H.: 2004, *Science* **305**, 218–222.

Christie-Blick, N., Sohl, L. E. and Kennedy, M. J.: 1999, *Science* **284**, 1087a.

Chyba, C. F.: 1993, *Geochim. Cosmochim. Acta* **57**, 3351–3358.

Conway Morris, S.: 1998. *The Crucible of Creation: The Burgess Shale and The Rise of Animals*, Oxford University Press, Oxford, 242 pp.

Conway Morris, S.: 2000, *Proc. Natl. Acad. Sci. USA* **97**, 4426–4429.

Croal, L. R., Johnson, C. M., Beard, B. L. and Newman, D. K.: 2004, *Geochim. Cosmochim. Acta* **68**, 1227–1242.

Crowley, T. J., Hyde, W. T. and Peltier, W. R.: 2001, *Geophys. Res. Lett.* **28**, 283–236.

Dauphas, N., van Zuilen, M. A., Wadhwa, M., Davis, A. M., Marty, B. and Janney, P. E.: 2004, *Science* **306**, 2077–2080.

de Ronde, C. E. J. and Ebbesen, T. W.: 1996, *Geology* **24**, 791–794.

de Wit, M. J., Hart, R., Martin, A. and Abbott, P.: 1982, *Econ. Geol.* **77**, 1783–1802.

de Wit, M. J. and Hart, R. A.: 1993, *Lithos* **30**, 309–335.

Des Marais, D. J. and Morre, J. G.: 1984, *Earth Planet. Sci. Lett.* **69**, 43–47.

Dimroth, E.: 1982, in A. V. Sidorenko (ed.), *Sedimentary Geology of the Highly Metamorphosed Precambrian Complexes*, pp. 16–27.

Doolittle, W. F.: 1998, *Trends Genet.* **14**, 307–311.

Douzery, E. J. P., Snell, E. A., Bapteste, E., Delsuc, F. and Philippe, H.: 2004, *Proc. Natl. Acad. Sci. USA* **101**, 15386–15391.

Durand, B.: 1980, *Kerogen: Insoluble Organic Matter from Sedimentary Rocks*.

Dymek, R. F. and Klein, C.: 1988, *Precambrian Res.* **39**, 247–302.

Evans, D. A. D.: 2000, *Am. J. Sci.* **300**, 347–433.

Fedo, C. M. and Whitehouse, M.: 2002a, *Science* **296**, 1448–1452.

Fedo, C. M. and Whitehouse, M. J.: 2002b, *Science* **298**, 917.

Feng, D. F., Cho, G. and Doolittle, R. F.: 1997, *Proc. Natl. Acad. Sci. USA* **94**, 13028–13033.

Foriel, J., Philippot, P., Rey, P., Somogyi, A., Banks, D. and Ménez, B.: 2004, *Earth Planet. Sci. Lett.* **228**, 451–463.

Forterre.: 2005, *Bichimie* **87**, 793–803.

Forterre, P. and Philippe, H.: 1999, *Biol. Bull.* **196**, 373–375.

French, B. M.: 1971, *Am. J. Sci.* **271**, 37–78.

Friend, C. R. L., Nutman, A. P. and Bennett, V. C.: 2002, *Science* **298**, 917.

Fuerst, J. A. and Webb, R.: 1991, *Proc. Natl. Acad. Sci. USA* **88**, 8184–8188.

Furnes, H., Banerjee, N. R., Muehlenbachs, K., Staudigel, H. and de Wit, M.: 2004, *Science* **304**, 578–581.

Garcia-Ruiz, J. M., Hyde, S. T., Carnerup, A. M., Van Kranendonk, M. J. and Welham, N. J.: 2003, *Science* **302**, 1194–1197.

Glockner, F. O., Kube, M., Bauer, M., Teeling, H., Lombardot, T., Ludwig, W., Gade, D., Beck, A., Borzym, K., Heitmann, K., Rabus, R., Schlesner, H., Amann, R. and Reinhardt, R.: 2003, *Proc. Natl. Acad. Sci. USA* **30**, 30.

Gogarten, J. P., Kibak, H., Dittrich, P., Taiz, L., Bowman, E. J., Bowman, B. J., Manolson, M. F., Poole, R. J., Date, T., Oshima, T., Konishi, J., Denda, K. and Yoshida, M.: 1989, *Proc. Natl. Acad. Sci. USA* **86**, 6661–5.

Gradstein, F. M., Ogg, J. G., Smith, A. G., Agterberg, F. P., Bleeker, W., Cooper, R. A., Davydov, V., Gibbard, P., Hinnov, L., M. R. H., Lourens, L., Luterbacher, H.-P., McArthur, J., Melchin, M. J., Robb, L. J., Shergold, J., Villeneuve, M., Wardlaw, B. R., Ali, J., Brinkhuis, H., Hilgen, F. J., Hooker, J., Howarth, R. J., Knoll, A. H., Laskar, J., Monechi, S., Powell, J., Plumb, K. A., Raffi, I., Röhl, U., Sanfilippo, A., Schmitz, B., Shackleton, N. J., Shields, G. A., Strauss, H., Dam, J. V., Veizer, J., Kolfschoten, T.v. and Wilson, D.: 2004, *A Geological Time Scale 2004*, Cambridge University, Cambridge, 610 pp.

Gray, M. W. and Doolittle, W. F.: 1982, *Microbiol. Rev.* **46**, 1–42.

Grotzinger, J. P. and Knoll, A. H.: 1999, *Annu. Rev Earth Planet. Sci.* **27**, 313–358.

Grotzinger, J. P. and Rothman, D. H.: 1996, *Nature* **383**, 423–425.

Gupta, R. S. and Golding, G. B.: 1996, *Trends Biochem. Sci.* **21**, 166–71.

Haeckel, E.: 1866. *Generelle Morphologie der Organismen: Allgemeine Grundzüge der organischen Formen-Wissenschaft, mechanisch begründet durch die von Charles Darwin reformirte Descendenz-Theorie*, Georg Reimer, Berlin.

Han, T. M. and Runnegar, B.: 1992, *Science* **257**, 232–235.

Hanor, J. S. and Duchac, K. C.: 1990, *J. Geol.* **98**, 863–877.

Hayes, J. M., Kaplan, I. R. and Wedeking, W.: 1983, in J.W. Schopf (ed.), *Earth's Earliest Biosphere, Its Origin and Evolution*, Princeton University Press, pp. 93–134.

Hedges, S. B., Blair, J. E., Venturi, M. L. and Shoe, J. L.: 2004, *BMC Evol. Biol.* **4**, 2.

Hedges, S. B., Chen, H., Kumar, S., Wang, D. Y., Thompson, A. S. and Watanabe, H.: 2001, *BMC Evol. Biol.* **1**, 4.

Hoffman, H. J., Grey, K., Hickman, A. H. and Thorpe, R.: 1999, *Geol. Soc. Am. Bull.* **111**, 1256–1262.

Hoffman, P. F., Kaufman, A. J., Halverson, G. P. and Schrag, D. P.: 1998a, *Science* **281**, 1342–1346.

Hoffman, P. F. and Schrag, D. P.: 1999, *Science* **284**, 1087a.

Hoffman, P. F., Schrag, D. P., Halverson, G. P. and Kaufman, J. A.: 1998b, *Science* **282**, 1644.

Hofmann, H. J., Narbonne, G. M. and Aitken, J. D.: 1990, *Geology* **18**, 1199–1202.

Hoffmann, K.-H., Condon, D. J., Bowring, S. A. and Crowley, J. L.: 2004, *Geology* **32**, 817–820.

Holm, N. G. and Charlou, J. L.: 2001, *Earth Planet. Sci. Lett.* **191**, 1–8.

House, C. H., Schopf, J. W., McKeegan, K. D., Coath, C. D., Harrison, T. M. and Stetter, K. O.: 2000, *Geology* **28**, 707–710.

House, C. H., Schopf, J. W. and Stetter, K. O.: 2003, *Org. Geochem.* **34**, 345–356.

Hyde, W. T., Crowley, T. J., Baum, S. K. and Peltier, W. R.: 2000, *Nature* **405**, 425–429.

Iwabe, N., Kuma, K., Hasegawa, M., Osawa, S. and Miyata, T.: 1989, *Proc. Natl. Acad. Sci. USA* **86**, 9355–9359.

Javaux, E. J., Knoll, A. H. and Walter, M.: 2003, *Orig. Life Evol. Biosph.* **33**, 75–94.

Javaux, E. J., Knoll, A. H. and Walter, M. R.: 2001, *Nature* **412**, 66–69.

Javaux, E. J., Knoll, A. H. and Walter, M. R.: 2004, *Geobiology* **2**, 121–132.

Jekely, G.: 2003, *Bioessays* **25**, 1129–1138.

Jenkins, G. S. and Scotese, C. R.: 1998, *Science* **282**, 1644.

Johnson, C. M., Beard, B. L., and Albarède, F.: 2004a, *Geochemistry of Non-traditional Stable Isotopes*, Mineralogical Society of America.

Johnson, C. M., Beard, B. L., Beukes, N. J., Klein, C. and O'Leary, J.: 2003, *Contr. Mineral. Petrol.* **144**, 523–547.

Johnson, C. M., Beard, B. L. Roden, E. E., Newman, D. K. and Nealson, K. H.: 2004b, in *Geochemistry of Non-traditional Stable Isotopes*, vol. 55, Mineralogical Society of America, pp. 359–402.

Johnson, C. M., Roden, E. E., Welch, S. A. and Beard, B. L.: 2005, *Geochim. Cosmochim. Acta* **69**, 963–993.

Kah, L. C. and Bartley, J. K.: 2001, *Precambrian Res.* **111**, 1–283.

Karl, D. M., Wirsen, C. O., and Jannasch, H. W.: 1980, *Science* **207**, 1345–1347.

Kimura, H. and Watanabe, Y.: 2001, *Geology* **29**, 995–998.

Kirschner, M. and Gerhart, J.: 1998, *Proc. Natl. Acad. Sci. USA* **95**, 8420–8427.

Kitchen, N. E. and Valley, J. W.: 1995, *J. Metamor. Geol.* **13**, 577–594.

Knauth, L. P. and Lowe, D. R.: 1978, *Earth Planet. Sci. Lett.* **41**, 209–222.

Knoll, A. H.: 1999, *Science* **285**, 1025–1026.

Knoll, A. H.: 2003. *Life on a Young Planet; The First Three Billion Years of Evolution on Earth*, Princeton University Press, Princeton, 277 pp.

Knoll, A. H. and Holland, H. D.: 1994, in Stanley, S. M. (eds.), *Proterozoic oxygen and evolution: an update. Biological Responses to Past Global Changes*, National Academy Press, Washington D.C., pp. 21–33.

Knoll, A. H., Walter, M. R., Narbonne, G. M. and Christie-Blick, N.: 2004, *Science* **305**, 621–622.

Konhauser, K. O., Hamade, T., Raiswell, R., Morris, R. C., Ferris, F. G., Southam, G. and Canfield, D. E.: 2002, *Geology* **30**, 1079–1082.

Kudryavtsev, A. B., Schopf, J. W., Agresti, D. G. and Wdowiak, T. J.: 2001, *Proc. Natl. Acad. Sci.* **98**, 823–826.

Lake, J. A. and Rivera, M. C.: 1994, *Proc. Natl. Acad. Sci. USA* **91**, 2880–2881.

Lancet, M. S. and Anders, E.: 1970, *Science* **170**, 980–982.

Lepland, A., Arrhenius, G. and Cornell, D.: 2002, *Precambrian Res.* **118**, 221–241.

Lepland, A., van Zuilen, M. A., Arrhenius, A., Whitehouse, M. and Fedo, C. M.: 2005, *Geology* **33**, 77–79.

Lindsay, J. F., Brasier, M. D., McLoughlin, N., Green, O. R., Fogel, M., Steele, A. and Mertzman, S. A.: 2005, *Precambrian Res.* **143**, 1–22.

López-García, P. and Moreira, D.: 1999, *Trends Biochem. Sci.* **24**, 88–93.

Lowe, D. R.: 1983, *Precambrian Res.* **19**, 239–283.

Lowe, D. R.: 1994, *Geology* **22**, 387–390.

Lowe, D. R. and Byerly, G. R.: 1986, *Nature* **324**, 245–248.

Lowe, D. R. and Byerly, G. R.: 1999, *Geologic Evolution of the Barberton Greenstone Belt, South Africa*, The Geological Society of America.

Lowe, D. R. and Byerly, G. R.: 2003, *Geology* **31**, 909–912.

Madigan, M. T., Martinko, J. M. and Parker, J.: 2003. *Brock Biology of Microorganisms*, Prentice Hall, Upper Saddle River, NJ, USA.

Margulis, L.: 1981. *Symbiosis in Cell Evolution*, W. H. Freeman, San Francisco, CA.

Margulis, L. and Fester, R.: 1993. *Symbiosis as a Source of Evolutionary Innovation*, MIT Press, Cambridge, MA.

Martin, W., Hoffmeister, M., Rotte, C. and Henze, K.: 2001, *Biol. Chem.* **382**, 1521–1539.

Martin, W. and Muller, M.: 1998, *Nature* **392**, 37–41.

McCaffrey, M. A., Moldowan, J. M., Lipton, Paul A., Summons, R. E., Peters, K. E., Jeganathan, A. and Watt, D. S.: 1994, *Geochim. Cosmochim. Acta* **58**, 529–532.

McCollom, T. M. and Seewald, J. S.: 2001, *Geochim. Cosmochim. Acta* **65**, 3769–3778.

McCollom, T. M. and Seewald, J. S.: 2006, *Earth. Planet. Sci. Lett* **243**, 74–84.

Mojzsis, S. J., Arrhenius, G., McKeegan, K. D., Harrison, T. M., Nutman, A. P. and Friend, C. R. L.: 1996, *Nature* **384**, 55–59.

Mojzsis, S. J., Coath, C. D., Greenwood, J. P., McKeegan, K. D. and Harrison, T. M.: 2003, *Geochim. Cosmochim. Acta* **67**, 1635–1658.
Mojzsis, S. J. and Harrison, T. M.: 2002a, *Earth Planet. Sci. Lett.* **202**, 563–576.
Mojzsis, S. J. and Harrison, T. M.: 2002b, *Science* **298**, 917.
Mojzsis, S. J., Harrison, T. M., Arrhenius, G., McKeegan, K. D. and Grove, M.: 1999, *Nature* **400**, 127–128.
Mojzsis, S. J., McKeegan, K. D. and Harrison, T. M.: 2005, Life on Earth before 3.83 Ga? Carbonaceous inclusions from Akilia (West Greenland). *NAI Annual Meeting* abstract.
Moreira, D. and López-García, P.: 1998, *J. Mol. Evol.* **47**, 517–530.
Narbonne, G. M.: 1998, *GSA Today* **8**, 1–6.
Narbonne, G. M.: 2005, *THE EDIACARA BIOTA: Neoproterozoic Origin of Animals and Their Ecosystems: Annual Review of Earth and Planetary Sciences*, vol. 33, pp. 421–442.
Nei, M., Xu, P. and Glazko, G.: 2001, *Proc. Natl. Acad. Sci. USA* **98**, 2497–2502.
Nier, A. O.: 1950, *Phys. Rev.* **77**, 789–793.
Nijman, W., de Bruijne, C. H. and Valkering, M. E.: 1999, *Precambrian Res.* **95**, 247–274.
Nisbet, E. G. and Sleep, N. H.: 2001, *Nature* **409**, 1083–1091.
Nutman, A. P., McGregor, V. R., Friend, C. R. L., Bennett, V. C. and Kinny, P. D.: 1996, *Precambrian Res.* **78**, 1–39.
Nutman, A. P. and Friend, C. R. L.: 2006, *Precambrian Res.*
Ohmoto, H., Kakegawa, T. and Lowe, D. R.: 1993, *Science* **262**, 555–557.
Ourisson, G., Rohmer, M. and Poralla, K.: 1987, *Annu. Rev. Microbiol.* **41**, 301–333.
Palin, J. M.: 2002, *Science* **298**, 961.
Papineau, D., Mojzsis, S. J., Karhu, J. A. and Marty, B.: 2005, *Chem. Geol.* **216**, 37–58.
Paris, I., Stanistreet, I. G. and Hughes, M. J.: 1985, *J. Geol.* **93**, 111–129.
Parkes, R. J., Cragg, B. A. and Wellsbury, P.: 2000, *Hydrogeol. J.* **8**, 11–28.
Pasteris, J. D. and Wopenka, B.: 2002, *Nature* **420**, 476–477.
Pasteris, J. D. and Wopenka, B.: 2003, *Astrobiology* **3**, 727–738.
Pearson, A., Budin, M. and Brocks, J. J.: 2003, *Proc. Natl. Acad. Sci. USA* **100**, 15352–15357.
Perry, E. C. Jr. and Ahmad, S. N.: 1977, *Earth Planet. Sci. Lett.* **36**, 280–284.
Pflug, H. D. and Jaeschke-Boyer, H.: 1979, *Nature* **280**, 483–486.
Philippe, H., Lopez, P., Brinkmann, H., Budin, K., Germot, A., Laurent, J., Moreira, D., Muller, M. and Le Guyader, H.: 2000, *Proc. R. Soc. Lond. B Biol. Sci.* **267**, 1213–1221.
Pinti, D. L., Hashizume, K. and Matsuda, J.: 2001, *Geochim. Cosmochim. Acta* **65**, 2301–2315.
Poole, A. M., Jeffares, D. C. and Penny, D.: 1998, *J. Mol. Evol.* **46**, 1–17.
Rasmussen, B.: 2000, *Nature* **405**, 676–679.
Reanney, D. C.: 1974, *Theor. Biol.* **48**, 243–251.
Reid, R. P., Visscher, P. T., Decho, A. W., Stolz, J. F., Bebout, B. M., Dupraz, C., Macintyre, I. G., Paerl, H. W., Pinckney, J. L., Prufert-Bebout, L., Steppe, T. F. and Des Marais, D. J.: 2000, *Nature* **406**, 989–992.
Rivera, M. C., Jain, R., Moore, J. E. and Lake, J. A.: 1998, *Proc. Natl. Acad. Sci. USA* **95**, 6239–6244.
Robert, F.: 1988, *Geochim. Cosmochim. Acta* **52**, 1473–1478.
Robert, F., Selo, M., Hillion, F. and Skrzypczak, 2005, *Lunar Planet. Sci. XXXVI*, 2 pp.
Rose, N. M., Rosing, M. T. and Bridgwater, D.: 1996, *Am. J. Sci.* **296**, 1004–1044.
Rosing, M. and Frei, R.: 2003, *Earth Planet. Sci. Lett.* **217**, 237–244.
Rosing, M. T.: 1999, *Science* **283**, 674–676.
Rosing, M. T., Rose, N. M., Bridgwater, D. and Thomsen, H. S.: 1996, *Geology* **24**(1), 43–46.
Rouxel, O., Dobbek, N., Ludden, J. and Fouquet, Y.: 2003, *Chem. Geol.* **202**, 155–182.
Sano, Y., Terada, K., Takahashi, Y. and Nutman, A. P.: 1999, *Nature* **400**, 127.
Sapp, J.: 2005, *Microbiol. Mol. Biol. Rev.* **69**, 92–305.

Schidlowski, M.: 1988, *Nature* **333**, 988.

Schidlowski, M.: 2001, *Precambrian Res.* **106**, 117–134.

Schidlowski, M., Appel, P. W. U., Eichmann, R. and Junge, C. E.: 1979, *Geochim. Cosmochim. Acta* **43**, 89–199.

Schopf, J. W.: 1983, *Earth's Earliest Biosphere, Its Origin and Evolution*, Princeton University Press.

Schopf, J. W.: 1993, *Science* **260**, 640–646.

Schopf, J. W.: 2000, *Proc. Natl. Acad. Sci.* **97**, 6947–6953.

Schopf, J. W. and Klein, C.: 1992, *The Proterozoic Biosphere: A Multidisciplinary Study*, Cambridge University Press.

Schopf, J. W. and Kudryavtsev, A.: 2005, *Geobiology* **3**, 1–12.

Schopf, J. W., Kudryavtsev, A., Agresti, D. G., Wdowiak, T. J. and Czaja, A. D.: 2002, *Nature* **416**, 73–76.

Searcy, D. G.: 1992, in H. Hartman and K. Matsuno (eds.), *The Origin and Evolution of the Cell*, World Scientific, Singapore, pp. 47–78.

Seilacher, A.: 1992, *J. Geol. Soc. Lon.* **149**, 607–613.

Sheehan, P. M., Coorough, P. J. and Fastovski, D. E.: 1996, in G. Ryder, D. Fastovski, and S. Gartner (eds.), *The Cretaceous-Tertiary Event and Other Catastrophes in Earth History*, vol. 307, Geological Society of America Special Paper, pp. 477–490.

Shen, Y., Buick, R. and Canfield, D. E.: 2001, *Nature* **410**, 77–81.

Simpson, A. G. and Roger, A. J.: 2002, *Curr. Biol.* **12**, R691–693.

Siever, R.: 1994, *Geochim. Cosmochim. Acta* **56**, 3265–3272.

Sogin, M. L.: 1991, *Curr. Opin. Genet. Dev.* **1**, 457–463.

Stanier, R. Y. and Van Niel, C. B.: 1962, *Arch. Mikrobiol.* **42**, 17–35.

Stolz, J. F., Feinstein, T. N., Salsi, J., Visscher, P. T. and Reid, R. P.: 2001, *Am. Mineral.* **86**, 826–833.

Takemura, M.: 2001, *J. Mol. Evol.* **52**, 419–425.

Tice, M. M., Bostick, B. C. and Lowe, D. R.: 2004, *Geology* **32**, 37–40.

Tice, M. M. and Lowe, D. R.: 2004, *Nature* **431**, 549–552.

Ueno, Y. and Isozaki, Y.: 2001, *Int. Geol. Rev.* **43**, 196–212.

Ueno, Y., Yoshioka, H., Maruyama, S. and Isozaki, Y.: 2004, *Geochim. Cosmochim. Acta* **68**, 573–589.

Ueno, Y., Yurimoto, H., Yoshioka, H., Komiya, T. and Maruyama, S.: 2002, *Geochim. Cosmochim. Acta* **66**, 1257–1268.

Van Kranendonk, M. J. and Pirajno, F.: 2004, *Geochem.: Explor. Environ. Anal.* **4**, 253–278.

Van Kranendonk, M. J., Webb, G. E. and Kamber, B. S.: 2003, *Geobiology* **1**, 91–108.

van Zuilen, M. A., Lepland, A. and Arrhenius, G.: 2002, *Nature* **418**, 627–630.

van Zuilen, M. A., Lepland, A., Teranes, J. L., Finarelli, J., Wahlen, M. and Arrhenius, A.: 2003, *Precambrian Res.* **126**, 331–348.

van Zuilen, M. A., Mathew, K., Wopenka, B., Lepland, A., Marti, K. and Arrhenius, A.: 2005, *Geochim. Cosmochim. Acta* **69**, 1241–1252.

Villarreal, L. P. (ed.): 2005, *Viruses and the Evolution of Life*, ASM press, Washington.

Walsh, M.: 1992, *Precambrian Res.* **54**, 271–293.

Walter, M. R., Buick, R. and Dunlop, J. S. R.: 1980, *Nature* **284**, 443–445.

Wang, D. Y., Kumar, S. and Hedges, S. B.: 1999, *Proc. R. Soc. Lond. B, Biol. Sci.* **266**, 163–171.

Westall, F., de Wit, M. J., Dann, J., van der Gaast, S., de Ronde, C. E. J. and Gerneke, D.: 2001, *Precambrian Res.* **106**, 93–116.

Westall, F. and Folk, R. L.: 2003, *Precambrian Res.* **126**, 313–330.

Whitehouse, M., Kamber, B. S. and Moorbath, S.: 1999, *Chem. Geol.* **160**, 204–221.

Whitehouse, M. J., Kamber, B. S., Fedo, C. M. and Lepland, A.: 2005, *Chem. Geol.* **222**, 112–131.

Woese, C. R. and Fox, G. E.: 1977, *Proc. Natl. Acad. Sci. USA* **74**, 5088–5090.

Woese, C. R., Kandler, O. and Wheelis, M. L.: 1990, *Proc. Natl. Acad. Sci. USA* **87**, 4576–4579.

Xiao, S., Zhang, J. and Knoll, A. H.: 1998, *Nature* **391**, 553–558.

Yamaguchi, K. E., Johnson, C. M., Beard, B. L. and Ohmoto, H.: 2005, *Chem. Geol.* **218**, 135–169.

Yoon, H. S., Hackett, J. D., Ciniglia, C., Pinto, G. and Bhattacharya, D.: 2004, *Mol. Biol. Evol.* **21**, 809–818.

Zillig, W.: 1991, *Curr. Opin. Genet. Dev.* **1**, 544–551.

Zuckerkandl, E. and Pauling, L.: 1965, *J. Theor. Biol.* **8**, 357–366.

Earth, Moon, and Planets (2006) 98: 291–297
DOI 10.1007/s11038-006-9092-8

© Springer 2006

8. A Synthetic Interdisciplinary "Chronological Frieze": an Attempt

DIDIER DESPOIS and MURIEL GARGAUD

Observatoire Aquitain des Sciences de l'Univers, Université Bordeaux 1, Bordeaux, France
(E-mail: despois@obs.u-bordeaux1.fr, gargaud@obs.u-bordeaux1.fr)

(Received 1 February 2006; Accepted 4 April 2006)

Abstract. This chapter introduces the chronological and interdisciplinary "frieze" which presents the main events relevant (in our opinion) to the problem of the emergence of life on Earth. This selection of events is directly connected to the previous chapters of this book.

Keywords: Origin of life, chronology, time scales, hadean, archean, proterozoic

8.1. General Description

The frieze enclosed at the end of this book (Figure 8.1) presents a table of various events considered by the authors as relevant to the origins of life on Earth. These events have been tentatively chronologically ordered in accordance with actual knowledge within all the scientific disciplines involved in astrobiology.

When isotopic dating is available, as is the case for most geological events or processes as well as for some astronomical events, an absolute age is obtained which corresponds to the time elapsed between the event and present. For other astronomical events or processes, durations or time intervals are estimated either from theoretical models or observational arguments. In all cases time can be expressed either with respect to present (BP) or with respect to a reference time, t_0[1]. A separate timescale, starting at $(t_0{}^*)$[2], has however been introduced for the early phases of the Sun formation, due to the present uncertainty on the precise timing of these events with respect to planetary formation. In other cases (prebiotic chemistry and early biological evolution) when the dating remains relative, the chronology is qualitatively presented as a logical sequence of events (taking into consideration that this sequence is not unique).

---

[1] t_0 corresponds to the age of the oldest solids in the Solar System which have been dated until now

[2] $t_0{}^*$ is defined as the beginning of the collapse of the interstellar cloud core which gave birth to the Solar System

In our initial project, it was considered to give a rough estimate of the duration of events or reactions for chemical and early biological processes; such a target rapidly appeared inaccessible as it consists in an extremely complex or even impossible task. The only possibility consisted in establishing some time-beacons. The opportunity to assign an absolute age (by radiochronology) to an indisputable microfossil identified in a sedimentary sequence constitutes a reliable time marker for biological evolution. Unfortunately, such absolute markers remain so far extremely sporadic, and in their absence only a relative succession of biological events can be suggested (even if still hotly debated).

An attempt was made to systematically refer to all available data sources and in case of controversies, to present alternative models and views and to cite the appropriate references. Unfortunately, this was not always possible: for example, too many scenarios are currently proposed for prebiotic chemistry and early biological evolution; likewise dating of the "first" microfossils is subject to many controversies impossible to report in a synthetic form.

As already discussed in the previous chapters, this *"chronological frieze"* does not present a complete overview of the Earth evolution but rather is limited to the period ranging between t_0^* and the so-called explosion of life at the beginning of the Cambrian period. The selected period includes three geological eras: the Hadean, Archaean and Proterozoic, which are classically regrouped under the Precambrian. The fact that the Precambrian is the period during which the most important events relative to the emergence and early evolution of life took place justifies our selection.

The data inserted in the "chronological frieze" were subject to long interdisciplinary discussions in order to resolve the constraints linked to the different types of chronometers used by each discipline concerned with the origins of life. When possible an interpretation relevant for the origin of life is proposed, and tentatively inserted in an unique and coherent interdisciplinary timescale. This theoretical *modus operandi* is limited due to the huge lack of chronological data relative to prebiotic and early biological evolution, which did not permit their insertion into a reliable time scale. Consequently, it was decided to insert the part of the "chronological frieze" relative to the prebiotic chemistry and early biological evolution between 4.4 and 2.7 Ga. The date of 3.8 Ga was chosen because it represents the age of the oldest sedimentary rocks so far recognized.

Whatever the dating method is, age is determined within an uncertainty domain; consequently, when available, an error bar on the date/age is given. It must be noted that this error bar gives the analytical uncertainty on age determination, and is totally independent of the interpretation of the age. For instance, in an old magmatic rock, the age measured on a zircon crystal can reflect the age of the source of the magma, the age of magma crystallization or even the age of a subsequent thermal event (called metamorphism). The discussion relative to these aspects is presented in a column called "reliability." In some critical cases, discussion has been extended further as "notes."

8.2. How to Interpret the Data of the "Chronological Frieze"?

8.2.1. EON AND PHASE

Eon and phase refer to periods of time. Eons are officially defined by the International Union of Geological Sciences (IUGS). However, since Hadean is not official in the latest version of the international stratigraphic chart, it was decided to place the Hadean–Archaean boundary at 4.0 Ga, which is roughly the age of the oldest known rocks. In addition to this official international nomenclature, Earth history is subdivided in several phases, which refer to periods important in terms of life apparition and development; these phases are closely related to the various chapters of this book.

8.2.2. AGE, ERROR BARS, TIME AND DURATION

They are expressed in mega years (Ma or Myr) or giga years (Ga or Gyr). The choice of "anno" or "year" reflects the convention differences between scientific domains: Geology and Astronomy respectively. Ages refer to present time, and are thus expressed as "age before present" (BP), more precisely defined as 1950 AD(see chapter 1).

8.2.3. REFERENCE TIMES

Excepted for the formation of the Sun, and to some extent, for the formation of the protoplanetary disk, all relative times presented in the "chronological frieze" refer to an absolute time t_0, that corresponds to the oldest solids in the Solar System which have been dated until now; these solid particles are the CAIs (Calcium–Aluminium Inclusions) of the Allende meteorite, dated at 4.5685 Ga BP. For the first 1–10 Myr, corresponding to the stellar formation and early protoplanetary evolution, a t_0^* time has been introduced, arbitrarily chosen as the start of the collapse of the initial molecular cloud core. This t_0^* time can be linked to the absolute time t_0 if the durations of the various stellar successive phases (e.g. protostar, T Tauri) are estimated; it must be clear that these durations are model-dependent.

8.2.4. EVENTS

The column "events" lists what we consider the most relevant phenomenons for our understanding of emergence of life. They are presented under a generic title. In case of complex events made of a succession of composite steps, the column labelled "Comments about these events" attempts to describe these different steps as clearly as possible. Their importance or their role in the emergence of life is presented in the column "Consequences for life."

8.2.5. DATING METHODS

The tools used to determine the age of an event are reported in this column. Chapter 2 provides additional information about methods.

8.2.6. RELEVANT OBSERVATIONS

This column gives additional information about the dated objects, the nature of observations, and/or the model used.

8.2.7. RELIABILITY

This column reports the problems linked to the data interpretation. In some cases, it points to the vivid controversies in the scientific community.

8.2.8. CONSEQUENCES FOR LIFE'S ORIGIN AND EVOLUTION

This column attemps to present the potential or possible link between a particular event and the emergence of life.

8.2.9. REFERENCES

Only the number of the relevant chapter of this book is reported here; detailed references can be found in the chapter itself.

8.2.10. NOTES

They refer to the following further details on data interpretation and possible controversies:

(#N1) t_0^* must be distinguished from t_0, which is the age of the oldest dated sample of primitive material in the solar system (see below: CAIs, dated 4.568 ± 1 Gyr).

(#N2) The date corresponding to t_0 (4568.5 Ma BP) is now rather well established : (1) Amelin (2002) and Bouvier et al. (2005) show that Allende CAI ages are consistent with this age within an error bar of ± 0.4 Ma (2)

Lugmair and Shukolyukov (1998) get the same age by adding Pb–Pb age of Ste Marguerite (H4) with ^{26}Al–^{26}Mg age (3) The same age is obtained if we add the Pb–Pb age of 6 angrites (Baker et al., 2005) and the ^{26}Al–^{26}Mg internal isochrone age of 99555 (their total rocky age are obviously 1.2 Ma too old)

(#N3) Hence possible role in the existence of Europa-like bodies

(#N4) A last metal-silicate equilibrium is recorded at 30 Ma with an error bar 11–50 Ma. Moreover N-body models predict a rather rapid first phase in telluric planet formation up to (roughly) the size of Mars ("planetary embryos") followed by a slower one (build up of Earth-size bodies by embryo encounters). The age of 30 Ma for the Earth could correspond to a collision with a Mars size body and indeed, a recent work propose an ^{182}Hf–^{182}W age of 45 Ma for the Moon crustal differentiation, which could correspond to the end of magmatic oceans on the Moon.

(#N5) The magmatic ocean stage could have lasted longer on Earth than on Moon, may be 60–70 Ma (^{129}I–^{129}Xe chronometers coupled to ^{129}I–^{129}Xe). The different ^{142}Nd isotopic anomalies between chondrites and the Earth (Boyet and Carlson, 2005) show that the effective formation of Earth magmatic ocean was achieved 30 Ma after accretion. In addition, due to their different gravities, less plagioclase crystallized in Earth magma ocean when compared to Moon where plagioclase accumulation formed the anorthositic crust.

(#N6) Before this shock, a H_2-rich primitive atmosphere may have existed; it subsequently disappeared by hydrodynamic escape or as the consequence of impacts.

(#N7) 100 Ma correspond to the epoch at which the atmosphere begins to retain ^{129}Xe produced from ^{129}I. However the atmosphere may have stayed open during a much longer period (in order to allow the escape of the ^{136}Xe produced by ^{244}Pu radioactivity). About 100 Ma is only an average closure time, because of problems linked to the measure of the age of reservoirs.

(#N8) The isotopic systems ^{146}Sm–^{142}Nd et ^{244}Pu–^{136}Xe indicate a very active Earth during the Hadean, allowing for a global differentiation during the first 100–400 Ma. The oldest dated zircon (4.4 Ga) does not have high $\delta\ ^{18}$O typical of interaction with water at low temperature. This $\delta\ ^{18}$O alone, could be interpreted as mantle value, however, zircon crystal also contains quartz and plagioclase inclusions; which demonstrates that it crystallized in a continental crustal magmatic rock.

(#N9) Other authors consider that most, if not all, halogens and H, C, N, O, S, P, were already present and available on Earth surface at 4.4 Ga when first oceans formed.

(#N10) On Earth this episode is not certain but only extremely probable; indeed, on the Moon, the Imbrian event is recognized only on the visible face of the Moon (which could then be considered as a local event)

(#N11)A similar isotopic deviation can be produced abiotically. The sedimentary nature of the graphite was analysed, and even its age has been questioned.

(#N12) It is impossible to say whether these metabolisms appeared simultaneously or successively. A parallel evolution appears possible, and even supported (e.g. phylogenetic radiation within the bacteria and prior divergence of the archaea) for most respiration types, photosynthesis and methanogenesis.

(#N13) The interpretation of these observations is strongly debated. If macroscopic laminar structures do correspond to fossil stromatolites, this would prove the occurrence of laminated microbial communities (microbial mats) implying the presence of metabolically diverse microorganisms. If the presence of microfossils associated to these structures is confirmed, it would constitute evidence for the existence of cellular life. If negative $\delta\,^{13}C$ values are actually biogenic and syngenetic with the rock, some type of autotrophic metabolism (photosynthesis and/or chemoautotrophy) would have already developed.

(#N14) Eukaryotic morphologies characterized by extent surface decoration are widely accepted

The frieze may be found in print as an insert at the back of the print publication. This supplementary material is also available in the online version of this article at http://dx.doi.org/10.1007/s11038-006-9092-8 and is accesssible for authorized users.

Acknowledgements

We'd like to thank warmly all the authors for their help during the "making of" of this "chronological frieze." We are especially very grateful to the coordinators of the various chapters of this review for fruitfull discussions, comments and critical reading of the (numerous) "proto-frieze" which precede this one: Philippe Claeys, Purification Lopez-Garcia, Hervé Martin, Thierry Montmerle, Robert Pascal and Jacques Reisse. We also thank especially Francis Albarede, Daniele Pinti and Franck Selsis for fruitfull discussions on oldest dated sample, oceans and atmosphere formation. Finally, we also thank Judith Pargamin for her help in the editorial presentation of this frieze and Michel Viso for his constant support to this project.

References

Amelin Y., Grossman L., Krot A. N., Pestaj T., Simon S. B., and Ulyanov A. A. (2002), *Lunar and Planetary Institute Conference Abstracts* **33**, 1151.

Baker, J., Bizzarro, M., Wittig, N., Connelly, J., and Haack, H.: 2005, *Nature* **436**, 1127–1131.

Bouvier, A., Blichert-Toft, J., Vervoort, J. D., McClelland, W., and Albarede, F.: 2005, *Geochim. Cosmochim. Acta* **69**(Suppl.), 384 .

Boyet, M. and Carlson, R. W.: 2005, *Science* **309**, 576–581.

Lugmair, G. W. and Shukolyukov, A.: 1998, *Geochim. Cosmochim. Acta* **62**, 2863–2886.

Earth, Moon, and Planets (2006) 98: 299–312
DOI 10.1007/s11038-006-9093-7

© Springer 2006

9. Life On Earth... And Elsewhere?

THIERRY MONTMERLE
Laboratoire d'Astrophysique de Grenoble, Université Joseph Fourier, Grenoble, France
(E-mail: montmerle@obs.ujf-grenoble.fr)

PHILIPPE CLAEYS
DGLG-WE, Vrije Universiteit Brussel, Brussels, Belgium
(E-mail: phclaeys@vub.ac.be)

MURIEL GARGAUD
Observatoire Aquitain des Sciences de l'Univers, Université Bordeaux 1, Bordeaux, France
(E-mail: gargaud@obs.u-bordeaux1.fr)

PURIFICATIÓN LÓPEZ-GARCÍA
Unité d'Ecologie, Systématique et Evolution, Université Paris-Sud, Orsay, France
(E-mail: puri.lopez@ese.u-psud.fr)

HERVÉ MARTIN
Laboratoire Magmas et Volcans, Université Blaise Pascal, Clermont-Ferrand, France
(E-mail: H.Martin@opgc.univ-bpclermont.fr)

ROBERT PASCAL
Départment de Chimie, Université Montpellier 2, Montpellier, France
(E-mail: rpascal@univ-montp2.fr)

JACQUES REISSE
Faculté des Sciences Appliquées (CP 165/64), Université Libre de Bruxelles, Brussels, Belgium
(E-mail: jreisse@ulb.ac.be)

FRANCK SELSIS
Centre de Recherche Astronomique de Lyon and Ecole Normale Supérieure de Lyon, Lyon, France
(E-mail: franck.selsis@ens-lyon.fr)

(Received 1 February 2006; Accepted 4 April 2006)

Abstract. This concluding chapter is divided into two main parts. The first part is a summary of the main facts and events which constitute the present body of knowledge of the chronology of life in the solar system, in the form of "highlights" in astronomy, geology, chemistry and biology. The second part raises the interrogation "Is life universal?", and tries to provide answers based on these facts and events. These answers turn out to differ widely among the various disciplines, depending on how far they feel able to extrapolate their current knowledge.

Keywords: Astrobiology, chemical evolution, early Earth, early Mars, early Venus, eukaryotes, exoplanets, habitable planets, Late Heavy Bombardment, microfossils, Moon, origin of life, prebiotic molecules, solar system, star planet information, supernovae, universality of life

9.1. The Chronology of Life in the Solar System: Highlights

9.1.1. FROM THE BIRTH OF THE SUN TO THE BIRTH OF THE EARTH

According to recent estimates, the Earth was assembled in less than 100 Myr.[1] Since its very existence is a pre-requisite for the emergence and/or evolution of life as we know it, a key problem for astronomers consists in trying to reconstitute its origin, and to establish whether the formation of terrestrial planets among solar-like stars is a common or an exceptional phenomenon. Unfortunately, until now this question remains unanswered; however, the observations of nearby solar-like stars and star-forming regions, lead to establish at least five important facts:

(i) Molecular clouds, out of which stars form, contain numerous organic molecules. They also contain dust grains. The grain mantles are formed during the pre-collapse phase. Once a protostar is formed, the heating of the central object and/or the shocks created by the outflowing material inject back into the gas phase the mantle components. Further reactions take place in the central regions of protostars, forming new complex molecules.

(ii) Circumstellar disks, composed of gas (hydrogen, helium and heavier elements) and dust grains (mostly carbon compounds, silicates and ices), having masses equivalent to many solar systems, are present around all young solar-like stars. Molecules are frozen onto the grain mantles in the plane of the protoplanetary disks. Planetesimals coagulate and form the bricks of planets, comets and asteroids, possibly incorporating the molecule-rich mantle components.

(iii) To date, nearly 200 "exoplanets" (with masses currently ranging from five Earth masses to 12 Jupiter masses)[2] have been discovered. The probability for nearby solar-like stars to host a planet we can detect with our present technology is about 5%.

(iv) Mostly theoretical considerations show that the presence of giant planets influences the formation of terrestrial (i.e., rocky) planets from circumstellar disks. Indeed, numerical simulations without giant planets tend to form smaller and more numerous terrestrial planets than simulations including Jupiter-like planets.

---

[1] For comments on the units of time adopted by astronomers (Myr or Gyr) vs. those adopted by geologists and biologists or chemists (Ma or Ga), and how they use them, see Chap. 2 on chronometers (Chap. 2.5).

[2] This is the limit, officially adopted by the IAU (International Astronomical Union), between a "planet" and a "brown dwarf" star. See (Chap. 2.1).

(v) On the other hand, the study of the oldest meteorites (chondrites) shows that the young solar system must have been filled very early with colliding planetary embryos known as "planetesimals".

In Chapter 3, we have attempted to describe in a simplified fashion the chronology of the evolution of the solar system, starting with the birth of its parent star, our Sun, and ending with the very young Earth, our planet. In analogy with a musical piece, this chronology can be divided in three (logarithmic) movements: *allegro*, for the formation of the Sun (inferred from the formation of solar-like stars), that takes about 1 Myr (we called it the "stellar era"); *andante*, for the evolution of circumstellar disks and the formation of giant planets (the "disk" era), which takes about 10 Myr, and *lento*, for the formation of terrestrial planets and the early evolution of the Earth (the "telluric era"), which is dominated by collisions between small bodies (*fortissimo!*) and takes about 100 Myr. It is in the course of the *andante*, that we start to combine circumstellar disk evolution with young solar system physics (the so-called meteoritic record). However, we have been forced to introduce an *intermezzo*, because our understanding of how dust grains, observable in circumstellar disks only up to a few mm in size, grow into solar-system planetesimals of a few km, suffers from a severe bottleneck. The highlights of this "musical piece" are as follows.

(1) *The "stellar era"* (1 Myr). Modern ideas on the formation of the Sun and solar-like stars stress that, as a rule, stars start forming inside dense, cold "molecular clouds", not in isolation but in clusters, ultimately comprising a few tens to a few thousand visible stars. This is thought to be the result of turbulence and magnetic fields threading these clouds. Because they are rotating, "protostars" form disks while collapsing under the pull of gravitation; they also form "bipolar jets" mediated by, again, magnetic fields. The so-called "primitive solar nebula" must have been at least part of such disks, which are observed around all stars except the most massive ones (i.e., above a few solar masses).

(2) *The "disk era"* (10 Myr). Observations show that circumstellar disks disappear, in a statistical sense, after a few million years only. This is interpreted as corresponding to giant planet formation, putting severe constraints on theoretical models for the transition from dust grains to planetesimals and their assembling into giant planets. On the other hand, X-ray observations show that while disks evolve, they must be irradiated by hard radiation (X-ray flares), for which there is now direct evidence, and by inference by energetic particles. In the young solar system, there is possible evidence for this energetic "internal" irradiation from the presence of so-called "extinct radioactivities" in the most primitive meteorites. However, some of these "extinct radioactivities" also point out to a supernova explosion in the

vicinity of the very young Sun (less than 1 Myr), which, contrary to the ubiquitous presence of circumstellar disks, may have been a relatively rare occurrence. Then, after the *intermezzo* mentioned above, theory takes over, subject to severe dynamical constraints, to explain the formation of planetary embryos and giant planets.

(3) *The "telluric era"* (100 Myr). Strongly influenced by gravitational interactions from giant planets, and also constrained by observations of "debris disks" around evolved young stars, a few terrestrial planets, and a primordial asteroid belt, form by accretion from planetesimals. This is the time of the last cataclysmic collisions, giving rise to the terrestrial planets. In particular, one such late (around 30 Myr) collision between a Mars-sized body and the young Earth is invoked to account for the formation of the Moon.

At this stage, the Earth starts to be differentiated (formation of a core), and an atmosphere appears. Another era begins, that of geology. The story continues in Chapter 4.

9.1.2. THE HISTORY OF THE EARLY EARTH AND GEOLOGICAL CONSTRAINTS ON THE ORIGIN OF LIFE

The succession of events that occurred on Earth surface since its formation mainly depends on both internal and external factors.

The main internal factor is the progressive cooling of our planet that started just after accretion and that still continues today. Most of internal heat (accretion energy, radioactive elements, latent heat of core crystallization, etc.) has been fixed 4.568 Ga ago and its budget slowly decreased until now. This cooling is progressive and accounts for the main changes in our planet dynamics (Magma Ocean, plate tectonic development, etc. – see Chapter 4.1). In addition, the greater heat production on the primitive Earth can have favoured the development of hydrothermalism and consequently created niches potentially favourable for the development of some forms of life (Chapter 6). As these changes are slow, life could have more easily accommodated and adapted to them.

Among external factors, some also changed progressively, such as the surface temperature of the Sun (Chapter 6.2), but others were more brutal, such as for instance the Late Heavy Bombardment or even glaciations (Chapters 4.4 and 6.3). These events may have had a more drastic and cat-astrophic effect on life evolution and development. Indeed, meteorites, micrometeorites or comets, not only can partly or totally vaporize oceans and consequently eradicate possible pre-existing life, but they are also able to bring on Earth chemical elements that modified the composition of the atmosphere and oceans (Chapters 4.2 and 6.2), as well as they were able to bring on the Earth surface the "bricks" of life.

Here too, the history of the Earth can be divided in few periods, each one being characterized by a specific thermodynamic conditions and geodynamic environments and consequently by different set of niches and conditions for life development.

(1) *The Early Hadean period (4.568–4.40 Ga)* [No life possible].

It is highly probable that this period was not favourable for the emergence of life and for its development on Earth. Indeed, most of the Earth mantle was at a very high temperature such that it was molten, giving rise to a magma ocean; continental crust did not exist and liquid water was not condensed on the Earth surface. In addition, it is also highly likely that the meteoritic bombardment intensity was still important. If this period was not favourable for the emergence of life, however it was a period of elaboration of conditions and building of niches and environments where subsequently life has been able to develop.

(2) *The Late Hadean period (4.40–4.00 Ga)* [Conditions favourable for potential emergence of life].

Recently, very old (4.3–4.4 Ga) zircon crystals have been discovered in Jack Hills (Australia; Chapter 4.3); their composition indicates that they formed in the continental crust and that the latter was already stable. In addition, their oxygen isotopic composition also shows that liquid water (ocean?) already existed at the surface of the Earth. Similarly, recent works on meteoritic bombardment showed that very probably the Hadean period was relatively cool ("Cool Early Earth") and that the meteoritic bombardment only lasted between 4.0 and 3.8 Ga (Chapter 4.4). Consequently, since the 4.4–4.3 Ga period, the conditions for prebiotic chemistry and appearance of life were already met (liquid water, continental crust, no strong meteoritic bombardment, etc...). This does not mean that life existed so early, but this demonstrates that some (all?) necessary conditions assumed for life development were already present on Earth.

(3) *The Archaean period (4.00–2.50 Ga)* [Conditions favourable for life evolution].

This period started with the Late Heavy Bombardment (LHB; 4.0–3.8 Ga; Chapter 4.4). Even if internal Earth heat production was lower than during Hadean times, it remained 2–4 times larger than today. The correlated high geothermal flux gave rise to rocks such as komatiites and TTG suites, which are no more generated on Earth since 2.5 Ga. The Archaean is a period of very intense continental crust differentiation. The greater heat production favoured the development of hydrothermalism and possibly created niches potentially favourable for the development of some forms of life (Chapter 6). The Isua greenstone belt in Greenland contains some of the oldest rocks known today on the Earth surface. Among them, outcrop detrital sediments demonstrate that ocean and emerged continents already existed 3.865 Ga ago (Chapter 4.3). These conclusions are corroborated by the younger (3.5–3.4 Ga) rocks in both Barberton (S. Africa) and Pilbara (Australia). Weak traces as well as isotopic signatures in these sediments (Chapter 7.1) are interpreted as possible indications of primitive life, but this interpretation is strongly debated and these markers cannot for the moment be actually considered as unequivocal evidence for the earliest forms of life. If specialists still hotly debate about the age of the oldest undisputable traces of life (Chapters 5.2–5.6 and 7.2), all agree that life appeared during part, if not all, of this period, which can be considered as having been extremely favourable for life development and diversification.

(4) *Proterozoic and Phanerozoic period (2.50 Ga – today)* [Conditions for diversification and explosion of life].

After Archaean times and almost since 2.50 Ga, the Earth had significantly cooled for its geodynamic regime to become similar to the modern plate tectonics as we know it today on Earth. Thermal and compositional conditions in oceans were not drastically different from the ones we know today and the solar bolometric luminosity (i.e., summing over all wavelengths) significantly increased since 4.4 Ga (Chapter 6.2.1.3). The main change appears to be a consequence of life development and consists in the increasing amount of dioxygen in the Earth atmosphere between 2.3 and 2.0 Ga (Chapter 6.2.1.2) as a consequence of photosynthesis. The Proterozoic–Phanerozoic boundary, about 540 Ma ago, consists only in the explosion and strong diversification of Metazoan life, which, due to the final increase of atmospheric O_2 (and correlated O_3), resulted in the conquest of emerged continent surfaces.

9.1.3. FACTS OF LIFE

That life appeared, or at least evolved, on Earth is a fact, but exactly when and how it did remains, and will most likely remain for a long time, unresolved. Life must have arisen after a period of active pre-biotic chemistry of unknown duration (Chapter 5.1). Life resulted from the merging of genetic systems capable to store, express, and transmit information, and also able to evolve, i.e., to accumulate changes (mutations) upon which natural selection could operate. Life resulted also from metabolic abilities for management of organic and inorganic sources for self-sustainment, including free energy production, storage and use. The merging of both systems was made possible by, or was associated with, their separation from the environment, while allowing metabolite exchange, by a membrane envelope defining a compartment (Chapters 5.2–5.6). The relative order of emergence of the different systems is controversial and highly speculative, as was forewarned in Chapter 5.1. They may even have evolved in parallel. Indeed, from a given (though unknown) moment in time (perhaps close to the theoretical limit between non-life and life), they did co-evolve. From the common properties occurring in extant terrestrial life, it is possible to conclude that a "last common ancestor", possessing those common traits to all living beings, existed (Chapter 5.7). However, although it must have been already quite complex, its precise nature is hotly debated, and when it thrived on Earth is impossible to tell. It must have lived some time between the emergence of the first living entities and the divergence between the different domains of life.

When did life appear in geological times? The answer is very uncertain. For many authors, this could be during the period comprised between 3.8 Ga (after the LHB) and 3.5 Ga. But a more cautious answer would be: some time between formation of the Earth and oceans (4.4 Ga) and the oldest unambiguous microfossils at 2.7 Ga (see Chapter 7). Will it be possible in the future to restrict further these limits? Looking at the farthest limit (4.4 Ga), if the LHB was actually sterilizing, life as we know it today must have developed only after 3.8 Ga. Life might have evolved and got extinct previously, but in that case, it will be impossible to know if such early life was based on similar molecules and mechanisms as the one that we know today. Looking at the closest limit, 2.7 Ga, it is quite possible that this threshold will eventually be pushed back in time. Claims for the occurrence of microfossils and isotopic traces of microbial activities at 3.5 Ga exist, although the evidence presented is not yet strong enough not to be criticized (Chapter 7.1). Hopefully, methodological advances and improvement in biomarker knowledge will help yielding more compelling answers about the nature of the putative oldest traces of life. However, it will be very difficult to distinguish early living entities from more derived bacteria or archaea at the oldest fossil-record level. Furthermore, the eventual recognition of a certain group of organisms in the fossil record does not preclude the possibility that it evolved much earlier without leaving traces in that fossil record. For instance, eukaryotes, whose oldest certified fossils date back to 1.5 Ga, may have appeared earlier without leaving easily identifiable morphological remains. It is clear, nevertheless, that eukaryotes appeared only after prokaryotes since their ancestor acquired mitochondria (derived from endosymbiotic bacteria). In addition to time constraints fixed by paleontology, many questions remain open in early microbial evolution regarding both the relative emergence of the different groups and their metabolic strategies, and the date at which they appeared (Chapters 5.2–5.6 and 7.2). Hopefully too, progress in molecular phylogenetic and comparative genomic analyses can help resolve some of these evolutionary questions by pointing out to enzymes or metabolic pathways that are more ancestral than others. But we need to assume as scientists that in most cases, we will be left with a bundle of appealing, unfortunately non-testable, historical hypotheses.

9.2. Our World in Perspective: Is Life Universal?

Life, as we know it, is present on only one of the terrestrial planets in the solar system. Perhaps life appeared on Mars and Venus at some stage as there was liquid water on Early Mars and possibly on Early Venus. Orbital distance was the critical factor that determined the divergent evolution of Venus and the Earth, as our planet was far enough from the Sun to avoid a runaway greenhouse effect. The different fates of Mars and the Earth are more related with their gravity, tectonic and volcanic activity, and magnetic field, three parameters determined by the planetary mass, which in the case of Mars is too small to sustain a dense atmosphere on geological timescales. Apart from having the right mass, composition and orbital distance, other less generic effects (such as the stabilization of the Earth rotation axis by the Moon, the early bombardment history) may have played an important role in the sustainment and development of life on Earth.

There is currently no observational evidence of living systems elsewhere in the Universe, which immediately raises the question: Are we alone? But in this review we have deliberately limited ourselves to the emergence and early evolution of life on Earth, i.e., the development of *microbial* life. The question of the development of *intelligent* life is related to the post-Cambrian evolution of species up to the appearance of Man and the development of technology.[3] We did not discuss this point, and therefore we discard it as being outside of the scope of this concluding chapter.

Even limited to the question of the universality of life in general, the answers the authors of this article are able to provide will ultimately rely only on the available data. As a matter of fact, these data (and their interpretation) differ widely according to the different fields of research, and reasonable scientific conclusions depend on how far (in time and space) they feel able to extrapolate from their current knowledge of the chronological history of the solar system and the Earth as described in this review.

More precisely, the "observable horizon" shrinks as we consider stars, planetary systems, pre-biotic chemistry, and finally life. In short, astronomers are now able to observe myriads of forming solar-like stars, and soon hundreds of planetary systems, out to a horizon hundreds of light-years away from the Earth. Thus astronomers can rely on a large statistical basis to draw conclusions, for instance about the "typical" time for the disappearance of

---

[3] In particular, this is the context of the famous so-called "Fermi paradox".

circumstellar disks, and about the diversity of exoplanets, such as the ubiquitous presence of "hot Jupiters". However, to date no planetary system similar to the solar system is known, so that, in spite of continuing discoveries, astronomers are currently unable to tell what fraction of planetary systems is harbouring habitable exoplanets. Astronomers, geologists and geochemists can, to some extent, reconstitute the early history of planets, such as gravitational interactions, collisions, accretion and internal differentiation, composition and early evolution of atmospheres and oceans, but they are able to do so in some detail only in the single planetary system one knows because it is ours: this horizon is about 10,000 times smaller than the previous one, illuminated by a single star of known evolution in luminosity and temperature, with no more than a few dozens of planets, satellites and minor bodies available for individual study (albeit spanning a tremendously large range of physical, chemical and magnetic conditions). At least for some time, chemists and biologists are confined to one object only, the Earth, to understand how life evolved, from the first (still undefined) "brick of life". They don't even know whether the path life took is the only one possible, since they have to rely on the Earth only, yet the Earth itself now contains millions of living species. And as is well known, "life destroys life", so to reconstitute its chronology back to its origin is a daunting, perhaps impossible, task.

At the end of this review, let us then try to be more specific and elaborate our contribution to the challenge of astrobiology. In other words, let's examine what each community has brought here through the chronology of the formation of the solar system and the Earth – a trip reminiscent of that of Micromégas, Voltaire's 18th century interplanetary traveler...

For astronomers, *all* solar-like stars (at least) are born surrounded by circumstellar disks. The conspicuous existence of exoplanetary systems, even if very different from ours, suggest that planet formation is commonplace, perhaps even universal, in spite of the fact that only 5% of the observed solar-like stars in the solar vicinity[4] are found to harbour giant planets relatively close to the parent star. If one admits that liquid water is a necessary condition for the emergence and development of life, it is possible to define for each system a "radiative habitable zone" where the stellar irradiation is such that liquid water may be present at the surface of terrestrial planets (or perhaps the large satellites of giant planets) having orbits confined within this zone. The fraction of Sun-like stars that host at least one 0.5–5 Earth-mass terrestrial planet in their radiative habitable zone is still unknown but will hopefully be determined in the solar vicinity

---

[4] Their average distance is about 30 pc, or 100 light-years. For more details on exoplanets and their statistics, see http://vo.obspm.fr/exoplanetes/encyclo/catalog-main.php?mdAff = stats#tc.

by the next generation of space observatories.[5] But recalling that our galaxy, the Milky Way, contains about 10^{11} stars, clearly a very large number of exoplanets are likely to be candidates for harbouring at least primitive forms of life.

Geologists and geochemists, on the other hand, would raise the concern whether planets in an *a priori* "habitable zone" would have oceans and atmospheres lasting long enough to develop and sustain life "in the open" to eventually lead to the Cambrian explosion of living species: we have evidence to the contrary in the solar system with Venus and Mars. On Earth, it took ~1–2 billion years before life emerged: this required that during all this time, the planet remained habitable and as we have seen, this necessary condition was perhaps the consequence of an impact between the young Earth and a body of the size of Mars leading to the Moon formation. Without this large satellite, it is not so sure that the physical conditions on Earth would have remained sufficiently stable for such a long time.

For chemists, life in the Universe may not be considered uniquely from this point of view, and seeking for indications of forms of life that evolved differently will be a major field of research following the discovery of extrasolar planets. Moreover, the presence of organic matter, in molecular clouds and protostellar envelopes, in meteorites, comets, and in various places in the solar system has shown that the building blocks of life including aminoacids are very common molecules. We are far from the clear-cut separation between organic chemistry and inorganic chemistry, which prevailed at the beginning of the 19th century when the organic world was supposed to be related to life. It simply indicates that the somewhat unexpected but common chemical processes that have supported the development of life on Earth may occur elsewhere. A logical next step is to look for spectroscopic indications for life-related chemistry in exoplanetary atmospheres. Although we do not know for sure what could be an unambiguous signature of life, the spectroscopic characterization of terrestrial exoplanets will allow us to identify potentially habitable planets and to search for atmospheric compounds of possible biogenic origin. The detection of an out-of-equilibrium atmosphere containing for example O_2 or O_3 together with some reduced molecules like CH_4 or NH_3 would be probably considered as a strong indication that some metabolic pathways took place on the planet. This is currently the formidable goal of future space missions like *Darwin* and *TPF* (see below, footnote 5).

For biologists, the situation is different. Astrophysicists, geologists and chemists find the objects of their disciplines in other planets. By contrast, life is known only on Earth, so that biologists generally feel uneasy in a field

[5] COROT launched by CNES in late 2006, *Kepler*, by NASA in 2009, *Darwin* by ESA and *Terrestrial Planet Finder (TPF)* by NASA in 2020?

where their object of study is purely hypothetical elsewhere. A great step was made in the 1970s due to the discovery of extremophilic microorganisms, whose study considerably extended the physico-chemical conditions in which life, terrestrial life, could occur. Some of these organisms might be able to thrive in other planets. In addition to microbiologists studying extreme environments, evolutionists trying to reconstruct the origin and early evolution of life are among the less skeptical biologists. Given the appropriate physico-chemical settings, some form of life based on a similar biochemistry to the terrestrial one could have developed. However, we unfortunately ignore how life evolved and the question of whether life was a rare unique event or a necessary outcome remains still unanswered.

Should we then envisage that the conditions on Earth (and in the young solar system) were so exceptional that life not only emerged, but also was able to evolve very quickly and survive major catastrophes such as, possibly, the Late Heavy Bombardment or later cosmic collisions, or major atmospheric changes?

As a matter of fact, one cannot be but surprised at some peculiarities of the Earth and the solar system, which may or may not have been important for life to emerge and survive. To list but a few, in chronological order:

- Is the possible presence of a supernova in the vicinity of the forming Sun, attested by the presence of ^{60}Fe in meteorites, important or circumstantial? Does it tell us that Sun-like planetary systems can be born only in "special", rare stellar associations?
- Has such a supernova anything to do with chirality? The compact remnant of a supernova is in general a neutron star[6] and, therefore, it may well be that a neutron star was located near the protosolar nebula. A neutron star can be described as a huge natural synchrotron source (a "pulsar") which means that above the poles of this fast rotating stellar object, a strong circularly polarized light is emitted at all wavelengths. Of course, the polarization is opposite above each pole but, for a particular position of the protosun with respect to the neutron star, it can be shown that if a mixture of chiral molecules pre-exists in the nebula, an enantioselective photolysis of these molecules could take place. This means that even in the case of a 50/50 mixture of molecules of opposite chirality (racemic mixture), an enantiomeric excess could be observed after photolysis. Interestingly enough, some aminoacids found in carbonaceous chondrites show enantiomeric excesses as large as 12%. Therefore, the origin of homochirality of the constituents of living matter could have been

---

[6] In reality, only a small fraction of young supernova remnants, known as "plerions" (like the famous Crab nebula, the remnant of a supernova having exploded in 1059 AD), are definitely known to host a neutron star, in the form of "pulsar". See also Chapter 2.

indirectly influenced by the explosion of a supernova in the vicinity of the forming solar system.

– Was the circumstellar disk around the young Sun a rare survivor of a UV or X-ray induced early evaporation in the vicinity of hot, massive stars?

– What was the role of the Moon? If one believes some large-scale mixing of pre-biotic material in water, like tides, or the stabilization and orientation of the Earth's rotation axis, was necessary, or even essential, for a large-scale spreading of protolife across its surface, then the Earth is the only planet in the solar system with a satellite massive enough and close enough to fulfill this condition.

– Was the exchange of ejected material between planets important for the preservation of life? During the early Hadean, the faint young Sun may have hosted three habitable planets (Venus, Earth and Mars), which were sporadically submitted to giant impacts. If life had already emerged on one of these planets, it could have survived these sterilizing impacts inside ejecta and reseeded its planet of origin or another one. In that case, odds might favour life emerging in systems with more than one habitable planet.

– What was the role of the Late Heavy Bombardment? Did it obliterate all forms of life possibly existing in oceans at that time, so that the process had to start all over again? Certainly, the present evidence is not inconsistent with the presence of bacterial life before the LHB, which could have survived even in small proportions, the near-vaporization of the oceans.

Lastly, many events in relation with the emergence of life and its early evolution remain totally unknown and are related to processes that are often dependent on the answers given to previous questions.

– Are the conditions for "starting" life (whatever that means: the first replication, for instance) extremely demanding and narrow, thus very rare, or on the contrary "inevitable", i.e., due to happen sooner or later?

– Once life "starts", how fast is its evolution, and how does it depend on the physical and chemical evolution of planetary atmospheres over time?

– Did life happen very early (before the Late Heavy Bombardment) with a reducing atmosphere favourable to pre-biotic synthesis? Then, how long did it take for life to settle on the whole planet and to bring about detectable traces?

– What was the first oxido-reduction reaction used by life for the necessary energy production and what was the first energy storage system?

– What was the composition of the atmosphere before and after the Late Heavy Bombardment and what is the relative importance of endogeneous and exogeneous origin for the building blocks of life? Did life on Earth start from a heterotrophic process depending on the abiotic production of organic molecules? Did it happen very early with a reducing atmosphere favourable to pre-biotic synthesis? Then, how long did it take for autotrophy to develop, for life to settle on the whole surface of the planet and to bring about detectable traces?

– Or, alternatively, was the emergence of life intimately associated with energy production pathways? Which means that it may have proceeded later in an environment less favourable to the formation of prebiotic molecules and that life may have developed faster.

Certainly, future multidisciplinary research in astronomy, geology, biology and chemistry, in the solar system and (partly) in young stars and exoplanets, along the lines developed in this review, will help us address these questions.

For the time being, we just hope to have provided readers from these various backgrounds with enough elements of knowledge and scientific curiosity, but also evidence of ignorance, that they may endeavour to try to answer these questions, and others, by themselves, and to participate to the development of this highly unconventional science called astrobiology. Indeed, astrobiology is unconventional because it is necessarily interdisciplinary or even supradisciplinary. It necessitates a strong collaboration between scientists with very different backgrounds and different scientific attitudes towards enigmatic facts or observations. People involved in astrobiology must necessarily be open-minded scientists, convinced that the number of things we don't know is incredibly much higher that the number of things we know, yet are stubborn enough not to be discouraged.

Further readings

Books

Gargaud, M., Barbier, B., Martin, H. and Reisse, J. (eds.): 2005, *Lectures in Astrobiology*, Vol. 1, Springer, Berlin, Heidelberg, New York.

Gargaud, M., Claeys, Ph. and Martin, H. (eds.): 2006, *Lectures in Astrobiology*, Vol. 2, Springer, Berlin, Heidelberg, New York (in press).

Lunine, J.: 1999. *Earth, Evolution of a Habitable World*, Cambridge University Press, Cambridge.

Norris, S. C.: 2004, *Life's Solution: Inevitable Humans in a Lonely Universe*, Cambridge University Press, Cambridge.

Mayor, M. and Frei, P. -Y.: 2001, *Les nouveaux mondes du Cosmos: à la découverte des exoplanètes*, Editions du Seuil, Paris.

Mayr, E.: 2004. *What Makes Biology Unique: Considerations on the Autonomy of a Scientific Discipline*, Cambridge University Press, Cambridge.

Review articles

Bada, J. L., and Lazcano, A.: 2002, *Science* **296**, 1982–1983.

Chyba, C. E., and Hand, K. P.: 2005, *Ann. Rev. Astr. Ap.* **43**, 31–74.

Ehrenfreund, P., and Charnley, S. B.: 2000, *Ann. Rev. Astr. Ap.,* **38**, 427–483.

Greenberg, J. M.: 1996, in D. B. Cline (ed.), *Physical Origin of Homochirality on Earth*, American Institute of Physics, Proc. 379, Woodbury New York, pp. 185–186.

Earth, Moon, and Planets (2006) 98: 313–317
DOI 10.1007/s11038-006-9126-2

ALBAREDE Francis
Professor
Ecole Normale Supérieure, Lyon, France
E-mail: albarede@ens-lyon.fr
Geochemistry: early Earth, meteorites

AUGEREAU Jean-Charles
Research scientist CNRS
Laboratoire d'Astrophysique de Grenoble
Université Joseph Fourier, Grenoble, France
E-mail : augereau@obs.ujf-grenoble.fr
Astrophysics: young stars, evolved disks, dynamics,
dust evolution, exoplanets

BOITEAU Laurent
Research Scientist CNRS
Group Dynamique des Systmes Biomolculaires Complexes,
Départment de Chimie
Université Montpellier II, France
E-mail: laurent.boiteau@univ-montp2.fr
Chemistry, physical organic chemistry: origin of life, origins
of amino acids and peptides

CHAUSSIDON Marc
Research Director CNRS
Centre de Recherches Pétrographiques et Géochimiques
(CRPG), Nancy
E-mail: chocho@crpg.cnrs-nancy.fr
Geochemistry/Cosmochemistry : Early Solar System evolution,
isotopic composition of meteorites and archeans rocks

CLAEYS Philippe
Professor
Vrije Universiteit Brussels, Brussels, Belgium
E-mail: phclaeys@vub.ac.be
Geology: comets and asteroids impacts and their consequences
for life evolution

314

DESPOIS Didier
Research Scientist CNRS
Observatoire Aquitain des Sciences de l'Univers,
Université Bordeaux1, Bordeaux, France
E-mail: despois@obs.u-bordeaux1.fr
Astrophysics/exobiology: comets, radioastronomy, stars and
solar system formation

DOUZERY Emmanuel J. P.
Professor
Institut des Sciences de l'Evolution, Montpellier,
Université Montpellier II, France
E-mail : douzery@isem.univ-montp2.fr
Biology, Molecular Phylogeny: mammals, tree reconstruction,
DNA and protein clocks, systematics.

FORTERRE Patrick
Professor
Institut de Génétique et Microbiologie, Orsay,
Université Paris-Sud, France
Head of the department of fundamental and medical micro-
biology at the Institut Pasteur, Paris
E-mail: forterre@igmors.u-psud.fr, forterre@pasteur.fr
Biology/exobiology: hyperthermophiles, archae,
phylogenomics, universal tree of life, LUCA, origin of DNA

GARGAUD Muriel
Research Scientist CNRS
Observatoire Aquitain des Sciences de l'Univers
Université Bordeaux1, Bordeaux, France
E-mail: muriel@obs.u-bordeaux1.fr
Astrophysics/astrobiology: interstellar medium
physico-chemistry, origins of life

GOUNELLE Mathieu
Assistant Professor
Muséum National d'Histoire Naturelle, Paris, France
E-mail: gounelle@mnhn.fr
Astrophysics, Meteoritics: origin of the solar system,
early solar system chronology, link between metetorites
and comets, micrometeorites

LAZCANO Antonio,
Professor,
President of the International Society for the Study of the
Origin of Life
Universidad Nacional Autónoma de México (UNAM),
Mexico City,
E-mail: alar@correo.unam.mx
Biology/microbiology: prebiotic evolution,
evolutionnary biology.

LÓPEZ-GARCÍA Purificación
Research Scientist CNRS
Unité d'Ecologie, Systématique et Evolution, Orsay
Université Paris-Sud, France
E-mail: puri.lopez@ese.u-psud.fr
Microbiology: microbial diversity, microbial evolution,
extreme environments

MARTIN Hervé
Professor
Laboratoire Magmas et Volcans,
Université Blaise Pascal, Clermont-Ferrand, France
E-mail : H.Martin@opgc.univ-bpclermont.fr
Geochemistry : geological and geochemical evolution of the
primitive Earth. Subduction zone magmatism.

MARTY Bernard
Professor
Ecole Nationale Supérieure de Géologie, Nancy, France,
Director of the Centre de Recherches Pétrographiques et
Géochimiques CRPG
Geochemistry: volatile elements; noble gases and nitrogen in
the solar system including the Sun and terrestrial planets;
formation of the atmosphere and the oceans; the recent
evolution of air upon industrial activity.

MAUREL Marie-Christine
Professor
Biochemistry of evolution and molecular adaptability
Institut Jacques Monod, Université Paris 6, France
E-mail: maurel@ijm.jussieu.fr
Biology-biochemistry: molecular evolution, origins of life,
etiology, activity and persistence of RNA

MONTMERLE Thierry
Research Director CNRS
Director of Laboratoire d'Astrophysique de Grenoble
Université Joseph Fourier, Grenoble, France
E-mail: montmerle@obs.ujf-grenoble.fr
Astrophysics: high-energy phenomena in the insterstellar
medium, formation and early evolution of stars and planets,
irradiation effects by hard radiation and energetic particles
in star-forming regions, stellar magnetic fields

MORBIDELLI Alessandro
Research Director CNRS
Observatoire de la Cte d'Azur, Nice, France
E-mail: morby@obs-nice.fr
Astrophysics: dynamic of asteroids and transneptunian
objects, origin and evolution of near-Earth objects, primordial
sculpting of the solar system

MOREIRA David
Research Scientist CNRS
Unité d'Ecologie, Systématique et Evolution, Orsay
Université Paris-Sud, France
E-mail: david.moreira@ese.u-psud.fr
Biology: molecular phylogeny, microbial evolution

PASCAL Robert
Research Scientist CNRS
Groupe "Dynamique des Systmes Biomoléculaires Com-
plexes", Chemistry Department, Université Montpellier 2
E-mail: rpascal@univ-montp2.fr
Chemistry: Organic reactivity, enzyme mimics, peptide and
amino acid chemistry, origin of life.

PERETÓ Juli
Associate Professor of Biochemistry and Molecular Biology
Institut Cavanilles de Biodiversitat i Biologia Evolutiva,
Universitat de Valencia, Spain
E-mail: pereto@uv.es
Biochemistry: origins of life, evolution of metabolic pathways

PINTI, Daniele L
Professor
GEOTOP-UQAM-McGill
Université du Québec Montréal, Qc, Canada
E-mail: pinti.daniele@uqam.ca
Geochemistry: Isotope geochemistry, biomarkers, noble gases.

PRIEUR Daniel
Professor
Laboratoire de Microbiologie des Environnements Extrmes,
Université de Bretagne Occidentale, Brest, France
E-mail: daniel.prieur@univ-brest.fr
Microbiology: oceanic hydrothermal sources,
extreme environments

REISSE Jacques
Professor Emeritus
Université Libre de Bruxelles, Belgium
Member of the Belgian Academy of Sciences
E-mail: jreisse@ulb.ac.be
Chemistry: stereochemistry and chirality, NMR,
sonochemistry, organic cosmochemistry,
study of the solid phase

SELSIS Franck
Research Scientist CNRS
Centre de Recherche Astronomique de Lyon,
Ecole Normale Supérieure de Lyon, France
E-mail: franck.selsis@ens-lyon.fr
Astrophysics, Planetary Science : Formation and evolution of
planetary atmospheres, extrasolar planets, biomarkers

VAN ZUILEN Mark
Research Scientist CNRS,
Equipe Géobiosphre Actuelle et Primitive,
Institut de Physique du Globe, Paris, France
E-mail: vanzuilen@ipgp.jussieu.fr
Geochemistry: Early life, early Earth environment,
isotope geochemistry, in situ analytical techniques
for microfossil characterization

Earth, Moon, and Planets 98: 319–370
DOI 10.1007/s11038-006-9127-1

GLOSSARY

The field of astrobiology is by definition multidisciplinary, therefore, any glossary devoted to it is likely to be incomplete. Nevertheless, the editors hope that the following glossary, which contains approximately 1000 terms commonly used in astrobiology, will help specialists in one field to obtain pertinent information or clarifications about other disciplines.

This glossary is also published In: M. Gargaud, Ph. Claeys, H. Martin (Eds), Lectures in Astrobiology (vol. 2) Springer, 2006.

M. Gargaud, Ph. Claeys, H. Martin

A

a *(astronomy)*: Orbital parameter, semi-major axis of elliptical orbit.

Abiotic: In absence of life.

Ablation *(astronomy)*: Loss of matter by fusion or vaporization during the entrance of an object into the atmosphere.

Absolute magnitude: Magnitude of a stellar object when seen from a distance of 10 parsecs (32.6 light-years) from Earth.

Acasta: Region of Northern Territories (Canada) where the oldest relicts of continental crust were discovered. These gneisses were dated at 4.03 Ga.

Accretion *(astronomy)*: Matter aggregation leading to the formation of larger objects (stars, planets, comets, asteroids). Traces of accretion of their parent body (asteroid) can be observed in some meteorites.

Accretion disk: Disk of matter around a star (or around a black hole) such that matter is attracted by the central object and contributes to its growth. Disks around protostars, new-born stars and T-Tauri stars are accretion disks.

Accretion rate *(astronomy)*: Mass accreted per time unit (typically 10^{-5} to 10^{-8} solar mass per year).

Accuracy: The accuracy of a method is its capacity to measure the exact value of a given quantity.

Acetaldehyde: CH_3CHO, ethanal.

Acetic acid: CH_3COOH, ethanoic acid.

Acetonitrile: CH_3CN, cyanomethane, methyl cyanide.

Achiral: Not chiral.

Acidophile: Organism which « likes » acidic media, which needs an acidic medium.

Actinolite: Mineral. Inosilicate (double chain silicate). [Ca_2 $(Fe,Mg)_5$ Si_8O_{22} $(OH,F)_2$]. It belongs to the amphibole group.

Activation energy *(E_a)*: Empirical parameter that permits to describe, via Arrhenius law, the temperature dependence of a reaction rate. In the case of a reaction $A + B \rightarrow C$, E_a can be described as the energy difference between the reactants $A + B$ and the activated complex (AB^*) on the reaction pathway to C. It corresponds to an energy barrier.

Active site *(of an enzyme)*: Part of an enzyme where substrate is specifically fixed, in a covalent or non-covalent way, and where catalysis takes place.

Activity *(solar or stellar)*: All physical phenomena which are time dependent and which are related to star life (like stellar wind, solar prominence, sun-spots). Their origin is mainly magnetic, they correspond to emissions of electromagnetic waves at different frequencies (from radio-waves to X-rays) and also to emissions of charged particles (protons, alpha particles and heavier particles). In the case of the Sun, the solar activity is cyclic. The shorter cycle is an eleven year cycle.

Activity: In non-ideal solution, activity plays the same role as mole fraction for ideal solution.

Adakite: Volcanic felsic rock generated in subduction zone, by partial melting of the subducted basaltic oceanic crust.

Adaptive optics: The goal of adaptive optics is to correct the effects of atmospheric turbulence on the wavefronts coming from stars. The basic idea consists of measuring the wavefront form each 10 ms (to follow the atmospheric turbulence), and to deform the mirror(s) accordingly to correct and stabilize the wavefront.

Adenine: Purine derivative which plays an important role in the living world as component of nucleotides.

Adenosine: Nucleoside which results from condensation of adenine with ribose.

ADP: Adenosine diphosphate (see also ATP).

Aerobian: Organism whose life requires free oxygen in its environment.

Aerobic respiration: Ensemble of reactions providing energy to a cell, oxygen being the ultimate oxidant of organic or inorganic electrons donors.

Aerosols: Liquid or solid sub-millimetric particles in suspension in a gas. Aerosols play an important role in atmospheric physics and chemistry.

Affinity constant: The efficiency of recognition of a target by a ligand is characterised by an affinity constant (in M^{-1}) which is the ratio of ligand-target concentration ($[\text{ligand-target}]_{eq}$) to the product of the free ligand and free target concentrations at equilibrium ($[\text{ligand}]_{eq}$ x $[\text{target}]_{eq}$) : K = ($[\text{ligand-target}]_{eq}/[\text{ligand}]_{eq}x[\text{target}]_{eq}$.

AIB: See aminoisobutyric acid.

Akilia: Region of Greenland near Isua where are exposed sediments and volcanic rocks similar in both composition and age (3.865 Ga) to the Isua gneisses.

Alanine *(Ala)*: Proteinic amino acid containing three carbon atoms.

Albedo: Fraction of the incident light which is reflected by a reflecting medium (i.e. atmosphere) or surface (i.e. ice cap). A total reflection corresponds to an albedo of 1.

Albite: Mineral. Tectosilicate (3D silicates). Sodic plagioclase feldspar Na-$AlSi_3O_8$.

Alcalophile: Organism, which « likes » alkaline (basic) media, or needs an alkaline medium

Alcohol: R-OH.

Aldehyde: R-CHO.

Aldose: Any monosaccharide that contains an aldehyde group (-CHO).

ALH84001: Martian meteorite found, in 1984, in the Alan Hills region (Antarctica). In 1996, the claim that it contains traces of metabolic activity and even, possibly, microfossils was the starting point of strong debates.

Aliphatic hydrocarbon: Hydrocarbon having an open chain of carbon atoms. This chain can be normal or forked.

Allende: Large carbonaceous chondrite (meteorite) of 2 tons, of the C3/CV type and found in Mexico in 1969.

Allochtonous sediment: Sediment that formed in a place different of its present day location (transported).

Alpha helix: A type of secondary helical structure frequently found in proteins. One helix step contains approximately 3.6 amino acid residues.

Alteration (*Weathering*): Modification of physical and chemical properties of rocks and minerals by atmospheric agents.

Amide: R-CO-NH2, R-CO-NH-R', R-CO-NR'R'' depending if the amide is primary, secondary or tertiary. The bond between the CO group and the N atom is generally called the amide bond but when it links two amino acid residues in a polypeptide or in a protein, it is called the peptide bond.

Amine: Derivatives of ammonia NH_3 in which one, two or three H atoms are substituted by an R group (a group containing only C and H atoms) to give primary, secondary or tertiary amines.

Amino acid (*AA*): Organic molecule containing a carboxylic acid function (COOH) and an amino function (generally but not always a NH_2 group). If the two functions are linked to the same carbon atom, the AA is an alpha amino acid. All proteinic AA are alpha amino acids.

Amino acids (*biological*): AA directly extracted from organisms, fossils, sedimentary rocks or obtained, after hydrolysis, from polypeptides or proteins found in the same sources.

Amino acids (*proteinic*): AA found as building blocks of proteins. All proteinic AA are homochiral and characterized by an L absolute configuration.

Amino isobutyric acid (*AIB*): Non proteinic alpha amino, alpha methyl AA. Detected in chondrites.

Amino nitrile: Molecule containing a CN group and an amine function. It could have played a role in the prebiotic synthesis of amino acids.

Amitsôq: Region of Greenland where are exposed the oldest huge outcrops (~3000 km^2) of continental crust, dated at 3.822 Ga (gneiss)

Amorphous: Solid state characterized by a lack of order at large distances

Amphibole: Mineral family. Inosilicate (water-bearing double chain silicate), including actinolite, hornblende, glaucophane, etc.

Amphibolite: Rock generated by metamorphism of basalt. It mainly consists in amphibole and plagioclase feldspar crystals, sometimes associated with garnet.

Amphiphile: Molecule with a hydrophilic part and a lipophylic part.

Amplification (*of DNA*): Production, in relatively large quantity, of fragments of DNA by in vitro replication, starting from a very small initial sample (see: PCR).

Amu (atomic mass unit): Atomic mass unit such that the atomic mass of the ^{12}C (carbon isotope) has exactly a mass equal to 12.0000.

Anabolism: General term to designate a group of biochemical reactions involved in the biosynthesis of different components of living organisms.

Anaerobian: Organism, which does not need free oxygen for his metabolism. In some cases, free oxygen is a poison for anaerobian organisms.

Anaerobic respiration: Ensemble of reactions providing energy to a cell, the ultimate oxidant being an inorganic molecule other than oxygen.

Anatexis: High degree metamorphism where rock begins to undergo partial melting.

Andesite: Effusive mafic magmatic rock (volcanic); it mainly consists in sodic plagioclase feldspar + amphibole ± pyroxene crystals. These magmas are abundant in subduction zones. Diorite is its plutonic equivalent.

Angular momentum: The angular momentum, L of a rigid body with a moment of inertia I rotating with an angular velocity ω is $L = I \times \omega$. In the absence of external torque, the angular momentum of a rotating rigid body is conserved. This constancy of angular momentum allows to fix constraints on star and planetary system genesis from huge interstellar gas clouds.

Anorthite: Mineral. Tectosilicate (3D silicates). Calcic plagioclase feldspar $CaAl_2Si_2O_8$.

Anorthosite: Plutonic magmatic rock only made up of plagioclase feldspar. Generally, it consists in an accumulation (by flotation) of plagioclase crystals. Relatively rare on Earth, anorthosites are widespread on the Moon where they form the "primitive" crust.

Anoxic: Environment (water, sediment, etc.) where oxygen is highly deficient or absent. Synonym: anaerobic.

Anthophyllite: Mineral. Inosilicate (double chain silicate). $[(Mg)_7 Si_8 O_{22} (OH,F)_2]$. It belongs to the amphibole group.

Antibody (immunoglobulin): Glycoprotein produced because of the introduction of an antigen into the body, and that possesses the remarkable ability to combine with the very antigen that triggered its production. Polyclonal antibodies are produced by different types of cells (clones), whereas monoclonal antibodies are exclusively produced by one type of cell by fusion between this cell and a cancerous cell.

Anticodon: Triplet of nucleotides of tRNA able to selectively recognize a triplet of nucleotides of the mRNA (codon).

Antigen: Foreign substance that in an organism, is able to induce the production of an antibody.

Antisense: Strand of DNA that is transcripted into a mRNA (messenger RNA).

Aphelia: In the case of an object in elliptical motion around a star, the point which corresponds to the largest distance with respect to the star.

Apollo: Ensemble of asteroids whose orbits intersect Earth orbit.

Aptamer: Synthetic polynucleotide molecule that binds to specific molecular targets, such as a protein oligonucleotide. Contrarily to nucleic acids, aptamers are able to fold back into a 3D structure, thus leading to the formation of binding cavities specific to the target.

Archaea: One of the three main domains of the living world. All the organisms of this domain are prokaryotes. Initially, these organisms were considered as the most primitive form of life but this is no longer accepted by most biologists. Most extremophiles such as hyperthermophiles and hyperacidophiles belong to archaea domain.

Archaean (*Aeon*): Period of time (Aeon) ranging from 4.0 to 2.5 Ga. Archaean aeon belongs to Precambrian. Unicellular life existed and possibly was already aerobe

Archebacteria: See Archaea.

Arginine: Proteinic alpha amino acid containing six carbon atoms and a guanido group in the side chain.

Aromatic hydrocarbon: Hydrocarbon that contains one or several benzene ring(s).

Aspartic acid *(Asp)*: Proteinic AA with an acidic side chain.

Assimilative metabolism: In a cell, reduction process by which an inorganic compound (C, S, N, etc) is reduced for use as a nutriment source. The reduction reaction necessitates an energetic supply (endothermic).

Asteroid belt: Ring-shaped belt between Mars and Jupiter where the majority of the asteroids of the Solar System are located

Asteroid: Small object of the Solar system with a diameter less than 1000 km. Many of them are orbiting around the Sun, between Mars and Jupiter (asteroid belt).

Asthenosphere: Layer of the Earth mantle, located under the lithosphere and having a ductile behaviour. It is affected by convective movements. Depending on geothermal gradient, its upper limit varies between 0 km under mid oceanic ridges and 250 km under continents.

Astrometry: Part of astrophysics that studies and measures the position and motion of objects in the sky. By extension, astrometry refers to all astrophysics phenomenons that can be documented or studied by observation and measurement of the motion of celestial object.

Asymmetric atom: Synonym for chirotropic carbon. See asymmetric.

Asymmetric: An atom is called asymmetric when it is surrounded by four ligands which are oriented in 3D space in such a way that they define an irregular tetrahedron. Such an atomic arrangement is chiral and can exist as two enantiomeric forms. These two enantiomers are called D or L depending on their absolute configuration. The D/L nomenclature is now replaced by the R/S nomenclature of Cahn, Ingold and Prelog but the D/L system is still accepted for amino acids and oses. The asymmetric carbon atom is a particular (but very important) example of asymmetric atom.

Atmosphere: Gaseous envelope around a star, a planet or a satellite. In absence of rigid crust or ocean, the atmosphere is defined as the most external part of the object. *The primary atmosphere* of a young planet corresponds to the first gaseous envelope directly generated from protostellar nebula. *The primitive atmosphere* of the Earth corresponds to the atmosphere in which prebiotic chemistry occurred, when free oxygen was in very low concentrations. In the second part of the past century, this primitive atmosphere was considered as being highly reductive. Today, it is considered as mainly consisting in carbon dioxide, nitrogen and water vapour.

ATP synthetase: Enzyme involved in ATP synthesis.

ATP: Adenosine triphosphate. Molecule which plays an important role in the living world for the energy transfers. Its hydrolysis in ADP and in inorganic phosphate is an exergonic reaction.

AU *(astronomical unit)*: Average Earth-Sun distance corresponding to 149.6×10^6 km (or approximately 8 minute light and 100 Sun diameters).

Authigenic minerals: In a sedimentary environment, minerals generated by direct local precipitation of dissolved ions.

Autocatalysis: Chemical reaction such that a reaction product acts as a catalyst for its own synthesis.

Autochthonous sediment: Sediment that formed in the place where it is now.

Automaton *(chemical automaton)*: As defined by A. Brack, chemical system able to promote its own synthesis.

Autotroph: Organism which is able to synthesize its own constituents from simple molecules like water and carbon dioxide.

B

Bacteria: One of the three domains of the living world. Organisms of this domain are all prokaryotes. They are the more abundant micro-organisms on Earth. One of their characteristics consists in the presence of a cell-wall containing muramic acid. Some of these organisms are extremophiles like *Aquifex* and *Thermoga*.

Bacteriochlorophylls: Pigment found in micro-organisms and containing a tetrapyrole acting as a ligand for an Mg cation (similarly to chlorophylls of green plants).

Barophile: Micro-organism that lives optimally (or can only grow) in high pressure environments such as deep sea environments.

Basalt: Effusive mafic magmatic rock (volcanic); it mainly consists in plagioclase feldspar + pyroxenes ± olivine crystals. It results of 20 to 25% melting of the mantle. Gabbro is its plutonic equivalent.

Bases in nucleotides: See nucleic base, purine bases and pyrimidine bases.

Benioff plane: In a subduction zone, interface between the subducted slab and the overlying mantle wedge.

Beta sheet: A secondary pleated structure frequently observed in proteins.

BIF (Banded Iron Formation): Sedimentary rocks widespread in Archaean terrains and no more generated today. They consist in alternation of black iron-rich (magnetite) and white amorphous silica-rich layers.

Bifurcation (thermodynamics): This term describes the behaviour of a system; which submitted to a very small variation in exchange conditions with its surroundings, jumps suddenly from one stationary state to another one. The characteristics of new stationary state cannot be predicted on the basis of a complete knowledge of the initial stationary state.

Binding energy (nuclear): In an atom the nuclear binding energy is the energy necessary to break the atomic nucleus into its elementary components. This is also the amount of energy released during the formation of an atomic nucleus.

Biocenosis: Group of interacting organisms living in a particular habitat and forming an ecological community (ecosystem).

Biochemical sediment: Sediment formed from chemical elements extracted from water by living organisms (for instance, limestone formed by an accumulation of shells).

Biofilm: Tiny film (few μm thick) consisting in colonies of micro-organisms fixed on an inorganic or organic surface.

Biogenic sediment: Sediment formed by precipitation of ions dissolved in water by the mean of living beings (e.g. calcium ion in shells).

BIOPAN: Experimental module made by ESA to be fixed on a Russian satellite of the Photon type, the aim of the system is to expose samples or dosimeters to space conditions as vacuum, microgravity or radiation.

Bioprecipitation: Precipitation of solid mineral phases linked to the metabolism of living organism. Through selective evolution, this process was used by organisms to reduce the toxic excess of some soluble compounds or elements (for example Ca), then to constitute new functionalities (exo- or endoskeleton, protective tests, teethes …). In some extent, the organisms can favour bioprecipitation while the surrounding physico-chemical conditions are unfavourable. This is for example the case of diatoms that precipitate an SiO_2 test under seawater conditions, which would normally dissolve SiO_2.

Biosensor: Device containing biological compounds on/in a membrane, which translates biological parameters such as electric potential, movement, chemical concentration, etc. into electrical signals.

Biosignature: observable considered as an evidence of the presence of life.

Biosphere: Ensemble of species living on Earth.

Biosynthesis: production of a chemical compound by a living organism, equivalent to anabolism.

Bio-tectonic: Environmental concept proposed to account for the situation in Lake Vostok (Antarctica), where crustal tectonics (faults) controls and activates hydrothermal circulation. The latter provides chemical elements allowing life to survive. In the ocean, as well as in the oceanic crust, life liked to magmatic activity should rather be called bio-magmatic.

Biotite: Mineral. Phyllosilicate (water-bearing sheet silicate). [$K(Fe,Mg)_3$ Si_3AlO_{10} $(OH)_2$]. It belongs to the mica group and is also called black mica.

Biotope: Smallest unit of habitat where all environment conditions and all type of organisms found within are the same throughout.

Bipolar flow *(astronomy)*: Flow of matter in two opposite directions, perpendicularly to the circumstellar disk associated to a new-born star. The components of the bipolar flow are molecules, atoms, ions and dust particles.

Birthline: Locus, in the Hertzsprung-Russell diagram where young stars become optically visible.

Black body *(radiation)*: Radiation emitted by a body at a temperature T and such that the coupling between matter and radiation is perfect. Such a body is black. The total power emitted by unit area and the power emitted at a well defined frequency depends only on the temperature (the T^4 dependence is given by the empirical Stefan law and by the theoretical Planck law). As a first approximation and on the basis of their emission properties, stars and planets can be described as black bodies.

Black smoker: Also called hydrothermal vent. Structure observed on the oceans floor generally associated with mid ocean ridges. There, hot hydrothermal fluid, rich in base metal sulphides, enters in contact with cold oceanic water. Polymetallic sulphides and calcium sulphate precipitate progressively building a columnar chimney around the vent.

Blast: Blast (Basic Local Alignment Search Tool) represents a powerful and rapid way to compare new sequences to an already existing database, which may either contain nucleotides (Blastn) or proteins (Blastp). Since the BLAST algorithm establishes local as well as global alignments, regions of similarity embedded in otherwise unrelated proteins could be detected. The Blast program gives a value for each of the high scoring results, together with the probability to find an identical score in a database of the same size by chance

Blue algae: See cyanobacteria. Old name for some procaryotic unicellular which are not algae

Bolometric light *(or bolometric magnitude)*: Total radiation output by time unit of a stellar object

Bootstrap: Statistical method with re-sampling, commonly used to measure the robustness and the reliability of phylogenic trees.

Braking radiation: Electromagnetic radiation emitted by high speed particle (electron, proton) when deviated by a magnetic or electric field. (Also see Bremsstrahlung).

Branching ratio: In the case of a chemical or of a nuclear reaction, such that different reaction paths exist, the branching ratio is the relative rate constants or probabilities of occurrence per unit of time for each different path.

Breccias (geology): Sedimentary or magmatic rocks consisting in an accumulation of angular fragments in a sedimentary or magmatic matrix.

Bremsstrahlung: German word also used in French and English to describe the electromagnetic radiation emitted by high speed particle (electron, proton) when deviated by a magnetic field.

Brown Dwarf: Space body born as a star, but whose mass is too small (< 80 Jupiter mass) to allow nuclear reaction: both core temperature and core pressure are insufficient to initiate hydrogen fusion.

C

C (alpha): Carbon atom linked to a chemical function we are interested in. More specifically, the carbon atom directly linked to the carboxylic function in amino acids. In alpha amino acids, the amino group is linked to the alpha carbon. The Greek letters alpha, beta, gamma ... are used to describe carbon atoms separated from the function by one, two, three... other carbon atoms.

CAI (Ca-Al Inclusions): Ca-and Al-rich inclusions abundant in some chondrite meteorites.

Caldera: Circular km-sized structure due to the collapse of superficial formation induced by the emptying of an underlying magma chamber.

Carbonaceous chondrite: Chondrite with high carbon content. The famous Murchison meteorite is a carbonaceous chondrite.

Carbonate compensation depth (CCD): Depth below which sedimentary carbonates are systematically dissolved in seawater. This depth depends on the physico-chemical conditions of the environment (amount of sinking particles due to surface biological production, deep-water circulation, etc.). These conditions can change from one ocean to another and may also change through time. The present Atlantic ocean CCD is at about 3500 m.

Carbonate: $(CO_3)^{2-}$-bearing mineral e.g. Calcite $=$ $CaCO3_;$ Dolomite $=$ $MgCa(CO_3)_2$.

Carbonation: In Ca-, Mg-, K-, Na- and Fe- bearing minerals, chemical reaction of alteration resulting in the formation of carbonates.

Carbonic anhydride: CO_2. Synonym for carbon dioxide. **Carbonyl:** -(CO)-, this chemical function is found in carboxylic acids, ketones, aldehydes, amides and many other organic molecules.

Carboxylic acid: Organic molecule containing a COOH group.

C-asteroid: Asteroid containing carbon. C-asteroids are the parent-bodies of carbonaceous chondrites.

Catabolism: Part of metabolism involving all degradation chemical reactions in the living cell resulting in energy production. Large polymeric molecules such as polysaccharides, nucleic acids, and proteins are split into their smallest size constituents, such as amino acids, after which the monomers themselves can be broken down into simple cellular metabolites. The small size constituents can be subsequently used to construct new polymeric molecules. These reactions involve a process of oxidation that releases chemical free energy, part of which is stored through synthesis of adenosine triphosphate (ATP). All these reactions require enzymatic catalysis.

Catalysis: Chemical process such that a substance (catalyst) increases reaction rate by changing reaction pathway but without being chemically modified during reaction. Enzymes are very efficient catalysts able to increase reaction rates by several orders of magnitude and also able to limit the number of secondary products and therefore to increase reaction selectivity.

CCD *(Charge-Coupled Device)*: Silicon photo electronic imaging device containing numerous photo-sensors (often at least 1000 X 1000). The most used astronomic detector in the visible wavelength domain.

Cell: Complex system surrounded by a semi-permeable membrane which can be considered as the basis unit of all living organisms.

Cenancestor: Last ancestor shared by all living beings. Synonym of LUCA.

Cenozoic *(Era)*: Period of time (Era) ranging from 65 Ma to today, it is also called Tertiary Era but in addition also it includes Quaternary area.

Chalcophile: Chemical element frequently associated with sulphur (i.e. Cu, Zn, Cd, Hg…).

Chandrasekhar mass: Also called Chandrasekhar limit. In astrophysics, the maximum possible mass of a white dwarf star. This maximum mass is about 1.44 times the mass of the Sun. After having burned all its nuclear fuel the star atmosphere collapses back on the core. If the star has a mass below Chandrasekhar limit, its collapse is limited by electron degeneracy pressure; if the mass is above the Chandrasekhar limit it collapses and becomes a neutron star and possibly a black hole.

Chemical derivatization: Typically, derivatization consists in the chemical modification of a compound in order to produce a new compound easier to detect due to its new physico-chemical properties (volatility, absorption in ultraviolet, fluorescence…). The derivatization of a molecule involves altering part of it slightly or adding a new part to the original compound.

Chemical sediment: Sediment formed by direct precipitation of ions dissolved in water.

Chemolithoautotroph: Chemotroph that uses CO_2 as only source of carbon.

Chemolithotroph: Chemotroph that takes its energy from the oxidation of inorganic molecules such as NH_3, H_2S, Fe^{++}. On Earth, the first living organisms could have been chemolithotrophs.

Chemolitotroph: Chemotroph that takes its energy from the oxidation of inorganic molecules. On Earth, the first living organisms could have been chemolitotrophs.

Chemoorganotroph: Chemotroph that takes its energy from oxidation of organic molecules.

Chemotroph: Organism that takes its free energy from the oxidation of chemicals.

Chert: Microcrystalline to cryptocrystalline sedimentary rock consisting of quartz crystals whose size in $< 30 \mu m$, or of amorphous silica. Synonym: flint.

Chicxulub: Large (180 km in diameter) impact crater, located in Mexico Gulf. It is assumed to result of the collision of a big (10 km in diameter) meteorite, 65 Ma ago. This impact is considered as the cause of the important biological crisis at the Cretaceous-Tertiary boundary that led to mass extinction of thousands of living species, such as ammonites, dinosaurs, etc.

Chirality: Property of an object (and therefore of a molecule) to be different from its mirror image in a plane mirror. A hand is an example of a chiral object (in Greek: *cheir* means hand). Any object is chiral or achiral and if it is chiral, it can exist as two enantiomorphous forms (called enantiomers for molecules).

328

Chlorophylls: Pigments of major importance in the oxygenic photosynthesis. The chemical structure of all chlorophylls is based on a porphyrin ring system chelating an Mg^{2+} cation. Chlorophylls are found in higher plants, algae and some micro-organisms.

Chloroplast: sub-Cellular structure that plays a fundamental role for photosynthesis in all photosynthetic eukaryotes. Chloroplasts have more probably an endosymbiotic origin.

Chondre: Small spherical aggregate of radiated silicate minerals (typically 1 mm in diameter) which is frequent in stony meteorites and especially in chondrites. Olivine is the main component of chondres.

Chondrite: Undifferentiated stony meteorite unmelted and frequently considered as a very primitive object. Chondrites have the same composition as the Sun except for volatile elements.

Chromatographic co-elution: Refers to organics compounds that cannot be separated by a given chromatographic method and that migrate at the same speed leading to a single detectable signal.

Chromatography: Preparative or analytical chromatography: experimental method based on the properties of all molecules to be absorbed more or less selectively by a solid phase (the stationary phase) and therefore to migrate at different rates when they are « pushed away » by a mobile phase which can be a gas (gas chromatography or GC) or a liquid (LC or HPLC for High Performance LC).

Chromophore: Chemical group that absorbs UV and visible wavelengths, thus colouring the molecules or objects that contain them.

Chromosome: Sub-cellular structure containing most of the genetic material of the cell. Prokaryotic cells generally contain only one chromosome made of a circular DNA molecule while eukaryotic cells generally have several chromosomes, each of them containing a linear DNA molecule.

CIP *(for Cahn-Ingold-Prelog)*: General nomenclature used in chemistry to describe chiral molecules and more generally stereoisomers. Following the International Union of Pure and Applied Chemistry (IUPAC), CIP nomenclature must replace all other nomenclatures including D, L Fisher nomenclature except for amino acids and sugars for which the D/L nomenclature is still accepted.

Circular dichroism: The absorption coefficient of right and left circularly polarized light are different if the absorbing medium is chiral. By plotting the difference between the absorption coefficients as a function of the light wavelength, the curve which is obtained corresponds to the circular dichroism curve of the medium.

Circumstellar disk: Disk of gas and dust particles around a star.

Class [0,I,II,III] *(astronomy)*: Classification of young stellar objects based on their electro-magnetic emission in the micro-wave and infra-red domains. It consists in an evolution sequence from the protostars (0 and I) to the T-Tauri stars (III).

Clast: Fragment of mineral or rock included in another rock.

Clay: Mineral family. Phyllosilicate (water-bearing sheet silicate), e.g. kaolinite, illite, smectite, montmorillonite.

CM matter: Pristine material of the solar system, analogue to constitutive matter of CM carbonaceous meteorites (M = Mighei) and very abundant in micrometeorites.

CNO cycle: Series of hydrogen burning reactions producing helium by using C,

N, O as catalysers. These reactions provide the energy of the main sequence stars with a mass $> 1.1\ M_{Sun}$.

Coacervat *(droplet)*: Protein and polysaccharides containing emulsion. According to Oparin, model of protocells.

Codon: Triplet of nucleotides in a mRNA molecule which corresponds to a specific amino acid of a protein synthesized in a ribosome or which corresponds to a punctuation signal in protein synthesis.

Coenzyme: Small molecule that binds with an enzyme and that is necessary to its activity. ATP, CoA, NADH, biotine are examples of coenzymes. Coenzymes are often derived from vitamins.

Collapse *(astronomy)*: Process that describes the formation of stars from dense cores. The process seems to be fast: less than 10^5 years.

Coma: Broadly spherical cloud of dust particles and gaseous molecules, atoms and ions surrounding cometary nucleus. It appears when nucleus becomes active, generally when approaching the Sun.

Combustion: Used to describe any exothermal chemical reaction involving dioxygen and organic reactants. Sometimes used in astrophysics to describe the thermonuclear fusion reactions taking place in the stars and which, obviously, are also exothermal.

Comet: Small body of the Solar system with an average size of 1 to 100 km, travelling generally on a strongly elliptic orbit around the Sun. Comets are constituted of ice and dust and are considered as the most primitive objects of the Solar system.

Cometary nucleus: Solid part of a comet (1-100 km diameter), made of ice (H_2O, CH_3OH, CO...) and of dust.

Cometary tail: Part of an active comet; three different cometary tails are known depending from their composition (dust particles and molecules, neutral sodium atoms, ions).

Complex molecule *(Astronomy)*: Molecule containing more than 3 atoms.

Complexation: Chemical term used to describe the non-covalent interaction of a molecule or, more frequently, an inorganic ion with other molecules (called ligands) to give a supramolecular system described as a complex.

Condensation: Chemical reaction involving two molecules and leading to the formation of a new chemical bond between the two subunits but also to the elimination of a small molecule (generally a water molecule). The formation of bonds between the subunits of many biochemical polymers are condensation reactions (examples: polynucleotides, polypeptides, polysaccharides).

Configuration: Term used in stereochemistry. Stereoisomers (except if they are conformers) have different configurations. As an example, butene can be cis or trans and it corresponds to two different configurations. In the case of chiral the D- and L-valine, the stereoisomers are enantiomers and they have « opposite » configurations. It is important to make a clear distinction between the relative configurations of two enantiomers and the absolute configuration of each of them.

Conglomerate *(chemistry)*: In the restricted case of crystallized chiral molecules, a crystalline state such that all molecules of the same chirality (homochiral) crystallize together giving a mixture of crystal which, themselves, are of opposite chirality.

Conglomerate *(geology)*: Detrital sedimentary rock consisting in an accumulation of rounded fragments in a sedimentary matrix.

Continental margin: Submarine part of a continent making the boundary with oceanic crust.

Continents: Emerged and associated shallow depth (< 300 m) parts of the Earth surface. Their average composition is that of a granitoids.

Continuum *(astronomy)*: in emission (or absorption) spectroscopy, emission (or absorption) background of a spectrum extending on a large frequency domain. Frequently, lines are superimposed on the continuous spectrum. Black body radiation corresponds to a continuous spectrum.

Cool Early Earth: Period of time between Earth accretion (4.55 Ga) and the late heavy bombardment (4.0–3.8 Ga). This model considers that period of early Earth as quiescent with respect to meteoritic impacts, thus being potentially favourable for life development.

Core *(geology)*: Central shell of a planet. On Earth core mainly consists in iron with minor amounts of nickel and some traces of sulphur; it represents 16% volume but 33% mass of the planet. It is subdivided in a solid inner core from 5155 to 6378 km depth and a liquid outer core from 2891 to 5155 km depth.

Coronography: Technique used in astronomy and that consists of masking one part of the observed field (for instance a bright star) to allow the observation of a lower luminosity object located in the immediate vicinity. This technique was first developed by Bernard Lyot, who masked the sun disk to observe its corona, thus leading to the given name of coronography.

COROT: French project for the search of extrasolar planets based on a 25 cm telescope and able to detect planets having a diameter equal to two times the Earth diameter and located at 0.5 AU from its star.

Cosmic rays: Highly accelerated ions coming from the Sun (solar wind, essentially protons) or coming from other and extrasolar sources (galactic cosmic rays).

Covalent bonds: Interatomic bonds such that two atoms share one, two or three electrons pairs leading to the formation of a single, a double or a triple bond. A covalent bond is described as a polarized bond if the two bonded atoms have different electronegativities.

CPT *(theorem)*: general theorem of physics which assumes that physical laws are unchanged if, simultaneously, space is reversed (parity operation), time is reversed (the sense of motion is reversed) and matter is replaced by anti-matter. In many cases but not all, the CP theorem alone is valid.

Craton: Huge block of old (often Precambrian in age) and very stable continental crust (see also shield)

Crust *(geology)*: The more superficial shell of the solid Earth. Its lower limit with the mantle is called Morohovicic (Moho) discontinuity. Together with the rigid part of the upper mantle it forms the lithosphere. Two main crusts exist 1) oceanic crust, basaltic in composition and about 7 km thick, it constitutes the ocean floor; 2) continental crust is granitic in composition, with a thickness ranging between 30 and 80 km, it constitutes the continents.

Cryosphere: Part of the Earth surface made of ice.

Cumulate: Igneous rock generated by accumulation of crystals extracted from magma.

Cyanamid: NH_2-CO-CN.

Cyanhydric acid: HCN, hydrogen cyanide. Triatomic molecule that during prebiotic period could have played the role of starting material for purine synthesis.

Cyanoacetylene or better cyanoethyne: H-CC-CN.

Cyanobacteria: Microorganism belonging to the Bacteria domain and able to perform oxygenic photosynthesis. In the past, these microorganisms were improperly called « blue algae ». The cyanobacteria could be the ancestors of chloroplasts.

Cyanogen: C_2N_2.

Cysteine *(Cys)*: Protinic amino acid containing three C atoms and a –SH group in its side chain. In a proteinic chain, two cysteine residues can be linked together by a –S-S- bond (disulfide bond) often used to stabilize protein conformation. Two cysteines linked together by a disulfide bond is a cystine molecule.

Cytidine: The ribonucleoside of cytosine. The corresponding deoxyribonucleoside is called deoxycytidine.

Cytochrome c: One particular example of the large cytochrome family. Cytochrome c is a protein involved in the electron transfers associated to aerobic respiration. In the eukaryotic cells, cytochrome c is localized in the mitochondrias.

Cytoplasm: Whole content of a cell (protoplasm) except the nucleus whose content is called nucleoplasm.

Cytoplasmic membrane: also called plasmatic membrane or cell membrane.

Cytosine *(C)*: One of the nucleic bases of pyrimidine type.

D

D/H ratio: D = Deuterium; H = Hydrogen. Due to their mass difference, molecules containing either H or D are able to fractionate. For instance, during vaporization, a molecule containing the light isotope (H) is more efficiently vaporized than its heavy equivalent (D). Chemical reactions, including biochemical ones are also able to fractionate these isotopes. Consequently the D/H ratio can provide information on the biotic or abiotic origin of some organic molecules. The D/H ratio can be very high in chondritic organic matter due to D-enrichment during reactions taking place in the interstellar clouds.

Dalton *(Da)*: Molecular mass unit equal to the sum of the atomic masses given in amu (atomic mass unit).

DAP: Abbreviation frequently used to design two different molecules, diamino propionic acid and diamino-pimelic acid.

Darwin: Research programme from the European Space Agency devoted to the search of extrasolar planets and the study of their atmosphere composition by spatial interferometry, in the infra-red spectral region. Five independent telescopes of 1,5 m each constitute the basis of this very sophisticated system.

Daughter molecule *(in comet)*: In the cometary coma, any molecule produced by photo-dissociation of a parent molecule coming from the nucleus.

Deamination: Reaction associated to the elimination of an amine group (NH_2, NHR or NRR').

Decarboxylation: Reaction associated to the elimination of a CO_2 molecule.

Decay constant (λ): Probability that over a given time interval, one radionuclide will decay. This probability is unique for each isotope and its value is independent of its chemical and mineralogical environment. The constant (λ) unit is s^{-1}.

Deccan *(Trapps)*: Voluminous stacking of basaltic flows emplaced in North-west India at the end of Cretaceous period. It constitutes the evidence of an extremely important volcanic event contemporaneous with the extinction of dinosaurs.

Delta *(isotopic)* (δ): Difference between the isotopic ratio of a sample (R_e) and that of a standard (R_s). $\delta = 1000 \cdot (R_e\text{–}R_s)/R_s$. e.g. $\delta^{18}O$.

Denaturation: Change of the native conformation of a biopolymer. More specifically and in the case of proteins, denaturation can be induced by increasing temperature and/or pressure or by adding a chemical reagent (like urea). Denaturation can be reversible or irreversible. It is generally associated with a loss of enzymatic properties

Denitrification: Capability of some organisms to transform dissolved nitrate (NO_3^-) into molecular nitrogen (N_2) or nitrogen oxide (N_2O). This metabolism requires anoxic or dysoxic conditions.

Dense clouds: See interstellar clouds.

Dense core *(Astronomy)*: Gravitationally bound substructure located inside a molecular cloud and surrounded by protoplanetary disks. Dense core collapsing leads to the formation of one or several stars.

Deoxyribose: Ose or monosaccharide having a structure identical to ribose except that the OH group in position 2' is replaced by an H atom. In living organisms, deoxyribose comes from ribose via a reduction reaction.

Depletion: Impoverishment of chemical element abundance when compared to a standard of reference composition.

D-Glucose: D enantiomer of glucose. It plays a primeval role in living cells where it produces energy through both anaerobic (fermentation) and aerobic

(associated with respiration) metabolism. D-glucose is also a precursor of several other molecules such as oses and nucleotides.

Diagenesis: Chemical and/or biochemical transformation of sediment after its deposition. This process, which generally consists in cementation and compaction, transforms a running sediment in a compact rock.

Diapirism: Gravity driven magma or rock ascent in the Earth. Generally low density materials rise up into greater density rocks.

Diastereoisomers *(or diastereomers)*: Stereoisomers that are not enantiomers. Diastereisomers are characterized by physical and chemical properties which can be as different as observed for isomers having different connectivity (constitutional isomers).

Diazotrophy: Capability of some organisms to assimilate nitrogen as N2. This assimilating process is also called nitrogen biofixation.

Differentiation (*Chemistry***):** Separation of an initially homogeneous compound into several chemically and physically distinct phases.

Differentiation (*Earth*): Separation from a homogeneous body of several components whose physical and chemical properties are contrasted. In the case of Earth these components are core, mantle, crusts, hydrosphere and atmosphere.

Diffuse clouds: See interstellar clouds.

Dinitrogen: N_2, also frequently called nitrogen.

Dioxygen: O_2, also frequently called oxygen.

Disk (of second generation): Disk around a star resulting from the breaking off of solid bodies previously formed like planetesimals, asteroids or comets. The beta Pictoris disk is probably a secondary disk.

Dismutation: Reaction that from a single reactant gives two products one of them being more oxidized than the reactant and the other one more reduced. A typical example is the transformation of an aldehyde into an alcohol and an acid.

Dissimilative metabolism: In a cell, mechanism that uses a compound in energy metabolism. The reduction reaction releases energy (exothermic).

Dissolve inorganic carbon equilibrium: Concentration proportions of the main chemical forms of the dissolved inorganic carbon in aqueous solution: $HCO_3^-/CO_3^{2-}/CO_2$. These proportions depend on the physico-chemical conditions, especially the pH.

Disulfide bond: Covalent bond between two S atoms. When the –SH groups of two cysteines react on each other in presence of an oxidant, it gives a cystine molecule, i.e. a cysteine dimer linked by a –S-S- bond. Disulfide bridges stabilize the ternary and quaternary structures of proteins. Some irreversible denaturazing of proteins can be associated to formation of disulfide bonds.

DNA: Desoxyribonucleic acid, long chain polymer of desoxynucleotides. Support of the genetic code in most living cells. Frequently observed as a double helix made by two complementary strands

Drake *(Equation of)*: Empirical formula (containing several adjustable parameters) which, following his author, gives a rough estimation of the number of « intelligent civilizations» in the galaxy.

Dust *(interstellar, cometary)*: Small solid particles (0.1 mm – 1 mm), generally made of silicates, metal ions and/ or carbonaceous matter.

Dysoxic: Environment (water, sediment, etc.) where oxygen is in limited amount. Synonym: dysaerobic.

E

e *(astronomy)*: Orbital parameter that measures the eccentricity of elliptical orbit.

Eccentricity: Parameter (e) characterizing the shape of an orbit. e is equal to 0 for a circle, equal to 1 for a parabola, higher than 1 for a hyperbola and between 0 and 1 for an ellipse.

Ecliptic: Geometric plane of the Earth orbit. More precisely, average planar of Earth-Moon barycentre orbit.

Eclogite: High degree metamorphosed basalt. It consists in an anhydrous rock made up of pyroxene and garnet.

Ecosystem: Community of organisms and their natural environment: Ecosystem = community + biotope.

Eddington: ESA project devoted to asteroseismology and search for extrasolar planets by the transits method. Kepler: similar project for exoplanetary science.

Electrophoresis: Analytical method used in chemistry and based on the difference between diffusion rates of ions when placed in an electric field. Initially the ions are adsorbed on a support or immersed into a viscous medium. Capillary electrophoresis is a technique adapted to the analysis of small samples.

Enantiomeric excess *(i.e. in percents)*: In the case of a mixture of two enantiomers whose respective concentrations are D and L with D greater than L, i.e. = (D – L / D + L). 100

Enantiomeric ratio: In the case of a mixture of two enantiomers, the enantiomeric ratio is the ratio of the two enantiomers.

Enantiomers: Two stereoisomers that differ only due to their opposite chirality (like an idealized left hand and an idealized right hand).

Enantiomorph: Two objects that differ only due to their opposite chirality (like an idealized left hand and an idealized right hand).

Enantioselectivity: A reaction leading to products which are chiral is said to be enantioselective if it gives an excess of one enantiomer. Enantioselectivity is generally induced by a chiral reactant or a chiral catalyst (like an enzyme). Enantioselectivity is also used to describe a reaction such that starting from a mixture of enantiomers; one of them reacts faster than the other. In this case, the reactant must be chiral.

Enantiotopic *(Chemistry)***:** A planar molecule like $H-CO-CH_3$ can be seen as a scalene triangle with summits of different colours. In the 2D space, such an object is chiral and the two faces are said to be enantiotopic. A chiral reagent is able to differentiate enantiotopic faces.

Endergonic: A chemical reaction or a physical change is endergonic if it requires a supply of free energy from its surroundings to succeed.

Endogenous *(Biochemistry)***):** That takes place inside. For instance an endogenous organic synthesis is a synthesis that occurs in a planetary atmosphere or at the bottom of an ocean. Antonym = exogenous.

Endogenous *(Geology)***:** Word that refers to petrogenetic mechanisms taking place inside the Earth, as well as to the rocks generated by such mechanisms. A magmatic or a metamorphic rock is an endogenous rock. Antonym = exogenous.

Endolithic *(biology)***:** Micro-organisms living in rocks.

Endosymbiosis: Process by which an eukaryotic cell lives in symbiosis with an other cell (generally a prokaryotic cell) located in its cytoplasm. Chloroplasts and mitochondria are considered as vestiges of endosymbiotic prokaryotes.

Endothermic: A chemical reaction or a physical change is endothermic if it needs a supply of energy from its surroundings to succeed.

Enstatite: Mineral. Inosilicate (simple chain silicate). Mg-rich Pyroxene [$MgSiO_3$].

Envelope *(circumstellar)***:** In astronomy, cloud of dust particles and gas surrounding a new-born star (protostar).

Enzyme: In biochemistry, a molecule which catalyses a reaction. Most enzymes are proteins but some are polynucleotides (ribozymes).

Epitope: Region on a macromolecule that is recognized by an antibody. It is generally about 5 to 12 amino acids long, which is the size of the antigen binding site on the antibody.

Equator *(celestial)***:** Plane perpendicular to the rotation axis of the Earth and corresponding to an extension in space of the terrestrial equator.

Escape velocity: Velocity required for a body to escape the planetary gravity field. On Earth, the escape velocity is 11 km/s.

Escherichia Coli: Bacteria found in the intestine and commonly used in experimental bacteriology. Its size is of the order of one micron. Its genome codes, approximately, for 3000 proteins.

Ester: Molecule that can be described as the result of condensation of an acid and an alcohol associated with elimination of a water molecule. (-O-CO-) is the ester bond.

Eubacteria: Sometimes used instead of bacteria, in order to point out the difference between Bacteria and Archaea. Archaeas themselves are sometimes called Archebacterias.

Eukaria: One of the three domains of life (together with bacteria and archaea). Eukaryotic cells are members of the Eukaria domain.

Eukaryote: Any organism from Eukarya domain and characterized by a nucleus (containing the genetic material) separated from the cytoplasm by a membrane. Eukaryotes can be unicellular or pluricellular.

Europe: Satellite of Jupiter with a diameter of about 3100 km (same size as the Moon). An ocean could exist beyond its icy surface.

Evaporite: Sedimentary rock generated by evaporation of huge volumes of water; chemical elements dissolved in water precipitate leading to the deposition of minerals such as halite (NaCl) or gypsum = $(CaSO_4, 2H_2O)$.

Exergonic: Chemical reaction or a physical change is exergonic when it provides free energy to its surroundings.

EXOCAM: Special reactor used to experimentally simulate exobiological processes.

Exogenous (*Biochemistry*): That takes place outside. For instance exogenous molecules are molecules synthesised in an extraterrestrial environment and imported on Earth by (micro) meteorites or interplanetary dust particles. Antonym = endogenous.

Exogenous (*Geology*): Word that refers to petrogenetic mechanisms taking place at the surface of the Earth (in atmosphere or hydrosphere), as well as to the rocks generated by such mechanisms. A sedimentary rock (sandstone, limestone, etc...) is an exogenous rock. Antonym = endogenous.

Exon: Sequence of transcripted nucleotides that is present in natural RNA and which corresponds to a DNA sequence. In DNA, exons are separated by introns (intervening sequences). Exons and introns are respectively coding and non-coding sequences for proteins.

Exoplanet: Planet orbiting around a star other than the Sun. The number of already discovered exoplanets is greater than 100. (More information on this fast developing field is available on www.obspm.fr/planets)

Exoplanetology: Part of astrophysics devoted to extrasolar planets.

Exosphere: The outermost region of a planet's atmosphere. This is the place where most of the photodissociation reactions of molecules take place. From the exosphere, the molecules with sufficient velocity can escape the Earth's gravitation.

Exothermic: A chemical reaction or a physical change is exothermic if it provides energy to its surroundings.

Explosive nucleosynthesis: Nucleosynthethic reactions taking place during star explosion. The characteristic time of these reactions is by far shorter than that of the same reaction occurring in a star at rest.

Extinction (*interstellar, atmospheric*): Decrease of the light intensity of a star due to light diffusion or absorption by a medium (planetary atmosphere, interstellar cloud). In an interstellar cloud, visible magnitude can be reduced by a factor of 1 while the reduction can be as large as 100 in a protostar dense nucleus.

Extremophile: Micro-organism that optimally lives (or optimally grows) in "extreme" physico-chemical environments (P, T, Ph...).

F

Fatty acid: Carboxylic acid R-COOH where R is a long chain containing only C and H atoms.

Fayalite: Mineral. Nesosilicate (isolated SiO_4 tetrahedrons). [$(Fe)_2SiO_4$]. This mineral is the ferrous end-member of olivine series (peridot family).

Feldspar: Mineral family. Tectosilicate (3D silicate). Feldspars are subdivided into two main chemical groups: 1) Alkali feldspars ($NaAlSi_3O_8$ = Albite and $KAlSi_3O_8$ = Orthoclase); 2) Plagioclase feldspars ($NaAlSi_3O_8$ = Albite and $CaAl_2Si_2O_8$ = Anorthite). These minerals constitute 52% of the continental and oceanic crusts.

Fermentation: Biochemical process such that complex organic molecules (i.e. glucose) are transformed into low molecular mass molecules (i.e. ethanol) by cells, in anaerobic conditions. Fermentation corresponds to an oxidation process but the final electron acceptor is an organic molecule instead of oxygen. During fermentation as during respiration, ATP is produced but less efficiently.

Ferrihydrite: Iron hydroxide, $5Fe_2O_3$ $9H_2O$.

Fisher: German chemist who was the first to introduce the D/L nomenclature to differentiate enantiomers and to characterize their absolute configurations.

Fisher-Tropsch *(reaction of)*: Reaction which gives hydrocarbons from a mixture of H_2 and CO. The FT reaction had and still has a great industrial importance but it could also have been important in prebiotic chemistry. The FT reaction requires metallic catalysts.

Flint (or flintstone): see chert.

Flint: Rock mainly made of amorphous silica and having a biogenic origin. It frequently appears as nodules in chalk or limestone.

Fluid inclusion: 1 to 100 μm-sized cavities in minerals that contain fluids trapped during mineral crystallization.

Fluorophore: Fluorescent molecule. This molecule absorbs light and its electrons are excited to higher energy states; their return to lower energy states is accompanied by light emission (fluorescence).

Formaldehyde or methanal: The simplest aldehyde (H-CO-H).

Formation *(geology)*: Group of terrains or rocks having the same characteristics.

Formic acid: HCOOH, methanoic acid.

Formose *(reaction)*: Starting from formaldehyde in water solution at high pH, this reaction leads to the formation of a large variety of sugars. Its importance in prebiotic chemistry remains an open question. It is also called Butlerow reaction.

Forsterite: Mineral. Nesosilicate (isolated SiO_4 tetrahedrons). [$(Mg)_2SiO_4$]. This mineral is the magnesium end-member of olivine series (peridot family).

Fossil *(geology)*: All kind of trace of passed life (bone, shell, cast, bio-molecule, track, footprint, etc....).

Fractionation *(Chemistry)*: Separation of chemical elements or isotopes by physical or chemical mechanisms.

Frasil: Ice disks with a diameter of few mm that are observed in water as soon as surfusion occurs. Frasil is common in Artic and Antarctic rivers but also below the huge ice platforms moving forward in the Antarctic Ocean.

Free-fall time: Time required for an object of mass m, initially at rest, to reach an object of mass M (M > m) under effect of gravitation alone. It gives a good approximation of the time required for an accretion disk to collapse during the protostar stage (typically 10^5 years).

FRET: (*Fluorescence Resonance Energy Transfer*). The energy of an excited electron in one molecule (the donor) can be passed on to an electron of another molecule in close proximity (the acceptor) through resonance, then released as a photon. If the donor is on a target and the acceptor on a ligand, a molecular recognition event can be detected by the photon emission resulting from the ligand-target complex formation.

Furanose: A "furanose ring" is a cyclic ose formed of 4 carbons and an oxygen atom.

G

Ga: Giga annum = one billion years (= Gy)

Gabbro: Plutonic magmatic rock. It has a granular texture and mainly consists in pyroxenes and plagioclase ($\pm$ olivine). Basalt is its effusive equivalent.

Gaia: Ambitious project of ESA to measure the position of one billion of stars with a precision of one micro arc second. Gaia is essentially devoted to the search for extraterrestrial planets.

Galactic "Open" cluster *(Astronomy)*: Cluster that can contain from a dozen to a few thousand of stars ("I" population) and are younger than globular clusters.

Garnet: Mineral. Nesosilicate (isolated SiO_4 tetrahedrons). Its general composition is $Y_2^{+++} X_3^{++}(SiO_4)_3$ (Y = Al, Fe^{+++}, Cr and X = Ca, Mg, Fe^{++}, Mn). Garnet is stable at high pressure (> 70 km depth). In the mantle, at depth, garnet is the only aluminium-bearing phase.

GC *(gas chromatography)*: Chromatographic method using gas as moving phase. It can be used in analytical and preparative chemistry.

Gene: Segment of DNA, containing hundreds to thousands nucleotides, found in a chromosome. A gene codes for a specific protein.

Genome: Ensemble of genes of an organism.

Genomic: Science that studies genomes. Genomic includes studying and sequencing of genomes as well as method for the analyses of mRNA (transcriptome) and proteins (proteome). Genomic also tries to index the genes of an organism, to localize them on chromosomes, to determine their sequence and to study their functionality.

Genotype: Ensemble of the genetic characters of an organism.

Geocruiser: Asteroid with an orbit that intersects that of Earth. Objects on such an orbit may eventually collide with Earth.

Geographic pole: Point where the rotation axis (instantaneous) of a planet intersects the globe surface.

Geomorphology: Branch of the geosciences that studies the characteristics, the configuration and the evolution of land forms. The methods developed on Earth to interpret the geomorphologic mechanisms (magmatism, erosion, sedimentation, etc...) are especially well adapted to the interpretation of images from other planets, where remote sensing is the only available source of information.

Geothermal gradient: Thermal gradient corresponding to the temperature increase with depth. In the Earth crust geothermal gradient is ~30°C.km^{-1}.

Giant planets: Large size planets of low density, such as Jupiter, Saturn, Uranus and Neptune. One can distinguish two groups: gaseous giants, Jupiter and

Saturn, mainly made up of gas (H_2, He) coming from the protosolar nebula and icy giants, Uranus and Neptune, rich in ice (H_2O, NH_3, and CH4). They all have a core made of heavy elements and were formed in the outer part of the solar nebula, beyond the ice line.

Glacial-interglacial alternation: Oscillatory change of ice volume, mainly due to the extension of ice caps located in high latitudes. The episodes of large ice cap extension are called glacial, while those of minimum extension are called interglacial. These oscillations are not a permanent feature of the Earth's climate; they require specific conditions, in term of plate tectonic configuration (position of the continents), and orbital parameters. The present-day climate is interglacial.

Glaciation: Cold period in the Earth history characterized by the presence of a large cryosphere (ice). Frozen water accumulates and forms ice caps on the continents; consequently, it becomes unable to return to ocean whose level decreases. The main glaciations occurred during Precambrian, Early Cambrian, End of the Ordovician, Carboniferous, and from the Oligocene to the Quaternary.

Glass inclusion: 1 to 100 μm-sized cavities in minerals that contain magma trapped during mineral crystallization.

Glass: Amorphous material. In volcanic rocks, it can result of the rapid cooling of the magma.

Global ecosystem: (= planetary ecosystem). The whole set of interactions between the living and non-living compartments of a life-bearing planet, which contribute to regulate its state far from the thermodynamic equilibrium (where no life is possible).

Globular cluster *(Astronomy)*: Large spherical cluster containing from a few thousands to several millions of old stars ("II" population)

Glucide: Name for sugar molecule, also called carbon hydrate, saccharide or ose. In living cells, they can be energy providers and components of nucleic acids or of wall proteins (glycoproteins).

Glucose: $CH_2OH(CHOH)_4COH$. D-glucose is the wider spread ose. Glucose is an aldohexose : the carbonyl group belongs to the aldehyde function (aldose); it contains 6 atoms of carbon (hexose). L-Glucose is also called levulose and D-Glucose is called dextrose.

Glutamic acid *(Glu)*: Alpha AA with a side chain containing an acidic COOH function. Described as a hydrophilic AA.

Glutamine *(Glu)*: Amino acid containing 5 C atoms with a NH_2 group in the side chain, it is considered as hydrophilic.

Glycan: Synonymous of polysaccharides.

Glyceraldehyde: ($HO-CH_2-CHOH-CHO$); the simplest aldose, containing only one chirotopic carbon atom. The two enantiomers play an historical role in stereochemistry because they are at the origin of the D, L nomenclature which describes absolute configuration (Fisher). Glyceraldehyde is the biochemical precursor of other oses.

Glycerol: ($HO-CH_2-CHOH-CH_2OH$); 1,2,3-propanetriol also called glycerine, component of many membrane phospholipids (which are esters of glycerol).

Glycine *(Gly)*: The simplest amino acid and the only one which is achiral.

Glycolic acid: $HO-CH_2-COOH$.

Gneiss: Metamorphic rock made up of quartz, feldspars and micas. All mica crystals show the same orientation thus defining a surface of preferential cutting up called « foliation plane ».

Gondwana (*Gondwanaland*): Palaeozoic super-continent formed by convergence and agglomeration of continents (Peninsular India, Madagascar, Africa, Australia, South America and Antarctica) due to plate tectonic activity. It was mainly located in the South hemisphere.

Gram +: Bacteria previously coloured during a Gram test and which does not loose the colour after a treatment with ethanol (« positive » response).

Granite: Plutonic magmatic rock. It has a granular texture and mainly consists of quartz, alkali feldspar and plagioclase feldspar; mica can be present whereas amphibole is rare. Rhyolite is its effusive equivalent.

Granitoid: Family of quartz-bearing plutonic magmatic rocks including granites, granodiorites, tonalites and trondhjemites.

Granodiorite: Plutonic magmatic rock. It is similar to granite but contains no more than 10% alkali feldspar.

Green bacteria: Micro-organisms of the Bacteria domain able to perform anoxic photosynthesis.

Greenhouse effect: Warming of a planet surface due to the trapping by planet atmosphere of the electro-magnetic waves received and radiated by the planet.

Greenstone belts: Volcanic (basalts and komatiites) and volcano-sedimentary formation widespread in Archaean terrains. It generally presents an elongated shape(~100 km long and few tens km wide). Its green colour is due to metamorphism of basalts and komatiites.

Guanine (*G*): Nucleic base with purine structure.

Gy: Giga year = one billion years (= Ga).

H

Habitable zone (HZ): Zone around a star where the physical conditions (temperature, presence of liquid water, etc.) on the orbiting planets are considered favourable for the birth and the development of an "Earth-like" life.

Hadean (*Aeon*): Period of time (Aeon) ranging from 4.55 Ga (Earth formation) to 4.0 Ga (oldest known rock: Acasta gneisses). Hadean aeon belongs to Precambrian.

Hadley cell: On Earth tropospheric convection cell; the air masses rise at the level of the equator and descend at about 30° latitude. At intermediate latitude, this cell is relayed by Ferrel cell and by Polar cell at high latitudes. Sometimes, the term Hadley circulation is used to describe the whole system. This global tropospheric circulation results in a redistribution of heat from the equator towards the poles. The concept remains true on Mars as on many other planets. In the case of Mars and Earth, the planet rotation strongly affects the circulation. For instance, the air moving toward the equator in the lower atmosphere is deflected by the Coriolis effect to create the easterlies trade winds in the tropics.

Half-life (T): Synonym of period. For a single radioactive decay process the half-life is the time needed for the number of radioactive atoms to decrease to half its original value. Half-life (T) is linked to the decay constant (λ) by the relation $T = \ln(2)/\lambda$. T unit is s.

Halogen (*Chemistry*): Fluorine (F), Chlorine (Cl), Bromine (Br), Iodine (I), Astatine (At). Chemical element belonging to the (VIIA) period in the Mendeleïcr periodic table.

Halophile: Micro-organism that lives optimally (or can only grow) in environments with high salt concentration ($1M$ NaCl).

Hamiltonian (operator): Mathematical entity describing the motion of a classic or quantic particle. In quantum mechanics, this mathematical entity allows one to write the wave equation for a stationary state whose solutions give the probability density of one or several electrons in any point in space.

Hapten: Can be considered as an isolated epitope. By definition a hapten is a small molecule (few 100's Da), not antigenic by itself: alone hapten cannot induce immune response; it stimulates production of antibodies only in combination with a specific protein called carrier or schlepper.

Harzburgite: Peridotite made up of olivine and orthopyroxene. It generally corresponds to residual mantle after lherzolite melting and extraction of basaltic magma.

HD 209458b: First exoplanet, whose previous detection by radial velocimetry method has been confirmed by the observation of a transit in front of its parent star (HD 209458).

Heavy element *(astronomy)*: Any element other than hydrogen and helium.

Helium: Rare gas whose ^{3}He isotope is used in geology as marker of a recent degassing process from the deep mantle.

Hertzsprung-Russel diagram *(HR diagram)*: In astronomy, a two-dimensional diagram with star temperature (or spectral type) as abscissa and star luminosity (or absolute magnitude) as coordinate. Temperatures decrease from left to right and the spectral types sequence is OBAFGKM.

Heterocycle: Cyclic organic molecule containing heteroatoms (i.e. atoms other than C.) as constituents of the cyclic structure.

Heterosphere: Part of an atmosphere located above the homopause and where each gas density distribution decreases according to its own scale height.

Heterotrophous: Organism which uses reduced organic molecules as principal carbon source for its biosynthesis. Nowadays, these reduced organic molecules are generally produced by other organisms. At the beginning of life, these reduced molecules were probably found in the environment.

Histidine *(His)*: Proteinic amino acid containing an imidazole group in its side chain. Histidine residues (hystidyl) are frequently found in active site of enzymes.

HMT: ($C_6H_{12}N_4$); hexamethylenetetramine, could be a minor component of comet nucleus.

Homeostasis *(Biology)*: In living organisms homeostasis is the regulation of a physical or of a chemical factor, which attempts to actively keep these factors at equilibrium, even in case of external environmental change (i.e. thermal regulation). This involves a feedback mechanism.

Homeostasis: Property of a living organism to maintain unchanged some of its physico-chemical characteristics even in presence of a change in the environment. Homeostasis requires autoregulation.

Homoacetogens: Microorganisms of the Bacteria domain producing acetate from H_2 and CO_2.

Homochirality: Of the same chirality. All proteinic amino acids are L while ribose in RNA or ATP is always D. The origin of homochirality for the large majority of the chiral constituents of organisms remains an active research subject.

Homogenization temperature: Temperature at which a fluid inclusion transforms from a multi-phase (heteregeneous, for instance gaz + seawater) to a one-phase (homogeneous) state. This temperature is considered as the minimal temperature of formation of the fluid inclusion.

Homology *(biology)*: Two structures in two different species are homologous (and therefore comparable) irrespective of their forms, if they are connected in the same way to identical structures.

Homolysis: Homolytic cleavage or homolytic fission. Breaking of a bond such that each molecule fragment retains the same amount of binding electrons; consequently, no electric charge is created due to the cleavage.

Homolytic (rupture): Rupture of a chemical bond in which the bonding electrons are equally partitioned between the two fragments; no charge appears during this process.

Homopause: Atmospheric boundary between the homosphere and the heterosphere.

Homosphere: Part of an atmosphere where the gases are uniformly mixed. Their densities decrease as the altitude increases following a single mean scale height, defined by the mean molecular weight and the temperature.

Hornblende: Mineral. Inosilicate (double chain silicate). [$Na_{0-1}Ca_2$ (Fe^{++},Mg)$_{3-5}$ (Al, Fe^{+++})$_{0-2}$ Si_{8-6} $Al_{0-2}O_{22}$ (OH,F)$_2$]. It belongs to the amphibole group.

Hot Jupiter: Jupiter massive-like exoplanet, orbiting close to a star. Most of the extrasolar planets so far discovered belong to this type.

Hot spot: see mantle plume.

HPLC *(High Performance Liquid Chromatography)*: Very efficient liquid phase chromatography performed under high pressure up to 100-400 bars.

Hydrocarbons: Molecules containing only C and H atoms. If the hydrocarbon contains an aromatic system, it is said aromatic. If the hydrocarbon contains only tetra-coordinated C atoms, the hydrocarbon is called aliphatic. Some hydrocarbons result from the polymerization of isoprene (2-methylbutene); they are called isoprenoïd hydrocarbons. Latex contains isoprenoïd hydrocarbons.

Hydrogen bond: Intermolecular but sometimes intramolecular low energy bond (about 20 kJ/mol) involving generally an H atom linked to an electronegative atom like O, N, S and an atom bearing non-bonding electron pairs such as O or N. The H-bond implies an H-donor and an H-acceptor.

Hydrogen cyanide: H-CN.

Hydrogen sulphide: (H_2S)

Hydrogenoid (function): Exact electronic wave function describing the motion of the electron in an atom with a single electron.

Hydrolysis: Cleavage of a molecule due to reaction with H_2O.

Hydrosphere: The whole water available on Earth surface; it includes, oceans, seas, lakes rivers, underground waters and atmospheric water vapour.

Hydrothermal vent: see black smoker.

Hydroxy acid: Carboxylic acid containing also an alcohol function. Glycolic acid is the simplest hydroxy acid.

Hydroxyl *(group)*: -OH or alcohol group

Hyperthermophile: Micro-organism that lives optimally (or can only grow) in high temperature environments (> 80°C).

Hypoxanthine *(6-hydroxypurine)*: Purine base and biological precursor of adenine and guanine.

I

Ice shelves: Ice platforms generated by glaciers whose ice progresses from Antarctica land over ocean. Ross shelf is about 1000km wide and extents over 600km in ocean. They are the source of very large tabular icebergs.

Ices *(astronomy)*: Solid form (crystalline or amorphous) of volatile molecules like water, carbon dioxide or ammonia.

IDP: Interplanetary Dust particle.

Igneous rocks: Magmatic rocks = due to magma crystallization.

Immunoaffinity chromatography: (IAC) All chromatographic methods based on a stationary phase that contains antibodies or fragments of antibodies.

Immunogenic: Substance able to induce an immune response.

Impact melt rock: Rock associated with impact craters (=impactite). It resembles a lava and consists of a mixture of solid or molten fragments from the local lithologies floating in a glassy microcrystalline or recrystallized matrix. It is generated by cooling of the magma produced by the melting of the target lithologies under the high temperatures and pressures induced by the collision.

Impactite: Heterogeneous breccia generated at depth by the impact of a large-size stellar body (> 10,000 tons). It is made up of fragments of the rock substrate included in a vitreous matrix.

Impactite: Lithologies generated during meteorite impact. Melt-rock, suevites and breccias found inside or at close proximity of an impact crater are called impactites.

Inclination (1) *(astronomy)*: Angle between the orbital plane of a solar object and the ecliptic plane in degrees (always lower than 90°).

Inclination (2) *(astronomy)*: Angle between the orbital plane of an interstellar object and the "sky plane" (plane perpendicular to the "line of sight" i.e. the straight line joining the observer and the stellar object).

Indels: Acronym for insertions / deletions. Phylogenetic analysis on several DNA or protein sequences requires sequences with same length. During the process of alignment, alignment gaps (indels) must be introduced in sequences that have undergone deletions or insertions.

Interferometry: Observation and astrophysical measurement technique based on the use of several disconnected telescopes spread over an area, and allowing reaching the angular resolution a single-dish telescope with the same area would give. This technique can be implemented different ways: visibility measurement, direct imaging by aperture synthesis, differential phase measurement, nulling interferometry...

Interplanetary dust: Small grains left behind by asteroids and comets, and dispersed in a cloud including the whole Solar system.

Interstellar cloud: Cloud of gas (98 %) and dust (2 %). The gas is mainly H (diffuse cloud) or H_2 (molecular cloud). Molecular clouds are called dense clouds ($n(H_2) > 10^3$ molecules.cm^{-3}) or dark clouds if dense and cold (10-20 K).

Intertidal environment: Environment between high and low tide. Also called tide range.

Intron: Non-coding sequence of nucleotides that separates exons. Introns are removed during the maturation processes of the three types of RNA by splicing.

Ion-molecule *(reaction)*: Kind of reaction between two gaseous reactants and initiated by ionizing cosmic rays,

X-rays or UV radiation. They are important in interstellar clouds and in planetary ionospheres.

IR: Infrared. Wavelength ranging between 1 and 300 μm.

Iridium (Ir): Element belonging to platinum element family. Its concentration in Earth crust is extremely low. A local Ir enrichment as at the Cretaceous-Tertiary (K/T) boundary is interpreted as a strong argument in favour of meteoritic impact.

ISM: Interstellar medium.

Isochron (geology): Rectangular diagram plotting isotopic ratio of a disintegration system (abscissa = parental isotope; ordinate = daughter isotope) (e.g. $^{87}Sr/^{86}Sr$ versus $^{87}Rb/^{86}Sr$). In this diagram, cogenetic rocks of the same age plot along a straight line whose slope is proportional to age. This method of age determination is widely used in geology.

Isocyanhydric acid: $H-N=C$.

Isocyanic acid: $HN=C=O$.

Isoleucine (Ile): Proteinic amino acid containing six carbon atoms and described as hydrophobic. Ile is considered as one of the prebiotic AA.

Isoprenoid hydrocarbon: Hydrocarbon formed by polymerization of isopren $eCH_2=C(CH_3)-CH=CH_2$.

Isostasy (Geology): Hydrostatic equilibrium in the Earth's crust such that the forces tending to elevate landmasses balance the forces tending to depress them. When this equilibrium is broken, the return to equilibrium takes place by vertical movements of the crust. For instance, in Scandinavia during the last ice age the weight of the ice caps pushed the continental crust into the mantle. As the ice melted, the load decreased and the lithosphere bounced back towards equilibrium level, (this process is still going on at a rate of about 1m per century).

Isotopic ratio: Concentration ratio of two isotopes or concentration ratio of two isotopomers of a molecule (like H_2O and D_2O). Isotopic ratio can provide information on the age of a sample (when used in isochron calculation) as well as on its origin and source.

Isovaline (Iva): Hydrophobic non-proteinic amino acid, isomer of valine and containing five carbon atoms.

Isovaline: Hydrophobic non proteinaceous amino acid. This constitutional isomer of valine contains five C atoms.

Isua: Region of Greenland where are exposed the oldest sediments so far recognized 3.865 Ga (gneiss). They contain carbon whose origin could be biogenic. (See also Akilia)

J

Jeans escape: Process leading to the escape of atomic or molecular species from a planet atmosphere. It happens when the thermal agitation rate is greater than escape rate. The lighter elements or molecules (like H, H_2 or He) escape faster than the heavier ones.

Jovian planets: Other name for giant planets.

Jupiter: The fifth and largest planet (1400 times the Earth volume, 320 times the Earth mass) of the Solar system. Jupiter is 5.2 AU away from the Sun. Its gaseous envelope mainly made of H_2 and He, surrounds a core of ice and rocks (10–20 Earth mass).

Juvenile gases: Gases produced by or trapped inside the Earth and which reach the surface of Earth for the first time. ^3He is an example of juvenile gas detected in sub-marines geothermal fluids.

K

K/T (Strata): Few centimetre-thick sedimentary layer located at the Cretaceous Cainozoic boundary. Its Iridium-enrichment is interpreted as due to a giant meteoritic impact.

Kepler: Spacecraft NASA mission devoted to detection of Earth-type exoplanets (equipped with a 1m telescope).

Keplerian rotation: Orbital motion that follows Kepler's laws.

Kerogen: Insoluble organic matter found in terrestrial sediments and in some types of meteorites like carbonaceous chondrites.

Kilo base or kilo base pair (kb): Unit used to measure the number of nucleotides in a gene or a genome: 1000 base pairs of DNA or 1000 bases of RNA.

Komatiite: Ultramafic high-Mg lava. It contains olivine and pyroxene; minerals which sometimes can have needle or dendritic shapes (spinifex texture). Komatiites were abundant before 2.5 Ga and extremely rare after.

Kuiper Belt or Edgeworth-Kuiper: A large ring-shaped reservoir of comets beyond Neptune at about 30 astronomical units (AU) from the Sun.

L

L/D (Ratio): For a chiral molecule, the ratio between the L and the D enantiomer concentrations.

Lactic acid: $HO-CH(CH_3)-COOH$.

Lagrange points: The five points determining the equilibrium position of a body of negligible mass in the plane of revolution of two bodies (ex: star-planet couple) in orbit around their common gravity centre.

Late Heavy Bombardment: Heavy bombardment of the Moon (and certainly also of the Earth and others telluric planets) which happened between 4 and 3.8 Ga ago. It could correspond to either the end of a long period of bombardment by asteroids, meteorites and comets or to a short time cataclysmic phenomenon.

Laurasia: Palaeozoic super-continent formed by convergence and agglomeration of continents (Europe, North America and Asia) due to plate tectonic activity. It was mainly located in the North hemisphere. Continent resulting of Pangaea broke in two parts at the end of Palaeozoic.

Lava: Magma emplaced as a flow at the surface of the Earth or any other planet.

Leaching: Dragging of soluble elements or particles of a soil by infiltrated water.

Leucine (Leu): Proteinic amino acid containing six C atoms and considered as one of the prebiotic AA.

Lherzolite: Peridotite made up of olivine and pyroxenes (ortho- and clinopyroxenes) as well as of an Al-bearing mineral. Its melting generates basaltic magmas leaving a harzburgite residue.

Ligand: Any atom or group of atoms bonded to the atom we are interested in. As an example, the four ligands of a chirotopic (asymmetric) carbon atom are necessarily different.

Ligase: Class of enzymes that catalyses the binding between two molecules (or two DNA fragments).

Light-year: Measure of distance used in astronomy, it corresponds to the distance that light runs in one year ($0.946 \ 10^{16}$ m).

Liquidus: Line which, in composition vs. temperature or pressure vs. temperature diagrams, separates the domain where crystals and liquid coexist from the field where only liquid exists.

Lithophile *(Geochemistry)*: Chemical element frequently associated with oxygen. A lithophile element has a greater free energy of oxidation per gram of oxygen than iron; it occurs as an oxide generally in silicate minerals (i.e. Si, Al, Na, K, Ca, etc... (Synonym = oxyphile).

Lithosphere: External rigid shell of the Earth. Its definition is based on rheological behaviour of rocks. It includes crusts (continental and oceanic) as well as the upper rigid part of the mantle its thickness varies between 0 and 250 km and more or less corresponds to the 800°C isotherm.

Lithotroph: See chemolithotroph. Living organism that takes its energy from the oxidation of inorganic molecules such as NH_3, H_2S, Fe^{++}.

Low mass star: Star with mass <2 M_{sun}.

L_{sun}. *(Astronomy)*: Sun luminosity ($3.826 \ 10^{24}$ W).

LUCA: Last Universal Common Ancestor. Hypothetical microorganism that stood at the root of all lines leading to the present day living beings. Appeared after a long evolution, it cannot be considered as a primitive form of live.

Lysine *(Lys)*: Proteinic amino acid containing six C atoms with an amino group in its side chain and which, therefore, is basic and hydrophilic. Lysyl residues are frequently found in the active site of enzymes.

M

Ma: Mega annum = Mega year = one million years (=My)

Macronutrient: Chemical element necessary in large amounts for the proper growth and metabolism of a living organism (i.e. C, O, H, N, P, K, S, Ca, Mg...). In seawater, the main macronutrients are water, CO_2 nitrates and phosphates (sources for H, O, C, N and P respectively). See also micronutrient.

Mag: Magnitude.

Magma: Molten rock which can be completely liquid or consist in a mixture of liquid and crystals. It is produced by high temperature ($>650°C$ for granite; $>$ 1200°C for basalt) melting of pre-existing rocks. Mantle melting generates basalts whereas oceanic crust fusion rather generates adakites or TTG and continental crust gives rise to granites.

Magnesiowurstite: Mineral, Oxide. [(Mg,Fe)O]. Magnesiowurstite together with perovskite is probably the main component of the lower terrestrial mantle (depth $>$ 660 km).

Magnetic anomaly: Difference between the measured and the theoretical value of the magnetic field intensity of Earth.

Magnetic pole: Point where the magnetic dipole axis of a planet intersects the globe surface.

Magnetite: Mineral: Iron oxide [$Fe^{++}Fe^{+++}_2O_4$]. Its ferromagnetic properties make it able to record past Earth magnetic field characteristics. It can also exist in some bacteria called "magneto-tactic".

Magnitude (Mag) (Astronomy): Measure of brightness of a stellar object on a logarithmic scale. The difference between two successive magnitudes is a factor 2.512. Mag = $- 2.5 \log_{10}$ (I/I0)); Less bright is a star, more is its magnitude. The magnitude is calculated on a chosen spectral interval (visible, IR) or on the total spectrum (bolometric magnitude).

Major half-axis: For an elliptic orbit, half of the distance aphelia-perihelia.

Mantle plume: Ascending column of hot mantle assumed to be generated near the mantle-core boundary or at the upper-lower mantle boundary, (= hot spot). Near surface, this column can melt giving rise to oceanic island magmatism (i.e. Hawaii; La Réunion, etc.)

Mars Express: ESA space mission towards Mars.

Mantle: In a planet, mantle is the shell comprised between crust and core. On Earth it represents 82% of the volume and 2/3 of the mass, it is divided into upper mantle (until 700~km depth) and lower mantle (until 2900 km depth).

Mass loss rate (astronomy): Mass ejected per time unit by a star during its formation. Ejection takes place through stellar winds and bipolar jets (typically 10^{-5} to 10^{-8} solar mass per year).

Massive star: Star with mass $> 2 \, M_{sun}$.

Matrix (Chemistry): Parent molecule that allows the pre-positioning of isolated elements thus making possible their polymerisation.

Maturation (Genetics): Transformation step of mRNAs leading to their functional form. It occurs by splicing "Introns sequences".

Megaton: Unit of energy equivalent to the energy released by 10^9 kg of TNT (trinitrotoluene). Corresponds to $4.2 \; 10^{15}$ J.

Mercaptans: Other name of thiols: sulphur analogues of alcohols.

Mesopause: Atmospheric boundary between mesosphere and thermosphere.

Mesophase: Matter state exhibiting characteristics of two phases. Liquid crystals have the fluidity of liquids but are characterized by an order at short range similar to what is observed in crystals.

Mesosphere: Atmospheric layer located above the stratosphere and below the thermosphere, between 45–50 km and 80km (in the case of the Earth) or, if the stratosphere is absent, directly above the tropopause (as in the case of Mars and Venus).

Mesotartric acid: HOOC-CHOH-CHOH-COOH, alpha-beta-di-hydroxysuccinic acid, molecule containing two asymmetric carbon of opposite chirality. This molecule is achiral by internal compensation.

Mesozoic (Era): Period of time (Era) ranging from 250 Ma to 65 Ma, it is also called Secondary Era.

Metabolism: All the reactions taking place in a cell or in an organism. Metabolism is divided into two subclasses: anabolism and catabolism. The very large majority of metabolic reactions are catalyzed by proteinic enzymes.

Metallicity: In a star, a galaxy or a gas cloud, the metallicity is the proportion of heavy elements (heavier than helium).

Metamorphism: Solid state transformation of a rock due to change in pressure and/or temperature conditions. New mineral assemblage, stable in new P-T conditions will appear. New minerals crystallise perpendicular to oriented pressure thus defining a new planar structure called foliation. Most often metamorphism corresponds to dehydration of the rock.

Metasediments: Sediments transformed by metamorphism.

Metasomatism: Change in rock composition due to fluid circulation. For instance, in a subduction zone, the fluids (mainly water) released by dehydration of the subducted slab, up-rise through the mantle wedge. These fluids which also contain dissolved elements, not only rehydrate the mantle peridotite, but also modify its composition.

Meteor Crater: Impact crater in Arizona. It is about 1.2 km in diameter and 170m deep. The impact which took place 50 000 years ago was due to an iron meteorite of about 25 m in diameter. Impacts of this kind generally occur every 25,000 year.

Meteorite: Extraterrestrial object, fragment of an asteroid, of a planet (like Mars) or of the Moon that falls on the Earth surface.

Methanogen: Archeobacteria producing methane CH_4 from CO_2 and H_2. Some methanogens are hyperthermophilic.

Methanogen: Methane-producing microorganism of the Archaea domain.

Methionine *(Met)*: Proteinic amino acid containing five carbon atoms with a $-SCH_3$ group in its side chain.

Methylalanine: Synonymous of alpha aminoisobutyric acid (alpha-AIB).

MGS *(Mars Global Surveyor)*: American probe that have carried out a complete cartography of Mars (from September 1997).

MHD: Magnetohydrodynamic.

Mica: Mineral family. Phyllosilicate (water-bearing sheet silicate): biotite = black mica $[K(Fe,Mg)_3Si_3AlO_{10} (OH)_2]$; muscovite = white mica $[KAl_2Si_3AlO_{10} (OH)_2]$

Microarray: Microarrays are small, solid supports on which thousands of different substances (antibodies- proteins- DNA) can be implanted in specific locations. The supports are usually glass microscope slides, but can also be silicon chips or nylon membranes. The substance is printed, spotted, or actually synthesized directly onto the support. Microarrays are used to test large number of targets quickly or when the only small quantities of the sample to be studied are available.

Micrometeorite: Very small meteorite (< 1 mm). The 50-400μm fraction is the most abundant found on Earth. Micrometeorites constitute more than 99% of the extraterrestrial material able to reach the Earth surface (major impacts excepted)

Micronutrient: Chemical element required in small amounts for the proper growth and metabolism of a living organism (i.e. B, Cu, Co, Fe, Mn, Mo, Zn...). See also macronutrient.

Microorganism: Organism invisible without a microscope. Includes prokaryotes and unicellular eukaryotes (i.e. yeasts).

Microspheres: Spherical clusters of organic molecules found in Precambrian rocks or produced in laboratory from amino acids polymers (proteinoids). Today, the Fox proteinoids microspheres are not still considered as plausible models of primitive cells.

Microsporidia: Parasitic unicellular eukaryotes that have been shown to be highly derived fungi from the fungi. Microsporidia were thought for some time to be primitive.

Migmatite: High-temperature metamorphic rock affected by partial melting.

Milankovitch *(theory of)*: Theory connecting the Earth climate variations to astronomic variations such as changes of Earth's orbit or obliquity with time.

Miller-Urey *(experiment of)*: One of the first experimental simulations of what was considered as atmospheric prebiotic chemistry (1953). Synthesis of a large variety of organic molecules including few amino acids from a very simple mixture containing reduced small molecules (H_2, CH_4, NH_3 and H_2O) submitted to an electric discharge.

Mineral: Solid material defined by both its chemical composition and crystalline structure.

Minimal protosolar nebula: Minimal mass of gas, necessary for the formation of the solar system planets (= mass of all planets + H + He ≈ 0.04 solar mass).

Minor planet: Asteroid or planetoid.

Mitochondria: Organelles in the cytoplasm of all eukaryotic cells where ATP synthesis takes place during aerobic respiration. Mitochondria have their own DNA and could have an endosymbiotic origin.

Mitosis: Nucleus division, cell division step including cytokinesis.

MM: Micrometeorites.

Moho *(Mohorovicic)*: Discontinuity in seismic waves that marks the crust-mantle boundary.

Mole: SI unit for amount of substance; it is defined as the number of atoms in exactly 0.012 kg of carbon-12 (1mole of atoms = 6.02×10^{23} atoms; 1 mole of molecules = 6.02×10^{23} molecules).

Molecular beacon: Molecular beacons are single-stranded oligonucleotide hybridization probes that form a stem-and-loop structure. The loop contains a probe sequence that is complementary to a target sequence, and the stem is formed by the annealing of complementary arm sequences that are located on either side of the probe sequence. A fluorophore is covalently linked to the end of one arm and a quencher is covalently linked to the end of the other arm. Molecular beacons do not fluoresce when they are free in solution. However, when they hybridize to a nucleic acid strand containing a target sequence they undergo a conformational change that enables them to fluoresce brightly.

Molecular clouds: See interstellar clouds.

Molecular flow: see bipolar flow.

Molecular recognition: Chemical term referring to processes in which a specific molecule (ligand) adhere in a highly specific way to another molecule (target), forming a large structure.

Monophyly: Term that describes a taxonomic group sharing a single ancestor and all its descendants (i.e. the mammals).

Monosaccharides: See oses.

Montmorillonite: Mineral. Phyllosilicate (water-bearing sheet silicate). Clay mineral belonging to the smectite group.

MORB *(Mid Ocean Ridge Basalt)*: Basalt generated in mid oceanic ridge systems where oceanic crust is created. Most of ocean floor has a MORB composition.

m-RNA: Messenger RNA. Obtained by transcription of a DNA segment and able to orient the synthesis of a specific protein in the ribosome.

MS *(Mass Spectrometry)*: Analytical method involving a preliminary ionization of atoms or ionization and fragmentation of molecules followed by measures of atomic or molecular masses. These measures can be carried out from precise study of ions trajectories or time of flight in an electric and/or magnetic field.

MSR *(Mars Sample Return)*: NASA-CNES space mission project for the return of Martian soil samples (~1 kg) extracted by automatic probes. Launch expected between 2009 and 2014, and samples return three to five years later.

Murchison: Carbonaceous chondrite (CM) felt in Australia in 1969. Fragments recovered immediately after the fall were (and still are) subjected to many analyses, mainly chemical. More than 500 organic compounds were identified, including amino acids and nucleic bases.

Muscovite: Mineral. Phyllosilicate (water-bearing sheet silicate). [$KAl_2Si_3AlO_{10}(OH)_2$]. It belongs to the mica group and is also called white mica.

Mutation: Any change of the genetic material, transmitted to the descendants.

My: Mega year = one million years (= Ma)

Mycoplasma: The simplest and the smallest known microorganisms, they live as parasites in animal or vegetal cells. They long be considered as possible analogues of the first cells; now considered as Gram + bacteria which lack their rigid cell wall and evolved by reduction.

N

Nanobacteria: Hypothetical bacteria, whose size could be around few nanometres, smaller than any known bacteria. Their existence is very much debated.

N-carbamoyl-amino acids: A molecule showing many similarities with aminoacids except that one of the H atoms of the amino group is substituted by the carbamoyl polyatomic group (-CO-NH$_2$). N-carbamoyl-amino acids could have been prebiotic precursors of some amino acids.

Neutral (mutations) (Genetic): Term coming from the neutral theory of molecular evolution proposed by Kimura. A neutral mutation is a mutation leading to sequences selectively and functionally equivalent. They are said neutral in regards of evolution.

Neutron star: Remnant of a dead star with an initial mass greater than ~8M$_{Sun}$. When electron degenerated pressure becomes too low; electrons penetrate into nucleus and are transformed into neutron by reacting with protons (neutralisation). Then they develop a neutron degenerated pressure.

NGST (Next Generation Space Telescope): NASA project of space telescope (4m), it must succeed to the HST (Hubble Space Telescope).

Nitrification: Microbial oxidation, autotrophic or heterotrophic, of ammonium to nitrate

Nitrile: R-CN where the CN group is the cyano group (cyano as prefix but nitrile as suffix).

Non reducing atmosphere: Atmosphere of CO_2, N_2, H_2O where hydrogen is absent or in low quantity, either in the form of free H_2 or hydrogen-containing compounds, such as methane or ammonia. Also named oxidized or neutral atmosphere according to its composition.

Non-sense (codon): When a codon (triplet of nucleotides) does not specify an amino acid but corresponds to a termination codon (Term.). In the « universal code », these codons are UAA, UAG and UGA.

Normative (rock composition): Rock mineralogical composition recalculated from its chemical composition.

Nuclear pores: Complex structures, highly specialized, embedded in the nuclear membrane. They allow the transfer of macromolecule between nucleoplasm and cytoplasm.

Nucleation (Astronomy): Mechanism of formation of solid bodies by accretion of planetesimals.

Nucleic acid: Long chain polymeric molecule obtained by condensation of nucleotides. DNA and RNAs are nucleic acids.

Nucleic base: Linked to ribose or desoxyribose by a hemi-acetal bond, it gives nucleosides. Nucleic bases are purine bases or pyrimidine bases. Nucleosides together with a phosphate group are the sub-units of nucleotides. By condensation, nucleotides give the polynucleotides (including DNA and all RNAs).

Nucleides: Constituents of atom nucleus, i.e. protons and neutrons.

Nucleon: Common name for proton or neutron.

Nucleoplasm: Protoplasm within the nucleus of eukaryotes.

Nucleotide: Molecule made by condensation of a base (purine or pyrimidine), an ose and a phosphate group linked to the ose. Nucleotides are ribonucleotides when the ose is ribose or deoxyribonucleotide when the ose is deoxyribose. DNA is a polydeoxyribonucleotide while RNAs are polyribonucleotides. The symbol of a nucleotide is determined by the base (A for adenine, C for cytosine, G for guanine, T for thymine, U for uracil).

Nucleus (Biology): Eukaryote cell substructure that contains the chromosomes.

O

Obduction: Mechanism leading to the thrusting of oceanic lithosphere onto continental crust.

Obliquity: Angle between the ecliptic and the celestial equator, actually, 23.3 degrees for the Earth. This angle is $> 90°$ if the planet has a retrograde rotation.

Ocean resurgence: See upwelling.

Oceanic rift: Central depression in a mid ocean ridge, this is the place where oceanic plates are created.

Oligomer: Small polymer, generally containing less than 25 monomeric units

Oligomerization: Polymerization involving a small number of monomers.

Oligopeptides: Small polypeptide (less than 25 AA residues even if the definition is not so strict).

Olivine: Mineral. Nesosilicate (isolated SiO_4 tetrahedrons). $[(Fe, Mg)_2SiO_4]$. This mineral which belongs to the peridot family, is silica-poor is one of the main components of the terrestrial upper mantle. It is also common in meteorites.

OMZ: See Oxygen Minimum Zone.

Oort cloud: Huge spherical collection of comets, orbiting the Sun between 10 000 and 100000 AU.

Ophiolite: Part of oceanic lithosphere tectonically emplaced (obducted) on a continental margin.

Optical activity: Orientation change of the linearly polarized plane of light after its passage through a chiral medium.

Optical rotatory dispersion: Change of the optical power of a chiral medium with the wavelength of the linearly polarized light.

Orbital Migration: Change, in course of time, of a planet-star distance; this hypothesis is proposed in order to explain the presence of massive planets close to their star.

Organic molecule: Until the beginning of the 19th century, organic molecules were molecules extracted from plants or animals. Today, any molecule containing carbon atoms is called organic. Carbonates, CO and CO_2 are borderline cases. Organic molecules generally contain C atoms with an oxidation number lower than 4.

Organometallic molecule: Molecule containing one or more metal atoms.

Organometallic: Organic molecule containing one or more metallic atoms bonded by covalent bonds or by coordination to the organic moiety of the molecules. Metallic salts of organic acids are not considered as organometallic compounds.

Orgueil: Large CI carbonaceous chondrite (very primitive), without chondres, which felt in France in1864, near Montauban.

Ornithine: NH_2-CH)-COOH-$(CH_2)_3$-NH_2. Non proteinic amino acid, precursor of arginine.

Orogenesis: Mountain chain genesis.

Orographic: Related to relief. In weather science, the presence of a relief can induce changes in atmospheric circulation resulting in weather changes (i.e. orographic rainfalls).

Orthose: Mineral. Tectosilicate (3D silicates). Alkali feldspar $KAlSi_3O_8$.

Oses *(or saccharides)*: Large group of molecules of primeval importance in the living world. Some of them are monomers like glucose $(C_6(H_2O)_6)$ or ribose $(C_5(H_2O)_5)$. Some of them are dimers like lactose or saccharose; some of them are polymers (polysaccharides) like starch or cellulose.

Outer membrane: membrane surrounding the plasma membrane in Gram negative bacteria (i.e. *Escherichia coli*).

Oxidative phosphorylation: In living cells, biochemical process that results in ATP synthesis; energy is provided by protons transfer across the cell membrane.

Oxygen minimum zone (OMZ): Part of the water column where the oxygen consumption due to respiration exceeds its supply by lateral circulation, giving rise to dysoxic or anaerobic conditions. The OMZ areas occur generally where the intermediate or deep circulation is weak, and/or when biological productivity at the surface is high.

Oxyphile: See lithophile.

P

P4: Certification for laboratories accredited to analyze high-risk infectious agents. Such laboratories must be protected against any risk of contamination by viruses and microorganisms, from inside to outside and from outside to inside. These two conditions must absolutely be realised for extraterrestrial sample analyses.

PAH *(Polycyclic Aromatic Hydrocarbons)*: Organic molecules like naphthalene or anthracene containing several fused aromatic rings.

Paleomagnetic scale: Relative stratigraphic scale based on the successive inversions of the Earth magnetic field through time

Paleo-soil: Fossil soil. These formations are able to have recorded O_2 and CO_2 concentration of the primitive terrestrial atmosphere.

Paleozoic *(Era)*: Period of time (Era) ranging from 540 Ma to 250 Ma, it is also called Primary Era.

Pangaea: Super-continent that existed at the end of Palaeozoic era (– 225 Ma) and which later broke in two parts: Laurasia (N) and Gondwana (S).

Paralogous: Paralogous genes originate in gene duplication events, in contrast to the standard orthologous genes, which originate via speciation events.

Parent body: Asteroid or planet from which an object has been extracted, such as for example a meteorite.

Parent molecule *(in comet)*: Molecule present in the comet nucleus.

Parity *(violation of)*: Characterizes any physical property that changes when space is inverted or, in other words, which is not the same in the mirror world. Parity violation is observed in several phenomena related to the weak intranuclear interactions like the beta radioactivity. Parity violation is at the origin of the very small energy difference between enantiomers (PEVD for Parity Violation Energy Difference).

Parsec *(secpar)*: Unit of astronomical length of 3×10^{18} cm (about 200,000 AU or 3.26 light-years); it is based on the distance from Earth at which stellar parallax is 1 second of arc. Average distance between stars in the Sun vicinity is around 1 parsec.

Pathfinder: American probe that landed on Mars on July 4th, 1997, formally named the Mars Environmental Survey (MESUR). It contained a rover "Sojourner" that explored Ares Vallis during several months. For instance, an "Alpha Proton X-ray Spectrometer" was used to analyse soil and rock samples in order to determine their mineralogy.

PCR *(Polymerase Chain Reaction)*: Experimental method which, by successive molecular duplications, leads to a dramatic increase of a small initial amount of DNA. This result is obtained with an enzyme called DNA-polymerase isolated from thermophile bacteria.

Peptide: Polymer obtained by condensation of amino acids. In a peptide (or polypeptide), the number of AA residues is generally lower than 60. With a higher number of residues, the polymer generally adopts a well defined conformation and is called a protein.

Peridot: Mineral family, olivine is a peridot.

Peridotite: Rock made up of olivine and pyroxenes (ortho- and clino-pyroxenes) as well as of an Al-bearing mineral (spinel at low pressure and garnet at high pressure. Earth mantle is made up of peridotite.

Perihelia: In the case of an object in elliptical motion around a star, the point which corresponds to the shortest distance with respect to the star.

Permafost: In arctic regions, permanently frozen soil or subsoil.

Perovskite: Mineral. $[(Mg,Fe)SiO_3]$. Perovskite together with magnesiowurstite is probably the main component of the lower terrestrial mantle (depth > 660 km).

Petrogenesis: Mechanism(s) of formation of rocks.

PGE or platinum group elements: Transition metal belonging to the Platinum group: Ruthenium (Ru), Rhodium (Rh), Palladium (Pd), Osmium (Os), Iridium (Ir) and Platinum (Pt). They possess similar chemical properties (siderophile). In Earth's crust their abundance is very low, whereas it can be high in undifferentiated meteorites. On Earth, their high concentration in sediments is used as an indication of a meteorite impact, for exmaple at the KT boundary. The elemental ratio between the PGE can provide information on the nature of the impacted meteorite.

pH: Measure of the acidity of an aqueous solution. $PH = -\log_{10} (H^+)$ where (H^+) is the molar concentration of hydroxonium ion in solution. $pH = 7$ corresponds to the neutrality while a solution with $pH > 7$ is basic and a solution with $pH < 7$ is acidic.

Phanerozoic *(Aeon)*: Period of time (Aeon) ranging from 0.54 Ga to today; it followed Precambrian and was characterized by metazoa development.

Phenetic: Taxonomic system for living beings based on overall or observable similarities (phenotype) rather than on their phylogenetic or evolutionary relationships (genotype).

Phenotype: The observable characteristics of an organism, i.e. the outward, physical manifestation of an organism.

Phenylalanine *(Phe)*: Amino acid containing an aromatic phenyl group in its side chain. Phenylalanine contains nine amino acid residues.

Phosphoric acid: H_3PO_4, dihydrogen phosphate. Molecule which plays an important role in the living world: nucleotides are esters of phosphoric acid.

Photochemistry: Chemistry involving energy supply coming from « light » (from IR to UV). When electromagnetic frequency is greater (X-rays, beta-rays or gamma rays), the term radiochemistry is generally used. Specific processes like photo-activation, photo-ionisation, photo-dissociation (including the production of free radicals), and photolysis are various aspects of photochemistry

Photodissociation: Dissociation of molecules due to the energy provided by electromagnetic radiation.

Photolysis: See photodissociation.

Photosynthesis: Synthesis using photons as energy supply. Photosynthesis can be performed at laboratory or industrial level as a sub-domain of photochemistry. In biochemistry, the term « photosynthesis » describes the different biosynthetical pathways leading to the synthesis of molecules under the influence of light. The photons are absorbed by cell pigments and their energy is converted into chemical energy stored in chemical bonds of complex molecules, the starting material for the synthesis of these complex molecules being small molecules like CO_2 and H_2O. It is important to make a clear distinction between the aerobic photosynthesis (also called oxygenic photosynthesis) and the anaerobic photosynthesis. In the first case, water is the reductive chemical species and O_2 is a by-product of the reaction. It must be kept in mind that atmospheric dioxygen as well as oxygen atoms of many oxidized molecules on Earth surface have a biosynthetical origin. Green plants, algae and cyanobacteria are able to perform aerobic photosynthesis; their pigments are chlorophylls. The anoxygenic photosynthesis is not based on H_2O but on reductive species like H_2S. Finally, some halophile archaeas contain a pigment called bacteriorhodopsin and are able to use light energy supply for ATP synthesis. So, light energy is converted into chemical energy or better, in chemical free energy.

Phototroph: Organism whose energy source is light (photosynthesis).

Phylogenesis: History of the evolution of a group of organisms.

Phylogenic tree: Schematic representation describing the evolutive relationships between organisms. It gives an image of the evolution pattern.

Phylotype: Environmental sequence, representing an organism. Sequence of a clone obtained from environment and representing an organism.

Phylum *(pl. phyla)*: Group of organism evolutionary connected at high taxonomic rank.

Pillow lava: Lava extruded under water (ocean, lake) which produces its typical rounded pillow shape. Pillow lava forms the upper part of oceanic crust.

PIXE *(Proton-Induced X-ray Emission)*: Device that allows the detection and identification of many elements (metals) in proteins.

Plagioclase: Mineral. Tectosilicate (3D silicates). Calco-sodic feldspar whose composition ranges between a sodic ($NaAlSi_3O_8$ = Albite) and a calcic ($CaAl_2Si_2O_8$ = Anorthite) poles. They represent about 40% of the Earth crust minerals.

Planet: Body formed in circumstellar disks by accretion of planetesimals and may be of gas.

Planetary nebula: This expression is not related to planets or to nebula. A planetary nebula is the low density and expanding gaseous envelope of a white dwarf. The last phase of evolution of a star whose mass is $< \sim 8M_{Sun}$ consists in an explosion (nova) and in the ejection of a gaseous envelope; the star relicts giving rise to the white dwarf itself.

Planetesimals: Small solid bodies ($\sim$1–100 km) formed in the protosolar nebula, probably similar to asteroids and comets. Their collision and accretion built the planets.

Plankton *(plankton)*: The whole organisms living in water and drifting along ocean and lake currents. It includes zooplankton (small animals) and phytoplankton (plants). Fishes and sea mammals, able to swim independently of current flow, constitute the nekton. Neston refers to organisms drifting at the air-water interface whereas benthos designates organisms living in or on the aquatic ground.

Plasma membrane *(or cell membrane)*: Semi-permeable membrane that surrounds contemporary cells. It consists of a double layer of amphiphilic molecules (hydrophobic tail with hydrophilic head), mainly phospholipids, with proteins embedded in it. It may also contain some molecules, such as cholesterol or triterpenes, able to rigidify the whole.

Plasmid: Extrachromosomic DNA found in bacteria and yeasts, able to replicate independently.

Plasmon: Electronic cloud of a metal.

Plasts: Organites found in photo-trophic Eukaryotes and containing photosynthetic pigments such as chlorophylls.

Plate *(lithospheric)*: Piece of the rigid lithosphere that moves over the ductile asthenosphere affected by convection.

Plate tectonic: Theory that describes and explains the rigid lithospheric plate motion.

Plutonic: Magmatic rock, resulting of the slow cooling and crystallisation of magma at depth. Its texture is granular (big crystals), e.g. granite.

PMS Star: New-born star such that the internal temperature is yet too low (< 10^7 K) to initiate nuclear fusion of hydrogen into helium (i.e. T-Tauri stars) and characterized by a low mass.

PNA *(Peptide Nucleic Acids)*: Synthetic analogues of nucleic acids such that the ose-phosphate strand is replaced by a polypeptide backbone to which the bases are linked.

Polarized light: It is important to differentiate two limit cases. A linearly polarized light (or plane polarized light) is an electromagnetic radiation, whatever is its frequency, such that the electric vector and thus also the magnetic vector oscillate in a plane. If this radiation travels through a chiral medium, the plane of polarization is deviated (optical rotation). A circularly polarized light is an electromagnetic radiation such that its electric vector and thus also its magnetic vector describe, in space, an helix around the propagation direction. This helix can be right or left corresponding to the two possible circularly polarized lights of a definite frequency. The linearly polarized light corresponds to the superposition of two circularly polarized lights of the same frequency.

Pole motion: Geographic pole motion at the Earth surface. This motion has low amplitude, since the pole only moves of few metres.

Polymerization: Chemical reaction such that molecules called monomers are covalently linked together and form long chain molecules, with or without branching. The polymerization can be the result of an addition of polymers like in polyethylene or polystyrene (industrial polymers) but the polymerization can also be the result of condensation reactions involving at each step, the elimination of a small molecule (generally water). Nylon is an industrial condensation polymer and most of biological macromolecules such as polypeptides, polynucleotides and polysaccharides are condensation polymers.

Polynucleotide: Polymer resulting from condensation of nucleotides.

Polypeptide: Polymer resulting from condensation of amino acids.

Polysaccharide: Polymer resulting from condensation of oses (in the past, oses were also called monosaccharides).

POM *(polyoxymethylene)*: Polymer of formaldehyde.

Population I stars: Stars enriched in heavy elements (O, C, N...). They have been formed from the interstellar gas enriched by the previous generation(s) of stars formed over the billions of years of lifetime of the Galaxy. They are predominantly born inside the galactic disk. The Sun belongs to population I stars. Massive, hot stars are necessarily young and are therefore always population I stars.

Population II stars: Stars poor in heavy elements (O, C, N...). They have been formed from low metallicity gas that has not been enriched by successive generations of stars. They are believed to be born in the early ages of the Galaxy. They are actually distributed predominantly in the halo of the Galaxy thus confirming that they are probable remnants of the infancy of our Galaxy.

Poynting-Robertson (effect): Effect of stellar light on a small orbiting particle. This causes the particle to fall slowly towards the star. Small particles (below 1cm) are more affected because the effect varies as the reciprocal of particle size.

p-p chain: Series of hydrogen burning reactions that produce helium. These reactions provide the energy of the main sequence stars whose mass is $< 1.1 \, M_{Sun}$.

ppb *(part per billion)*: Relative concentration in mass = nanogram/gram.

ppbv *(part per billion in volume)*: Relative concentration in volume = nanolitre/litre).

ppm *(part per million)*: Relative concentration in mass = microgram/gram.

ppmv *(part per million in volume)*: Relative concentration in volume = microlitre/litre)

Prebiotic chemistry: All chemical reactions which have contributed to the emergence of life.

Precambrian: Group of aeons ranging from Earth genesis (4.55 Ga) until the beginning of Palaeozoic era (0.54 Ga). It includes: Hadean, Achaean and Proterozoic aeons.

Precession *(of equinoxes)*: Motion of Earth's rotation axis with respect to the celestial sphere due to the other bodies of the Solar system and more particularly to the Moon. The terrestrial pole describes a circle on the celestial sphere in 26000 years.

Primary structure *(of a protein)*: Sequence of the amino acid residues in the proteinic chain.

Primitive atmosphere: The Earth primitive atmosphere refers to the atmosphere present when prebiotic chemistry occurred, free oxygen concentration was very low. This atmosphere was considered as highly reductive; today it is rather assumed to have been made up of carbon dioxide, nitrogen and water vapour (greenhouse).

Primitive Earth: The young Earth from its formation until 2.5 Ga.

Primitive nebula: See protosolar nebula.

Prion: Protein able to induce a pathological state because its conformation is modified.

p-RNA: Synthetic RNA molecule in which the sugar is a pyranose instead of a furanose.

Prokaryote: Microorganism in which chromosomes are not separated from the cytoplasm by a membrane. Bacteria and Archaea are prokaryotes.

Proline *(Pro)*: Amino acid containing six C atoms with a unique characteristic. Its side chain links together the alpha atom and the N atom to give a five membered ring containing one N atom and four C atoms ; the amino group is no longer a $-NH_2$ group but a $-NH-$ group. Proline is frequently observed in protein secondary structures called beta turns

Proper motion *(Astronomy)*: Apparent angular movement of a star on the celestial sphere during a year (perpendicular to the line of sight). Proper motion analysis can lead to detection of planets in orbit around this star.

Proper motion: Apparent angular movement of a star on the celestial sphere. The detailed observation and study of such motion can lead to the detection of planets orbiting the star.

Propionaldehyde: CH_3-CH_2-CHO or propanal.

Protein Scaffold: Protein whose main function is to bring other proteins together in order for them to interact. These proteins usually have many protein binding domains (like WD40 repeats).

Proteins: Long chain biological polymers obtained by condensation of amino acids. The degree of polymerization (number of residues) ranges between 60 and 4000. Some proteins form aggregates. For example, haemoglobin is a tetramer containing two proteinic chains of one type and two proteins chains of another type; in each chain, a heme molecule containing a ferrous cation is settled in without being covalently bonded. Structural proteins are components of muscles or flagella; most enzymes are proteins and proteins are also carriers of other molecules like dioxygen or carbon dioxide. The activity of proteins is extremely dependent on their molecular conformation.

Proteobacteria: Group of bacteria including *Escherichia coli* and purple bacteria (photosynthetic bacteria). Mithochondria are relics of proteobacteria.

Proteome: The complete collection of proteins encoded by the genome of an organism.

Proteomics: Part of science that studies proteome. Proteomics not only analyses the complete protein collections but also determines their precise location, modifications, interactions, activities and functions. Consequently, it implies the simultaneous analyse of a huge amount of proteins in a single sample.

Proterozoic *(Aeon)*: Period of time (Aeon) ranging from 2.5 to 0.54 Ga.

This aeon belongs to Precambrian and follows Achaean. Apparition of oxygen in atmosphere and of metazoa.

Proton motive force: Free energy difference (measured in volts) associated with proton translocation across a membrane. It depends on the electrical membrane potential and on the pH difference between the two reservoirs separated by the membrane. The proton motive force provides energy required for ATP synthesis.

Protoplanetary disks: Disk around a new-born star where accretion of planets is supposed to take place.

Protosolar nebula: Rotating disk of gas, dust and ice, from which the solar system is originated.

Protostar: several similar definitions exist. 1) new-born star such that half of its luminosity is due to accretion. 2) Body involved in an accretion process which will bring it on the main sequence. 3) Collapsing interstellar cloud. 4) Young object which is not yet optically visible. Protostars are rare because their time life is short (10^4 to 10^5 years): « Protostars are the Holy Grail of IR and submillimetric astronomy ».

Psychrophilic organism: Organism that lives optimally at a temperature lower than 10°C. Some psychrophilic organisms live at temperatures lower than 0°C.

Pulsar : Small neutron star in very fast rotation (one rotation in less than 1s) emitting a highly focalized radiation, circularly polarized and detected as very regular pulses. Pulsars are remnants of supernovae.

Purine bases: Guanine and adenine are examples of purine bases because their molecular skeleton corresponds to purine. These bases are found in DNA as in RNA's.

Purple bacteria: Micro-organisms of the bacteria domain able to perform anoxic photosynthesis.

PVED *(Parity Violation Energy Difference)*: Energy difference between enantiomers due to parity violation at the level of the weak forces.

Pyranose: A "pyranose ring" is a cyclic ose formed of 5 carbons and an oxygen atom. It is the more stable form of oses.

Pyrimidine bases: Thymine, cytosine and uracil are examples of pyrimidic bases because their molecular skeleton corresponds to pyrimidine. Cytosine is found in DNA as in the RNAs while thymine is specific of DNA and uracil is specific of RNA's.

Pyrite: Mineral. Sulphide [FeS_2].

Pyrolysis: Thermal degradation of a molecule.

Pyroxene: Mineral family. Inosilicate (simple chain silicate). Divided in two families: 1) Orthopyroxene [$(Fe,Mg)_2Si_2O_6$] (e.g. enstatite) and 2) Clinopyroxenes [$(Ca,Fe,Mg)_2 Si_2O_6$] (e.g. augite, diopside).

Q

Q: Orbital parameter of a planet orbiting a star; distance between planet aphelia and star.

q: Orbital parameter; distance between planet perihelia and star.

Quartz: Mineral. Tectosilicate (3D silicate). It crystallizes in hexagonal or rhombohedric systems. In magma it characterizes silica sur-saturation. The rhombohedric crystals are chiral: quartz can therefore exist as D- or L-quartz depending from the helicity of the –O-Si-O-Si- chain. A chiral quartz crystal can induce an enantioselectivity during a chemical reaction between

358

achiral reactants via a catalytic effect.

Quaternary structure *(of a protein)*: In the case of protein forming supramolecular aggregates like haemoglobins, the quaternary structure corresponds to the arrangements in space of the sub-units.

Quencher: Molecular entity that deactivates (quenches) a fluorophore.

R

R: solar radius ($0.69 \ 10^6$ km) or 1/200 AU; approximately 10 times the Jupiter radius and 100 times the Earth radius.

Racemate *(crystal)*: Crystalline form of a chiral substance such that each unit cell of the crystal contains an equal number of molecules of opposite chirality.

Racemic *(mixture)*: Mixture of enantiomers containing an equal number of the two enantiomers. Such a mixture is described as achiral by external compensation.

Racemisation: Diminution of the initial enantiomeric excess of a homochiral ensemble of molecules or of a non-racemic mixture of enantiomers. It corresponds to an equilibration reaction: the system evolves spontaneously towards a state characterized by higher entropy and therefore, lower free energy. The highest entropy and lower free energy corresponds to the racemic mixture. Racemisation can be a very slow process: this is why enantiomers of amino acids, sugars and many other components of living species can be separated in many cases.

Radial velocity measurement: The line-of-sight velocity of a star or other celestial object towards or away from an observer. This methods allows to determine a star movements through space and consequently to infer the existence (or not) of an orbiting object.

Radial velocity measurement: The line-of-sight velocity of a star or other celestial object towards or away from an observer. This methods allows to determine a star movements through space and consequently to infer the existence (or not) of an orbiting object.

Radial velocity: Star velocity component parallel to the view line. It causes the frequency shift observed in spectral emission lines (Doppler Effect). Periodical change in radial velocity can be an indirect proof that a planet orbits around the star (reflex motion).

Radiation pressure: Pressure applied by electromagnetic waves on any atom, molecule or particle.

Radioactivity *(long lived species)*: Radioactive elements with long period (10^9 to 10^{11} years); they are still present in the Solar system.

Radioactivity *(short lived species)*: Radioactive elements with short period ($< 10^7$ years). Nowadays they have totally disappeared in the Solar system.

Raman *(spectroscopy)*: Physical method used for molecular structural analysis. Raman spectroscopy is based on inelastic diffusion of visible or UV light and gives information about the vibration modes of diffusing molecules. Raman spectroscopy must be considered as a complementary method with respect to IR spectroscopy.

Rare Earth Elements *(REE)*: Chemical elements with very similar chemical properties. This family (lanthanides) ranges from Lanthanum ($Z = 57$) to Lutetium ($Z = 71$). In geochemistry, they are commonly used as geological tracers of magmatic processes. Indeed they are poorly sensitive to weathering or metamorphism, but on the opposite they are excellent markers of magmatic processes such as melting or crystallization.

Rare gases *(He, Ne, Kr, Ar, Xe, Rn)*: Monatomic gases corresponding to the (VIIIA) period of Mendeleyev periodic table (see "Astrobiological data"). Their isotopes can be used to trace some geological events of Earth differentiation (e.g. atmosphere and ocean formation).

Recovery ratio (%): A sample prepared by adding a known mass of target analyte to a specified amount of matrix sample for which an independent estimate of target analyte concentration is available. Spiked samples are used, for example, to determine the effect of the matrix on a method's recovery efficiency.

Red giant *(astronomy)*: Old star (spectral type K or M) still performing fusion of hydrogen but having already a helium core.

Reducing atmosphere: Atmosphere with high hydrogen content. Carbon, oxygen and nitrogen mainly exist as CH_4, H_2O and NH_3.

Reduction potential: Measure of the tendency for a molecule or an ion to give an electron to an electron acceptor that, itself, can be a molecule, an ion or an electrode. A conventional reduction potential scale for molecules in water allows to determine, a priori, what chemical species will be the electron acceptor and what will be the electron donor when they are mixed together at a well defined concentration.

Refractory inclusion: Aggregate of refractory minerals incorporated into a meteorite. See CAl.

Refractory: A refractory compound is a compound that remains solid over the whole temperature range undergone by the host object (i.e. dust in a comet). Antonym: volatile.

Refractory: Substance which remains solid in all temperature conditions available in a particular body of the solar system (ex. dust particles in a comet). If not refractory a substance is said volatile.

Region *(HII region)*: interstellar cloud such that H exists essentially as H^+. Ionization is due to intense UV radiation coming from OB stars. HII is the old name used by spectroscopists to describe lines coming from H^+ recombination.

Replication: biochemical process by which a DNA strand or a RNA molecule is copied into the complementary molecule. Replication is different from transcription and from translation.

Reservoir: In biogeochemistry or geochemistry, a reservoir is the available quantity of one chemical element (or species) within one specific compartment of the global ecosystem (ocean, atmosphere, biosphere, sediments, etc...). Any reservoir may evolve through exchanges with the other ones (biogeochemical or geochemical fluxes).

Residence time: The average time a chemical element or species spends in a geochemical or a biogeochemical reservoir (ocean, atmosphere, biosphere, sediments, etc...). . Assuming the reservoir is in steady state (input flux = output = dQ/dt) and contains an amount Q of one element, the residence time $\tau = Q/(dQ/dt)$.

Retrotransposons: DNA sequences able to move from one site to another along the chromosome. Retroposons belongs to a transposons family which requires RNA molecule as intermediate. The more frequent retrotransposons are the

Alu sequences; around one million of such sequences have been identified in the human genome.

Ribose: Aldopentose of major importance in the living world; part of RNA nucleotides

Ribosomes: Intracellular structures containing many rRNA (ribosomal RNA) molecules together with a complex of 60–80 proteins. Synthesis of proteins takes place in ribosomes by condensation of amino acids; information about the correct sequence is given by a mRNA (messenger RNA) while each amino acid is linked to a specific tRNA (transfer RNA).

Ribozyme: RNA molecule acting as an enzyme.

Ridge *(geology)*: Submarine mountain chain located at divergent lithospheric plate margins. On the Earth the total length of ridges is 60000 km.

Rift: Rift valley limited by faults = graben.

Rigid rotator (hypothesis): Molecular system frozen at its equilibrium geometry. Interatomic distances and bond angles do not vary during rotation.

Ringwoodite: Mineral. Nesosilicate (isolated SiO_4 tetrahedrons). $[(Mg,Fe)_2SiO_4]$. Also called olivine γ phase. In Earth mantle, ringwoodite is stable between 520 and 660 km.

RNA world: Often considered as an early hypothetical stage of life evolution, based on a life without protein and such that RNAs would have played a double role: catalysis and support of genetic information. This theory is mainly based on the discovery of catalytic RNAs (ribozymes) and on an increasing knowledge about the importance of RNAs in contemporary life. For other scientists, the RNA world is the stage of evolution that preceded the DNA emergence.

RNA: Ribonucleic acid, a class of nucleic acids containing ribose as a building block of its nucleotides. RNAs themselves are divided into sub-groups like messenger-RNA (m-RNA), transfer-RNA (t-RNA), ribosomal-RNA (r-RNA).

Rock: Solid material made up of mineral assemblage. It constitutes telluric planets, asteroids and probably the core of giant planets.

Rocky planets: Other name for telluric planets.

Rodinia: Super continent that formed at about 750 Ma ago.

Root *(Genetics)*: for a particular genetic tree, the last common ancestor of all the organisms of the tree.

Rosetta: ESA mission launched at the beginning of 2004; it will reach the P/Cheryumov-Gerasimenko comet after a 10 years trip. An orbiter will follow the comet during one year, and a lander will perform « in situ » analyses on comet nucleus surface.

Rotatory power: Deviation of the polarization plane of a radiation of well-defined frequency by a chiral medium.

Rovibrational (level): Energy level of a molecule expressed as a rotational quantum number J associated to a given vibration level v.

r-process (rapid): Nucleosynthethic mechanism that takes place during explosion events (i.e. Supernova) and leads to the synthesis of neutron-rich nucleus with atomic number > 26 (Iron). During this process, the atomic nucleus captures a huge amount of neutrons and is immediately dissociated through β^- decay.

RRKM: Method of molecular dynamics for the calculation of rate constants of chemical reactions due to Rice, Ramsperger, Kassel & Marcus.

r-RNA: Ribosomal RNA. r-RNAs are the major components of ribosomes. Some of them have catalytic properties.

Rubisco *(or RuBisCo)*: Ribulose-1,5 biphosphate carboxylase-oxygenase; enzyme which catalyses CO_2 metabolism. It is the most abundant enzyme on Earth.

Runaway greenhouse effect: Amplification of a greenhouse effect due to vaporization of molecules able to absorb infrared radiation emitted or received by the planet and which, themselves, contribute to greenhouse effect (i.e. Venus).

Runaway growth *(astronomy)*: Increasing rate of planetary accretion with planet growth, it leads to an increasing size difference between the small and large bodies.

R_{uni}: Sun radius (0.69×10^6 km); 1/200 AU; ~10 times Jupiter radius, ~100 time Earth radius.

Runoff channels: Kind of channels observed on Mars and which seems to originate by large water flows over a long period of time.

S

Sagduction: Gravity driven rock deformation. When high density rocks (e.g. komatiites) emplace over low density rocks (e.g. TTG), they create an inverse density gradient that results in the vertical sinking of dense rocks in the lighter ones. Sagduction was widespread before 2.5 Ga.

Sarcosine: $(CH_3)NH-CH_2-COOH$, non-proteinic N-methyl amino acid.

Schist: Fine grain sedimentary rock characterized by a cleavage (slate). It results of fine sediment (i.e. mud) metamorphism.

Secondary structure *(of a protein)*: Spatial arrangement of the main chain of a protein. Alpha-helices, beta-sheets, beta-turns are examples of secondary structures

Sedimentary rock: Rock generated on Earth surface. It can consist in the accumulation of rock particles (detrital) or of organic matter (oil, petrol, coal). It can also be produced by physico-chemical or biogenic precipitation of dissolved ions. Detrital particles and ions derived of alteration and erosion of pre-existing rocks.

Selenocysteine: Frequently described as the 21st proteinic amino acid because sometimes coded by the genetic code. It has the structure of the cysteine with a Se atom replacing the S atom.

SELEX: *(Systematic Evolution of Ligands by Exponential Enrichment)* Method that generates high-affinity RNA or DNA ligands through successive *in-vitro* rounds of directed molecular evolution. This process allows to identify aptamers by iterative enrichment for molecules capable of binding a target.

Semi major axis: Orbital parameter. Half of the major axis of an elliptical orbit.

Semi minor axis: Orbital parameter. Half of the minor axis of an elliptical orbit.

Semi-minor axis: Orbital parameter. Small axis of orbital parameter.

Sense: DNA strand that is not copied into mRNA. Therefore, the sequence of the sense strand corresponds to the mRNA sequence which, itself, corresponds to the transcription of the antisense strand. Sense and antisense DNA sequences are strictly complementary.

Sequence *(genetics)*: Series of directly linked nucleotides in a DNA or a RNA strand; series of directly linked amino acids residues in proteins.

Sequence (main sequence) *(astronomy)*: Stage in star evolution when it performs hydrogen fusion in its core. In a HR diagram, main sequence stars draw a straight line.

Sequencing: Experimental determination of a DNA, RNA or protein sequence.

Serine *(Ser)*: Proteinic amino acid with a -CH_2-OH side chain. Being hydrophilic, serine is generally present in the external part of the skin and can be used as the proof for human contamination of meteoritic samples. Serine is produced in very small amounts during simulation experiments considered as experimental models of interstellar chemistry.

Serpentine: Mineral family. Phyllosilicate (water-bearing sheet silicate) (i.e. Antigorite $[Si_4O_{10}Mg_6(OH)_8]$). They are generated by olivine (and sometimes pyroxene) alteration. They play an important role in internal water cycle.

SETH *(Search for Extraterrestrial Homochirality)*: Search for proofs of enantiomeric excesses in extraterrestrial objects.

SETI *(Search for Extraterrestrial Intelligence)*: Search for electromagnetic signals (essentially in the radio-wave domain) that should be intentionally emitted by some living organisms and coming from sources located outside the Solar system.

Shield *(geology)*: Huge block of old (often Precambrian in age) and very stable continental crust, e.g. Baltic shield (includes Finland, Sweden, Norway, and Western part of Russia).

Shocked quartz: Quartz crystal whose structure contains defaults characteristic of the high pressures realised by meteorite impacts.

Shocked quartz: Quartz showing characteristic microscopic defects of its crystalline structure that were generated by the passage of a high pressure (> 5 Gpa) shock wave. The most common modification is the formation of planar deformation features (PDF) consisting of fine lamellae of amorphous SiO_2 oriented parallel to two (or more) specific crystallographic planes. Under even higher pressure the whole quartz crystal can be transformed to glass. Shocked quartz are a diagnostic criteria to recognize meteorite impacts as no other natural process is capable of producing the required high pressure dynamic shock wave.

Siderite: Mineral, iron carbonate $[FeCO_3]$.

Siderophile: Elements frequently associated to iron (like Au, Pt, Pd, Ni...).

SIDP *(Stratospheric IDP)*: IDP collected in the stratosphere by captors placed on airplanes.

Silicate: Wide mineral family of silicium oxides. The structure is based on a tetrahedron $(SiO_4)^{4-}$

Simple sugar: Old name for monosaccharide.

Site *(active site)*: For an enzyme, specific locus where the substrate is fixed, ready to react with the reactant.

SL *(Astronomy)*: Solar luminosity (3.826×10^{24} W) or 3.826×10^{33} erg s^{-1}

SM: Solar mass (2×10^{30} kg).

Small bodies: Comets and asteroids.

SMOW *(Standard Mean Ocean Water)*: Standard reference sample for H and O isotopic abundance measurements.

SNC: Family of about 30 meteorites which, based on several experimental observations, are considered as having a Martian origin. SNC is the abbreviation for Shergotty, Nakhla and Chassigny, three meteorites of this family.

Snow line: Limit between the regions where H_2O is gaseous and the region where it is solid. In a protoplanetary disks, the snow line marks the boundary between the region where telluric planets form from silicates, and the region where giant planets, made of an icy core surrounded by protoplanetary gases, are generated.

Snowball effect (*Astronomy*): During planetary accretion the larger objects grow more rapidly than the smaller ones, leading to an increasing difference in size between small and large bodies. This process accelerates (like a rolling snowball, the growing rate is proportional to its size), until the moment when the larger objects control the whole dynamic. See also Runaway growth.

Solar constant: Total energy delivered each second by the Sun and measured in $W.m^{-2}$, the surface being placed at 1AU of the Sun and perpendicularly to the light rays. The value of the solar constant is equal to 1360 $W.m^{-2}$.

Solar type star: Star of G type, similar to the Sun.

Solidus: Line which, in composition vs. temperature or pressure vs. temperature diagrams, separates the domain where crystals and liquid coexist from the field where only crystals (solid) exists.

Spallation: Atomic nuclei breaking due to the collision between two atomic nuclei with energy greater than columbic barrier; it leads to the formation of new elements, stable or radioactive.

Speckle interferometry: Acquisition and statistical treatment of images technique that consists of the recovery of part of the information included in astrophysical images and lost because of the atmosphere imperfect transmission. This technique, developed during the 80ies is not much used any longer and replaced by adaptive optics techniques that allow a real time partial correction of atmospheric turbulence effects.

Spectral type: Star classification procedure based on electromagnetic spectrum which, itself, depends on star surface temperature. The OBAFGKM classification ranges from surface temperature of about 50000 K (O type) until 3500 K (M type = Sun).

Spinel: Mineral. Oxide [$MgAl_2O_4$]. Stable at low pressure (depth < 70 km).). In the mantle, at shallow depth, spinel is the only aluminium-bearing phase.

Spore (*biology*): a) Resistant, dormant, encapsulated body formed by certain bacteria in response to adverse environmental conditions. b) Small, usually single-cell body, highly resistant to desiccation and heat and able to develop into a new organism.

s-process (slow): Nucleosynthetic mechanism that takes place over long periods of time in environments with moderated neutron flux (i.e. Red giant). This mechanism leads to the synthesis of atoms with atomic number > 26 (Iron) that are located in the nuclear stability valley.

Star: Celestial object, generally spherical, where thermonuclear reactions take place (e.g. the Sun) or will take place in the future (e.g. the PMS stars) or has taken place in the past (neutron star)

Stellar cluster: Group of few hundreds to several millions of stars. The smallest groups are named associations. Most of the stars are formed in open clusters.

Stereoisomers (*or stereomers*): Isomers (molecules with identical atomic composition, i.e. the same number of the same atoms) such that the atoms are identically interconnected by covalent

364

bonds. If two stereoisomers are different in the same way that a left hand is different from a right hand, the stereoisomers are enantiomers. In all other cases, stereoisomers are called diastereoisomers (or diastereomers).

Stereoselectivity: When a chemical reaction leads to the formation of stereoisomers and when these latter are not exactly produced in the same amount, the reaction is called stereoselective. Similarly, when two stereoisomers react at a different rate with a reactant, the reaction is stereoselective. In the chemical literature, some authors have introduced subtle differences between stereoselectivity and stereospecificity.

Steric effect: One of the multiple « effects » introduced by organic chemists to explain the relative stability or the relative reactivity inside a group of molecules. The steric effect takes into account the size of each atom or groups of atoms. For instance, the steric effect of a $-C(CH_3)_3$ group is larger than the steric effect of a CH_3 group. The steric effect can be explained on the basis of repulsive term in the Van der Waals forces.

Stony meteorites: Mainly made up of silicates, they can contain from 0 to 30 % of metal grains and several percents of sulphides. They can be differentiated (achondrite) or undifferentiated (chondrite) even if CI undifferentiated chondrites do not contain chondrules.

Stop codon *(Genetic)*: Codon that does not code for amino acids but that indicates the translation end. For the Universal code, these codons are UAA, UAG and UGA (synonym: termination codons).

Stratopause: Atmospheric boundary between stratosphere and mesosphere.

Stratosphere: Atmospheric layer located above the troposphere and below the mesosphere, between 9–17 km and 50 km. In stratosphere, temperature slightly increases with altitude which prevents it of convective movements.

Strecker *(synthesis)*: Synthetic method producing amino acids from aldehyde, HCN, NH_3 and water. Frequently considered as important prebiotic reaction.

Stromatolite: Sedimentary structure consisting of laminated carbonate or silicate rocks and formed by the trapping, binding, or precipitating of sediment by colonies of microorganisms (bacteria and cyanobacteria).

Strong force: One of the four fundamental forces in physics which contribute to the stability of the atomic nucleus.

Subduction: Plate tectonic mechanism, where an oceanic plate sinks under an other lithospheric plate (generally a continental plate, but sometimes also an oceanic plate).

Sublimation: Direct phase change from solid to gas state.

Succinic acid: $HOOC-CH_2-CH_2-COOH$.

Suevite: Rock associated with impact craters (=impactite). Defined in the Ries crater (Bavaria), this brecciated rock containing melt material (glass) consists of fragments of local impacted lithologies and basement rocks, mixed in a fine-grained clastic matrix.

Sun: Star belonging to the main sequence (in H-R diagram); it is 4.56 Ga old. Sun-Earth distance corresponds to 8 light-minutes or 1 astronomic unit (1 AU).

Supernova: Exploding star which, before explosion, was either a binary star (type I) or a massive star (type II). After explosion, the remnant becomes a neutron star.

Surface plasmon resonance (SPR): A biosensor system used for analyzing ligand binding and kinetics of specific molecules within complex mixtures without prior purification. Binding of a ligand to a biomolecule immobilized on a membrane (metallic film) results in changes in membrane surface plasmon resonance.

Symbiosis: Prolonged association between two (or more) organisms that may benefit each other. In the case of endosymbiosis, one organism lives into another one; their relationship is irreversible and implies a complete interdependence such as the two become a single functional organism. Mitochondria and chloroplasts are remnants of the endosymbiosis of photosynthetic bacteria.

Synchrotron radiation: Electromagnetic waves covering a large frequency domain and emitted under vacuum by high velocity electrons or ions, when their trajectory is altered, as by a magnetic field. Synchrotron radiation is naturally polarized.

Synonymous *(codons)*: Different codons that specify the same amino acid; e.g. AAA and AAG specify lysine.

T

T Tauri star: New-born low mass star (lower than two solar masses) which starts to become optically observable. Classical T-Tauri are very young (less than 1 Ma old) and are not yet on the main sequence (PMS stars). They are surrounded by a disk, their luminosity is variable with an excess of IR with

respect to UV. T-Tauri stars we observe today are probably similar to the young Sun.

Talc: Mineral. Phyllosilicate (water-bearing sheet silicate) [Mg_3 (Si_4O_{10}) $(OH)_2$]. Talc frequently results of hydrous alteration of magnesian minerals (i.e. olivine).

Taxonomy: Science that classifies living species. Similar species belong to the same taxon.

Tectonic: Science of rock and crustal deformation.

Tektites: Natural glasses formed at very high temperature ($> 2000°C$) by meteoritic impacts.

Telluric planets: Small and dense rocky planets (density: 3 to 5.5 g. cm^{-3}). These planets (namely Mercury, Venus, Earth and Mars) are silicate rich and were formed in the inner part of the proto-solar nebula, beyond the dust line.

Terraforming: Voluntary transformation of a planetary atmosphere in order to allow colonization by plants and animals (including man).

Tertiary structure *(of a protein)*: Arrangement of the side chains of the residues in space, overall conformation of a protein.

Theia: Name given by some authors to Mars-sized object, which, after impacting the young Earth, led to the Moon formation. This cataclysmic event took place 4.5 to 4.45 years ago.

Thermolysis: Thermal degradation of a molecule into smaller fragments (atoms, radicals, molecules and, rarely, ions).

Thermonatrite: Mineral. Water bearing sodium carbonate [$Na_2CO_3·H_2O$]. This mineral is water-soluble and primarily forms in evaporite deposits and in desert soils where it may occur as a surface deposit, or in volcanic fumaroles

Thermonuclear reaction: Reaction leading to formation of heavier nuclei by fusion of lighter nuclei. Thermonuclear reactions require high T and high P conditions. In natural conditions, these reactions occur spontaneously in cores of main sequence stars or of heavier stars like giant stars. It occurs also during supernovae explosions.

Thermopause: Atmospheric boundary between thermosphere and exosphere.

Thermophile: Organism living optimally at high temperatures. They can be divided in "moderate thermophiles", living optimally between 40 and 60°C, extreme thermophiles, between 60 and 80°C, and hyperthermophiles, living optimally at temperatures higher than 80°C (and up to ~120°C).

Thermosphere: Atmospheric layer located above the mesosphere and below the exosphere. At the top of the thermosphere, at an altitude of about 700 km (on Earth) the temperature can be > 1300 K.

Thin section *(geology)*: Rock slice generally 30 μm-thick. At such a thickness most minerals are transparent and can be observed by transmitted light with a polarizing microscope.

Thiocyanate: R-S-CN.

Thioester: R-S-CO-R'.

Thioester: R-S-CO-R'.

Thiol: R-SH.

Tholeiite: Relatively silica-rich and alkali-poor basalt.

Tholins: Solid mixture of complex organic molecules obtained by irradiation of reduced gases like CH_4, NH_3. Could be present on Titan.

Threonine *(Thr)*: Proteinic amino acid containing four C atoms and a –OH group in its side chain. Threonine is described as a hydrophilic amino acid.

Thymine *(T)*: Nucleic base belonging to purine and specific of DNA.

Titan: The biggest satellite of Saturn (~5000 km in diameter; which is approximately the same size as Mercury). In the solar system, this is the only satellite which possesses a dense atmosphere. Very probably, organic reactions could have taken place in its multi-components hydrocarbon-bearing atmosphere.

Titus-Bode law: Empirical law giving approximately the planet-Sun distance d as a function of the planet ranking n (d = $(4 + 3.2^{(n-2)}/10)$. Mercury is an exception to this law (d = 0.4 AU).

Tonalite: Plutonic magmatic rock (granitoid), made up of quartz and calcic plagioclase feldspar; biotite and sometimes amphibole are minor mineral phases. Tonalite does not contain alkali feldspar. Dacite is its effusive equivalent.

TPF *(Terrestrial Planet Finder)*: NASA project with a similar goal as the ESA Darwin project i.e. discovery of extrasolar terrestrial planets.

Transcription: Synthesis of an m-RNA as a copy of an anti-sense DNA single strand.

Transduction: Transfer of genetic material from one bacteria to another through viral infection.

Transfer of genes *(horizontal transfer)*: Transfer of a gene from one organism to another which does not belong to the same species. Such transfer can occur through viral infection or by direct inclusion of genetic material present in the external medium. Horizontal transfer is different from vertical transfer from parents to children.

Transferase: Enzyme that catalyses transfer of a chemical group from one substrate to another.

Transform fault: Boundary between two lithospheric plates which slide without any crust creation or destruction.

Transit: Motion of a planet in front of the disk of its star.

Translation: Sequence enzymatic reactions such that the genetic information coded into a messenger RNA (m-RNA) leads to the synthesis of a specific protein.

Trans-neptunian object: (TNO) Solar-system body located beyond the orbit of Neptune, in the Kuiper belt or beyond.

Triangular diagram: Diagram classically used in geology to plot the chemical or mineralogical composition of a sample. Each apex of the triangle represents 100% of one component whereas the opposite side corresponds to 0% of the same component.

Triple point: In a P-T phase diagram of a pure compound, it corresponds to the unique P, T value where the three phases (gas, liquid, solid) coexist at equilibrium. For water, P = 6.11 mbar and T = 273.16K.

t-RNA synthetase: Enzyme which catalyzes, in a very specific way, bond formation between an amino acid and its t-RNA.

t-RNA: Transfer RNA.

t-RNA: Transfer RNA. Polymer containing 70 to 80 ribonucleotides and specific of each amino acid to which it is linked. Able to recognize a triplet of nucleotides of m-RNA (codon) by a specific molecular recognition process involving a triplet of nucleotides of the t-RNA (anti-codon). The t-RNA's play a fundamental role for proteinic synthesis.

Trojans: Family of asteroids located at the Lagrange point on the Jupiter orbit. Their position together with the Sun and Jupiter positions determines an equilateral triangle.

Trondhjemite: Plutonic magmatic rock (granitoid), made up of quartz and sodic plagioclase feldspar; biotite is a minor mineral phase. Tonalite and trondhjemite are similar rocks except that in tonalite plagioclase is calcic whereas it is sodic in trondhjemite).

Tropopause: Atmospheric boundary between troposphere and stratosphere.

Troposphere: Lowest part of Earth atmosphere, as the temperature at its basis is greater than at its top, it is the place of active convection. On Earth, troposphere thickness ranges between 9 km (pole) and 17 km (equator). Most meteorological phenomena take place in troposphere.

Tryptophane *(Trp)*: Proteinic amino acid containing eleven C atoms; tryptophane contains a heterocycle in its side chain. It is described as an aromatic amino acid.

TTG: Tonalite, Trondhjemite, Granodiorite. Rock association typical of the continental crust generated during the first half of Earth history.

Tunguska event: Explosion which took place in 1908 (June 30th) in Siberia and devastated 2000 km^2 of forest. It was probably due to the explosion in the atmosphere of small (< 50 m) asteroid or comet.

Tunnel effect: Description of tunnel effect requires the use of quantum mechanics because it is a direct consequence of the wave properties of particles. When a system A gives another system B while the internal energy of A is lower than the energy barrier (activation energy) required to cross the barrier from A to B, it can be said that

A gives B by passing « through the barrier » by a tunnel effect.

Turbidite: Sedimentary deposit deposited by a turbidity current according to a characteristic fining upward grain size sequence in deep water at the base of the slope and in the abyssal plain.

Turbidity current: Also called density current. Viscous mass of mixed water and sediment that propagate downward along the continental slope (of an ocean or a lake) due to its greater density. It may reach high speeds and has a high erosive power. Such current are set in motion by earthquakes for example.

Turbulence (parameter) (*Astronomy*): In hydrodynamics, the « α model empirically describes the turbulent viscosity of a flow. It is based on a parameter α: $v_t = \frac{\alpha C_s^2}{\Omega}$ with C_s = local sound speed and Ω, the keplerian rotational frequency.

Tyrosine *(Tyr)*: Proteinic amino acid containing nine C atoms. Its side chain contains a phenolic group.

U

Upwelling: Upward movement of cool and nutrient-rich sub-surface seawaters towards the surface. Upwelling is mainly controlled by local atmosphere dynamics (pressure, wind), frontal contact between water masses with different densities (oceanographic fronts) or by Eckman transport.

Uracil *(U)*: Nucleic base belonging to pyrimidine family and specific of RNAs

Urea ($H_2N-CO-NH_2$): First organic molecule that has been synthesised from a mixture of inorganic molecules (Wohler).

UV radiation: Electromagnetic radiation characterized by wavelengths ranging from 0.01 to 0.4 microns (energies from 124 to 3.1 eV).

V

Valine (Val): Hydrophobic proteinic amino acid containing five carbon atoms.

Van der Waals forces *(V.d.W.)*: Interatomic forces acting between non-bonded atoms at the intramolecular level but also at the intermolecular levels. Repulsive V.d.W. forces are responsible for the no-infinite compressibility of matter and for the fact that atoms have sizes. Attractive V.d.W. forces are responsible for matter cohesion. V.d.W. forces play a major role in biochemistry: together with H-bonds and electrostatic interactions, they determine the preferred conformations of molecules and they contribute to molecular recognition phenomena.

Vernal point: Sun location on the celestial sphere at the vernal equinox (spring equinox). It is the origin of coordinates in the equatorial system.

Vertical tectonic: See sagduction

Vesicle (geology): Bubble-shaped cavities in volcanic rocks formed by expansion of gas dissolved in the magma.

Vesicle: Small sac, made of hydrophobic or amphiphile molecules, whose content is isolated from the surrounding environment.

Viking: NASA mission to Mars which started in 1976. The two landers (Chryse Planitia and Utopia Planitia) performed a series of very ambitious experiments to detect the presence of life on Mars. Unfortunately, many results were ambiguous.

Virus: System containing DNA or RNA surrounded by a proteinic envelope (capside). When introduced in a living cell, a virus is able to replicate its genetic material by using the host cell machinery.

VLBI *(Very Long Baseline Interferometry)*: Technique that allows a very accurate determination (50 microarcsec) of the position of astronomical sources of radiowaves. This method is based on interferometry measurements using very distant radiotelescopes (large base) located on the same continent or on different continents or even on Earth and on a satellite.

VLT *(Very Large Telescope)*: Group of four large telescopes (4 to 8 metres) and several smaller telescopes, able to work as interferometers and located in Chile. VLT is managed by ESO.

Volatile (volatile substance): Molecule or atom that sublimates at relatively low temperature (i.e. cometary ices).

Volatile: see refractory.

Volcanoclastic sediment: Sediment due to sedimentation in the sea or in a lake of volcanic products (i.e. ashes).

W

Wadsleyite: Mineral. Nesosilicate (isolated SiO_4 tetrahedrons). $[(Mg,Fe)_2 SiO_4]$. Also called olivine β phase. In Earth mantle, wadsleyite is stable between 410 and 520 km.

Wall *(of a cell)*: Extracellular membrane. In bacteria, cell wall structure is complex: the walls of gram-positive and gram-negative bacteria are different.

Water triple point: See triple point.

Watson-Crick: Canonical model of DNA (double helix) involving the pairing of two polynucleotide strands via H-bonds between A and T or G and C.

RNA is generally single-stranded but within a single strand Watson-Crick pairing can occur locally. When happens, it involves A—U and G—C pairing.

Weak bonds: Intermolecular or intramolecular bonds involving non-bonded atoms (atoms not bonded by covalent, coordination or electrostatic bonds). H-bonds are well known examples of weak bonds but Van der Waals forces and electrostatic interactions also contribute to weak bonds. The weak bonds play a fundamental role in the living world: they determine the conformation of molecules and more particularly the conformation(s) of biopolymers; they are responsible for the specificity of molecular recognition. The intermolecular association due to weak bonds is generally reversible

Weak force: One of the four fundamental forces of physics. Parity is violated for these forces. The coupling between weak forces and electromagnetic forces is at the origin of the very small energy difference between enantiomers (PVED for Parity Violation Energy Difference).

Weathering: See alteration.

White dwarf: Relict of a dead star with an initial mass < $\sim 8M_{Sun}$. Its gravity collapse is limited by electron degeneracy pressure.

Wind *(solar or stellar)*: Flow of ionized matter ejected at high velocity (around 400 km/s) by a star. Solar wind mainly contains protons).

Wobble *(genetics)*: Describes imprecision in base pairing between codons and anti-codons. It always involves position 3 of codon, mainly when the base is U.

X

Xenolith: Inclusion or enclave of foreign rock or mineral (xenocrystal), in a magmatic rock.

Y

Young sun paradox: Apparent contradiction between the lower brightness of the young Sun, (70% of the present-day intensity in the visible spectral range) and the early presence of liquid water (−4.4 Ga) on Earth. A strong greenhouse effect due to high concentrations of atmospheric CO_2 could account for this apparent paradox.

YSO *(Young Stellar Object)*: Star which has not yet completed the process of star formation. YSO includes objects ranging from dense cores, (that can be detected in the submillimetric IR frequency range) to pre-main sequence stars (T Tauri, Herbig AeBe) and HII regions.

YSO: Young Stellar Object.

Z

Z *(astronomy)*: Abundance of « heavy elements » i.e. all elements except H and He.

ZAMS *(Zero Age Main Sequence)*: Ensemble of new-born stars in which H fusion has just started.

Zircon: Mineral. Nesosilicate (isolated SiO_4 tetrahedrons). [$ZrSiO_4$]. This mineral which also contains traces of Th and U is extremely resistant to weathering and alteration. These are the reason why it is commonly used to determine rock ages. The oldest zircon crystals so far dated gave an age of 4.4 Ga. They represent the oldest known terrestrial material.

Zodiacal light: Diffuse faint light observed in a clear sky close to the ecliptic. It is due to the diffusion of the solar radiation by the electrons and the interplanetary dusts. Also used for any light diffused by dust particles in a planetary system.

Printed in the United States
By Bookmasters